In: Intrinsically Disordered Proteins

MYELIN BASIC PROTEIN

Intrinsically Disordered Proteins

Series Editors: Vladimir N. N. Uversky and A. K. Dunker

Measles Virus Nucleoprotein
Sonia Longhi
2008. ISBN-13: 978-1-60021-629-9

Myelin Basic Protein
Joan M. Boggs
2008. ISBN 978-1-60456-699-4

In: Intrinsically Disordered Proteins

MYELIN BASIC PROTEIN

JOAN M. BOGGS
EDITOR

Nova Science Publishers, Inc.
New York

For permission to use material from this book please contact us:
Telephone 631-231-7269; Fax 631-231-8175
Web Site: http://www.novapublishers.com

LIBRARY OF CONGRESS CATALOGING-IN-PUBLICATION DATA

Myelin basic protein / Joan M. Boggs (editor).

p. ; cm.

Includes bibliographical references and index.

ISBN 978-1-60456-699-4 (hardcover)

1. Myelin basic protein. 2. Myelin sheath. I. Boggs, Joan M.

[DNLM: 1. Myelin Basic Proteins--physiology. 2. Multiple Sclerosis--physiopathology. QU 55.7 M996 2008]

QP552.M88.M94 2008

572'.633--dc22 *2008017062*

Published by Nova Science Publishers, Inc. ✦ New York

CONTENTS

PREFACE

The compact myelin sheath formed around nerve axons speeds up nerve conduction and also nurtures the axon. Destruction of this sheath in demyelinating diseases such as multiple sclerosis (MS) results in nerve conduction failure and neurodegeneration. Myelin basic protein (MBP) is the second most abundant protein of central nervous system (CNS) myelin (after the proteolipid protein), representing about 30% of the total myelin protein and about 10% of myelin by weight. It is also present in peripheral nervous system (PNS) myelin but as a lower percentage of the total protein. In the CNS, myelin is formed by oligodendrocytes which extend membrane processes that wrap around the axon (Figure).

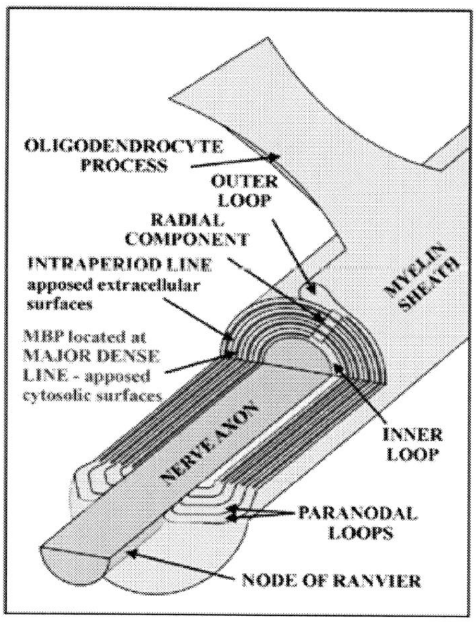

Figure. Diagram of the myelin sheath formed by an oligodendrocyte process around a nerve axon. The extracellular surfaces are apposed at the intraperiod line (thin black line) in compact myelin. MBP mediates adhesion at the cytosolic surfaces forming the major dense line (heavy red line). The paranodal loops, outer loop and inner loop contain cytosol, including cytoskeletal proteins. The radial component consists of a series of tight junctions and may also contain actin, and tubulin (see Chapter VIII). The nerve axon is bare at the node of Ranvier. Adapted from Boggs et al. (2008) and republished with permission from Elsevier. I thank George Harauz for providing an earlier version of this figure.

MBP binds to negatively-charged lipids on the cytosolic surfaces of the processes and is responsible for adhesion of these surfaces of myelin in the CNS, thus forming the major dense line observed by electron microscopy. It is thus a structural protein that has been shown to be essential for formation of compact CNS myelin; a naturally occurring shiverer mutant mouse, which has a deletion of the major part of the gene encoding MBP, produces only small amounts of uncompacted myelin. However, MBP is not essential for formation of PNS myelin, due to the presence of other proteins specific to PNS myelin, that may compensate for its absence.

MBP was first discovered in the 1960s by Dr. Marian Kies as a result of efforts to determine the factor from brain which induced experimental allergic encephalomyelitis (EAE) (Kies et al., 1961; Laatsch et al., 1962). The most abundant 18.5 kDa isoforms from bovine and human brains were sequenced in 1971 by Eylar et al. (1971) and Carnegie (1971), respectively. MBP is now known to be the product of a gene which has three different transcription start sites and yields two major families of proteins, the "classic" MBP family, which is expressed only in myelin and myelin-producing cells, and the golli proteins, which are also expressed in other cells, including immune tissue. The MBP gene and the functions of the golli proteins are described in Chapter I by Campagnoni and Campagnoni. Classic MBP also exists as a number of size isoforms due to differential splicing, and can be post-translationally modified in a number of ways, resulting in a diverse family of proteins (see Chapter II by Zand, and Chapter III by Mastronardi and Moscarello).

Although it is now known that other proteins from myelin can also induce EAE, and that determinant spreading occurs during autoimmune responses, the encephalitogenic properties of MBP have attracted wide interest in attempts to understand its role in the demyelinating disease multiple sclerosis (MS). Both T cell and B cell-mediated immune responses to MBP occur in MS, and its antigenic epitopes have been characterized (see Chapter IV by Ji and Goverman). The structures of immunogenic MBP peptides complexed with class II MHC proteins have been determined in order to aid in the design of synthetic peptides which will be useful for suppression of this immune response (see Chapter V by Katsara, Ramsland, Tselios, Matsoukas, and Apostolopoulos). In addition to inducing autoimmune response, a particular variant of MBP, with significantly reduced net positive charge, may be further involved in demyelination in MS. This charge isomer, in which a number of arginines are deiminated by the enzyme peptidyl arginine deiminase (PAD) to give uncharged citrulline residues, occurs normally in higher amounts in children than in adults. However, both it and the PAD enzyme are found in increased amounts in brains of adults with MS (see chapter III). This less-charged MBP variant has decreased ability to cause adhesion of negatively-charged lipid membranes, which may destabilize myelin. It may also elicit an increased immune response, as discussed below.

Despite the early amino acid sequencing of MBP and the characterization of its gene structure, attempts to determine its tertiary structure by crystallography, solution NMR spectroscopy, and other methods, have been unsuccessful. The reason is the large number of charged residues throughout the protein's sequence, and its low overall hydrophobicity, which maximize intramolecular electrostatic repulsion, resulting in an extended, natively-unfolded structure (Harauz et al., 2004). Such proteins have sufficient flexibility to bind to various charged surfaces and ligands, and to acquire whatever local conformation is necessary to optimize specific binding to several different targets (Dyson and Wright, 2002). Site-directed spin labeling and EPR spectroscopy are now being applied to determine the structure of MBP, when bound to lipids and other proteins, from the environment of its spin-labeled residues (see Chapter VI by Boggs, Bates, Musse, and Harauz). These studies have revealed that when MBP is bound to a lipid surface, an immunodominant epitope of MBP forms an amphipathic alpha-helix with its hydrophobic surface embedded in the bilayer, and its hydrophilic surface exposed. They further showed that deimination results in dissociation of much of the C-terminal half of MBP from the membrane surface, shortens the length of the amphipathic alpha-helix formed by the immunodominant epitope, and increases its susceptibility to proteolytic digestion. Greater exposure to proteolytic enzymes can cause release of this immunodominant epitope of MBP, which may initiate or sustain immune response to this epitope, as discussed in Chapter VI.

Studies of other intrinsically disordered proteins have shown that they are often multifunctional regulatory proteins (Tompa et al., 2005) involved in multiple interactions, which may integrate hubs of various biological activities (Uversky et al., 2005; Patil and Nakamura, 2006). In this regard, MBP also interacts with several other proteins in addition to binding to negatively-charged lipids, and has also been suggested to be a multifunctional protein (Boggs, 2006). It binds to actin, tubulin, tropomyosin, calmodulin, and clathrin and assembles microfilaments, microtubules, and clathrin baskets *in vitro*. MBP has been shown to be one of only two proteins isolated from the brain with the ability to stabilize microtubules from depolymerizing (STOP activity) in the cold, and also has this activity in oligodendrocytes (see Chapter VII by Galiano, Sambrooks, and Hallak). It is also able to bind bundles of actin filaments to lipid surfaces and may be able to serve as a membrane actin-binding protein (see Chapter VIII by Boggs). MBP's binding to actin and tubulin can be regulated by Ca^{2+}-calmodulin, and is influenced by its various post-translational modifications. The MBP-mediated binding of actin filaments to the membrane can also be regulated by changes in membrane surface potential, which could be achieved through local changes in lipid composition during signal transduction. Chapters VII and VIII describe studies suggesting that MBP is involved in transmitting extracellular signals received on the oligodendrocyte surface to the cytoskeleton.

In addition to interactions with actin and tubulin, which may be primarily electrostatic, MBP has a domain that has been predicted to be a PXXP SH3-target consensus sequence (Moscarello, 1997). This domain has recently been shown to form a poly-proline helix, and to bind SH3 domains of several proteins (see Chapter IX by Polverini). The interactions of MBP with both Ca^{2+}-calmodulin and the SH3 domain of Fyn tyrosine kinase, and the effects of post-translational modifications to MBP on these interactions, have been modeled *in silico* and are described in Chapter IX. Since the binding of MBP to different ligands can induce or stabilize a particular conformation, its structure when bound to lipids, Ca^{2+}-calmodulin, and other proteins is now being studied by solution and solid-state NMR spectroscopy (see Chapter X by Harauz and Ladizhansky). Further study of the structure of members of the MBP family when bound to different ligands and in different environments will help to understand its role in signaling in oligodendrocytes and myelin, and its involvement in the pathogenesis of MS.

I wish to take this opportunity to thank all of the authors for their chapters in this book and to Dr. George Harauz in particular, for helpful advice during its preparation. I also apologize to those whose work could not be included or was improperly cited because of space and time limitations. I hope that this book will be a valuable contribution to the study of myelin basic protein and myelin.

Joan M. Boggs, Editor

REFERENCES

Boggs, J.M. (2006). Myelin basic protein: a multifunctional protein. *Cell. Mol. Life Sci*, 63, 1945-1961.

Boggs, J.M., Gao, W., and Hirahara, Y. (2008). Myelin glycosphingolipids, galactosylceramide and sulfatide, participate in carbohydrate-carbohydrate interactions between apposed membranes and may form glycosynapses between oligodendrocyte or myelin membranes. *Biochim. Biophys. Acta*, 1780, 445-455.

Carnegie, P.R. (1971). Amino acid sequence of the encephalitogenic basic protein from human myelin. *Biochem. J*, 123, 57-67.

Dyson, H.J., and Wright, P.E. (2002). Coupling of folding and binding for unstructured proteins. *Curr. Opin. Struct. Biol*, 12, 54-60.

Eylar, E.H., Brostoff, S., Hashim, G., Caccam, J., and Burnett, P. (1971). Basic A_1 protein of the myelin membrane: The complete amino acid sequence. *J. Biol. Chem*, 246, 5770-5784.

Harauz, G., Ishiyama, N., Hill, C.M.D., Bates, I.R., Libich, D.S., and Farès, C. (2004). Myelin basic protein - diverse conformational states of an intrinsically unstructured protein and its roles in myelin assembly and multiple sclerosis. *Micron*, 35, 503-542.

Kies, M.W., Murphy, J.B., and Alvord, E.C., Jr. (1961). *Studies of the encephalitogenic factor in guinea pig central nervous sytems.* In J. Folch-Pi (Ed.), Chemical Pathology of the Nervous System (pp. 197). Elmsford, NY, USA: Pergamon Press.

Laatsch, R.H., Kies, M.W., Gordon, S., and Alvord, E.C., Jr. (1962). The encephalitogenic activity of myelin isolated by ultracentrifugation. *J. Exp. Med*, 115, 77-88.

Moscarello, M.A. (1997). *Myelin basic protein, the 'executive' molecule of the myelin membrane.* In B.H.J. Juurlink, R.M. Devon, J.R. Doucette, A. J. Nazarali, D.J. Schreyer, V.M.K. Verge (Eds.), Cell Biology and Pathology of Myelin: Evolving Biological Concepts and Therapeutic Approaches (pp. 13-25). New York, NY, USA: Plenum Press.

Patil, A., and Nakamura, H. (2006). Disordered domains and high surface charge confer hubs with the ability to interact with multiple proteins in interactions networks. *FEBS Lett*, 580, 2041-2045.

Tompa, P., Szasz, C., and Buday, L. (2005). Structural disorder throws new light on moonlighting. *Trends Biochem. Sci*, 30, 484-489.

Uversky, V.N., Oldfield, C.J., and Dunker, A.K. (2005). Showing your ID: intrinsic disorder as an ID for recognition, regulation and cell signaling. *J. Mol. Recog*, 18, 343-384.

In: Myelin Basic Protein
Editor: Joan M. Boggs

ISBN: 978-1-60456-699-4

Chapter I

THE PROPERTIES AND FUNCTIONS OF THE GOLLI MYELIN BASIC PROTEINS

Anthony T. Campagnoni and Celia W. Campagnoni**

ABSTRACT

Like the classic myelin basic proteins, the golli-MBPs are intrinsically unstructured protein products of the MBP gene. They are expressed in numerous cell types throughout the immune and nervous systems, and their function(s) are beginning to be understood. Since their discovery over a decade ago a number of approaches have been taken to elucidate their structure, including searches for binding partners, selective ablation of their expression in knock-out mice, and their overexpression in transgenic animals and in cell models. They appear to have an, as yet, undetermined role in the nucleus, but they have now been clearly shown to regulate Ca^{++} homeostasis in T-cells through modulation of CRAC channels, and in oligodendrocytes through voltage-gated Ca^{++} channels as well as through store-operated and ligand-gated Ca^{++} channels. Thus, they appear to play a significant role in oligodendrocyte development and in Ca^{++}-dependent processes, such as process extension/retraction and migration, as well as T-cell activation. Several studies have also shown that expression of the golli-MBPs is altered in a number of human and animal neuropathological conditions, providing further evidence for their importance in cellular function.

* Semel Institute for Neuroscience and Human Behavior, Geffen School of Medicine, University of California at Los Angeles, Neuroscience Research Building, 635 Charles Young Drive, Los Angeles, CA 90095-7332. Telephone: 1-310-825-5006; Fax: 1-310-206-5050; E-mail: acampagnoni@mednet.ucla.edu; ccampagnoni@mednet.ucla.edu

INTRODUCTION

About twenty years ago reports of the isolation of unexpected alternatively spliced products of the MBP gene began to appear in the literature (Newman et al., 1987; Campagnoni et al., 1993; Zelenika et al., 1993), but a clear relationship of these products to the MBP gene was not firmly established until the entire gene structure was elucidated (Campagnoni et al., 1993). The publication of the revised MBP gene structure and the golli products of the gene was met initially with some confusion and also some skepticism. Relative to the classic MBPs, among the most abundant proteins in the nervous system, the levels of the golli proteins seemed too low to be of real significance. Furthermore, the introduction of a second family of proteins that was immunologically similar to the known MBPs and expressed in both the immune and nervous systems, raised questions about prior concepts of "tolerance" in EAE, which had been studied for decades as a model for the inflammatory component of MS (Huseby and Goverman, 2000; Maverakis et al., 2000; 2003). Furthermore, the exact relationship between the classic MBPs, the golli-MBPs and their relationship to the structure of the MBP gene was confusing to some because the golli transcription unit was initially referred to as "overlapping" the classic MBP gene, when, in fact, the entire complex should be viewed as a single gene (see section on gene structure).

In the years since these proteins were identified, the importance of the golli proteins to the biology of T-cells and oligodendrocytes has become more clearly established. We are now beginning to understand the function of the golli proteins in Ca^{++} homeostasis in these cells and we have identified other potential roles for these proteins in the cell deserving of further investigation. In this review we summarize findings about the structural, molecular and cell biological properties of the golli proteins and our present understanding of the function of these proteins in T-cells and in oligodendrocytes.

The MBP gene encodes the "classic" and golli family of proteins

The structure of the MBP gene in mouse and human and their major golli splice products are diagrammed in Figure 1. The mouse MBP gene is ~ 105Kb in length (Campagnoni et al., 1993) and the human gene is ~ 180 Kb (Pribyl et al., 1993). The MBP gene contains three independent promoters, and mRNA products from all three transcription initiation sites have been identified in the mouse (Campagnoni et al., 1993; Fritz and Kalvakolanu, 1995; Zelenika et al., 1993; Kitamura et al., 1990) and from the first and third sites in the human (Roth et al., 1987; Pribyl et al, 1993; Grima et al., 1994; Tosic et al., 2002). The presence of exon 4 and the second transcription initiation site in the human gene is inferred.

The classic MBP mRNAs are derived from transcription start sites 2 and 3, and constitute major protein constituents of the myelin membrane. The most downstream promoter, governing tss3, is the strongest of the three promoters and is very active in oligodendrocytes. As such they encode some of the most abundant proteins in the brain and they are expressed almost exclusively in myelin-forming cells, although expression of low levels of classic MBPs has been reported in the immune system (Liu et al., 2001).

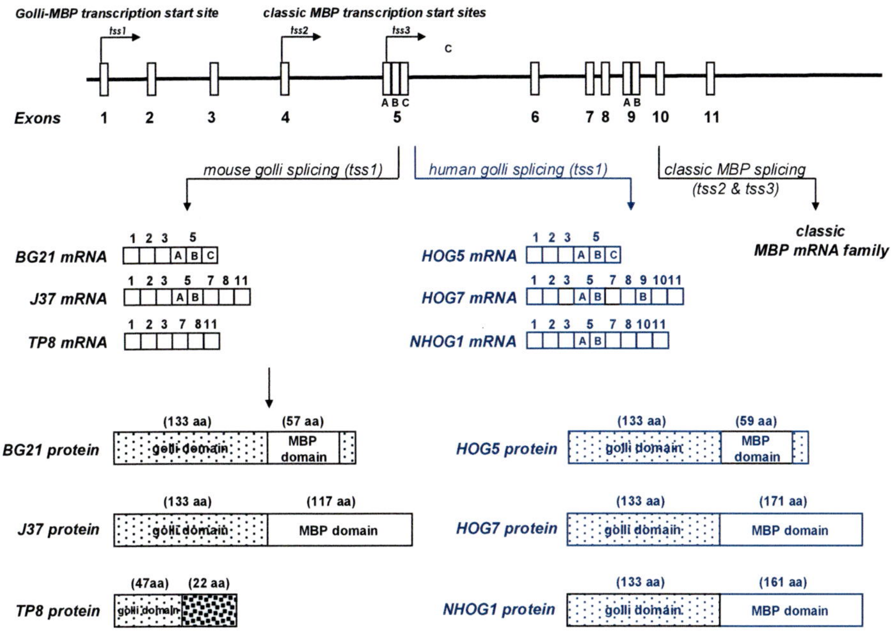

Figure 1. Diagram of the myelin basic protein gene showing the generation of the major golli-MBP products in human and mouse. The fourth exon has not been demonstrated in the human, but is inferred by analogy with the mouse. (See Fig. 1 in Chapter X for correspondence of former exon numbering of classic MBP with new numbering of all exons in total MBP gene).

The second family of proteins encoded by the gene is the golli proteins. The golli mRNAs are generated from the first transcription start site and they are expressed more ubiquitously than the classic MBPs. In contrast to the classic MBPs, they are expressed at similar levels in the thymus, spleen and brain (Feng et al., 2004). Many cell types, e.g., thymocytes, T-cells, B-cells and macrophages, as well as neurons and oligodendrocytes in the nervous system express the golli products.

The golli mRNAs from mouse and human all contain exons 1-3, and these encode a 133 aa golli domain with related primary sequences. Figure 1 shows the major golli cDNA products that have been cloned from mouse brain and human brain. Mouse golli TP8 and its unpublished human homologue are minor products. While they encode the golli peptide, they contain no classic MBP sequences, since the classic MBP exons are read out of frame. The major products of the golli gene also express MBP epitopes. Thus BG21 and its human ortholog, HOG5, contain the golli domain immediately upstream (i.e. on the N-terminal side) of the first 57 (mouse) or 59 (human) amino acids of classic MBP (Figure 2A). The mouse J37 consists of the 133 aa golli domain fused to a small MBP that does not correspond to one of the classic MBPs, however the human HOG 7 consists of the golli domain fused in frame to the human 18.5 kDa MBP. In mouse, BG21 and J37 are the major isoforms expressed throughout development; but in human brain the major isoforms change during development. For example, in fetal human brain the principal golli mRNAs appear to be HOG 5 and NHOG1, a golli isoform consisting of the golli domain fused to a 17 kDa classic MBP sequence (see Figure 2B).

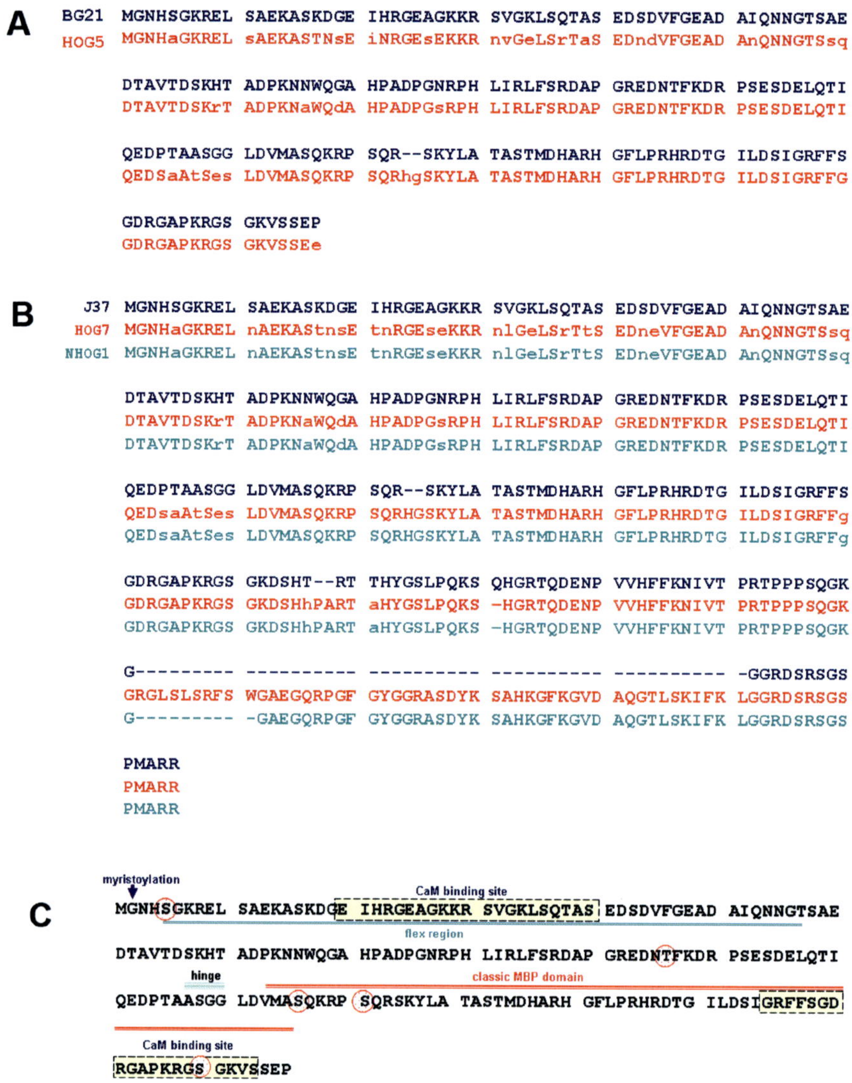

Figure 2. (A) Sequence comparison of the BG21 (mouse) and HOG5 (human) golli orthologs. (B) Sequence comparison of the mouse J37 and human HOG7 & NHOG1 golli isoforms, which contain longer classic MBP sequences than either BG21 or HOG5. (C) Illustration of identified and potentially significant residues and domains in the BG21 molecule. Potential phosphorylation sites are noted by speckled circles. Other sites and domains are shown.

In the more mature human brain the principal isoforms are HOG5 and HOG7. This developmental shift from NHOG to HOG 7 reflects the exon splicing pattern shift that normally occurs within the human classic MBPs with development (Roth et al 1987; Pribyl et al., 1996a). HOG 5/HOG 7 and HOG 5/NHOG1 constitute about 80% of the transcripts found in human tissue, although numerous other splice variants also have been identified (Pribyl et al., 1996a).

Features of the primary and higher ordered structure of the golli-MBPs

Unlike the classic MBPs, the golli proteins do not appear to be normal components of the myelin sheath, but are localized within the nuclei, cell bodies and primary processes of oligodendrocytes and neurons (Landry et al., 1996; Paez et al. 2007). As indicated, they are more ubiquitously expressed throughout the nervous system and the immune system than are the classic MBPs, which are primarily products of myelin forming OLs, and they have different developmental patterns of expression (Campagnoni et al., 1993; Pribyl et al., 1993; Landry et. al., 1996;1997). This, and their non-inclusion in the myelin sheath, suggested that they had some other biological function than the classic MBPs (Campagnoni & Skoff, 2001).

A clue to the function of proteins often comes from the physical properties of, and the presence of consensus sequences and domains in, the molecule. The J37 and HOG 7 isoforms contain significantly longer classic MBP sequences than BG21 or HOG 5; and, as might be expected, the pIs of the golli isoforms mirror the amount of (highly basic) classic MBP sequence found within the molecule. For example, BG21 and HOG 5 have pIs of ~6 and J37 and HOG7 have pIs from 9.6-9.8. All these golli isoforms have mean net charges and mean net hydrophobicity values that would categorize them as potential intrinsically unstructured proteins (Ahmed et al., 2007).

Primary sequence

A survey of the primary sequences of the mouse golli proteins predicts calmodulin-binding three calmodulin-binding motifs in J37, two in BG21, and four in HOG7 (Polverini et al., 2004) (see Chapter IX and see Figure 2C for BG21). Experimentally, both recombinant J37 and BG21 bind calmodulin at a 1:1 ratio in the presence of calcium but the association appears to be weak, since even at high ratios of calmodulin: BG21, a significant amount of unbound BG21 remained. Titration curves of BG21 with calmodulin monitored either by fluorescence intensity of the single tryptophan at 346 or by fluorescence anisotrophy did not plateau so it was impossible to calculate a dissociation constant (Kaur et al., 2003; Bamm et al., 2007).

There is an essential myristoylation motif at the amino terminus of the golli proteins. Elimination of this site by site-directed mutation of G^2 to A essentially eliminates golli function as a Ca^{++} regulator in both T cells and oligodendrocyte cell lines (Feng et al., 2006; Paez et al., 2007). Chemical confirmation of the presence of the myristoylated glycine residue has been obtained from LC-MS analysis (C. Campagnoni and K. Faull, unpublished)

Golli proteins are phosphorylated in vivo (Feng et al. 2004). BG21 transfected into Jurkat T cells grown in the presence of ^{32}P orthophosphate was immunoprecipitated as a phosphoprotein and the amount of phosphorylation increased when the cells were activated with the PKC stimulator, PMA. However, it is not yet clear what kinases or signaling pathways might be involved. Feng et al (2004) investigated two kinases and found that while both BG21 and J37 could be phosphorylated in vitro by PKC, only J37 could be phosphorylated by Erk 2 (MAPK). The 133 aa golli-specific peptide could be phosphorylated by neither kinase. As shown in Figure 2C, there are at least 5 potential PKC sites on the molecule, i.e. S^5 and T^{106} in the golli domain and $S^{136, 141, 188}$ in the MBP region common to both BG21 and J37.

Higher ordered structure of golli proteins

Harauz and his colleagues have extensively characterized the mouse recombinant golli proteins. While classic MBPs aggregate phospholipid vesicles, J37 (a golli isoform with significant classic MBP sequence) had no effect on a myelin-like preparation. Circular dichroism studies, however, showed that BG21 and J37, which are largely random coils in aqueous solution, each acquired more organized secondary structures in the presence of ganglioside GM1 or phospholipids, such as PI(4)P (Kaur et al., 2003; Bamm et al., 2007). Heteronuclear NMR measurements in 0.1M KCl confirmed that BG21 has little ordered secondary structure in solution and revealed that residues S^5-T^{69} were unusually flexible even for IUPs (see Chapter X). This region was postulated to be a candidate for protein-protein interactions, and another smaller mobile segment, A^{126} to G^{129} was postulated to be a hinge (Ahmed et al., 2007) (see Figure 2C).

The analyses of the primary and higher ordered structures of the protein provided evidence for a critical myristoylation site, suggesting a need for association of golli with a membrane in order for it to be functional. They also suggested a potential ability of golli to perform its functions through interactions with other proteins, a characteristic common to IUPs. From what is now known about the cell biology of golli proteins, these interactions could occur in complexes at the plasma membrane or in the nucleus (see below).

Nuclear localization sequence in golli proteins

There is evidence that sequences found within exon 6 of the (golli) MBP gene (please refer to Figure 1 for exon numbering) are responsible for targeting certain classic MBP isoforms, such as the 17kDa and 21.5kDa MBPs, to oligodendrocyte nuclei early in postnatal brain (Pedraza et al., 1997). In most cells, golli immunocytochemical analysis suggests that golli proteins are localized in both the nucleus and cytoplasm/processes. Interestingly, in certain cell types the golli MBPs appear to undergo rather dramatic subcellular localization shifts during development (Landry et al., 1996). This is particularly evident in cerebellar granular cells. During development, immature granule cells migrate from the outer layers of the cerebellum to a deeper layer where they form the internal granule cell layer in the mature cerebellum. Accompanying this transition is a significant shift of the localization of golli from the cell body and processes of these cells to the nucleus. Transfection studies with mutated golli cDNAs have identified a nuclear targeting element within a 36 amino acid region of the golli proteins, M^{134}-I^{169}, located in the MBP domain (Reyes and Campagnoni, 2002). The BG21 isoform of golli does not contain sequences derived from exon 6 (see Figure 1) found in the 17kDa and 21.5kDa classic MBPs, so its translocation to the nucleus cannot be attributed to the same NLS as that of the classic MBPs. Neither the regulation of nuclear targeting of golli proteins nor the function(s) of golli proteins in the nucleus have been resolved.

Approaches to defining the biological roles of golli proteins in cells

Several approaches have been used to identify the cellular function(s) of the golli proteins in the immune and nervous systems. These have included a search for binding partners using the yeast two-hybrid system with the golli domain as a "bait" (Fernandes et al., 2004), loss-of-function analyses through ablation of golli expression in knock-out (KO) mice (Jacobs et al., 2005; Feng et al., 2006) and gain-of-function analyses (a) through transfection of golli isoforms into cells in vitro (Reyes and Campagnoni, 2002; Paez et al., 2007) and (b) through cell-specific targeting of golli into oligodendrocytes in transgenic mice (Reyes et al., 2003; Martin et al., 2007).

Golli proteins bind to nuclear proteins involved in gene transcription

Fernandes et al. (2004) conducted a yeast two-hybrid screen of a rat PC12 library using the 133 aa golli domain as "bait". From this library a clone was isolated that encoded a golli-interacting protein (GIP) with a predicted molecular weight of 25kDa; and the rat clone was used to isolate the mouse homolog from a mouse oligodendrocyte library. Immunocytochemical analysis indicated that GIP was co-expressed with golli proteins in a wide variety of cells and that it was localized predominantly in the nuclei of these cells. Immunoprecipitation studies showed that GIP interacted with nuclear LIM interactor (NLI), a nuclear protein known to associate with LIM transcription factors, as well as the golli proteins; and that, in fact, all three could form a trimolecular complex with GIP serving as the intermediary.

GIP is identical to SCP-1 of a series of small carboxyl-terminal domain (CTD) phosphatases described by Gill and coworkers (Yeo et al., 2003; 2005). In eukaryotic cells, the transcriptional activity of RNA polymerase II is modulated in part by the phosphorylation status of S^2 and S^5 in a 26-52 tandem heptapeptide repeat at its carboxyl terminus. SCP-1 preferentially dephosphorylates S^5 in the consensus sequence $Y^1S^2P^3T^4S^5P^6S^7$. It co-immunoprecipitates with a number of transcription factors, among them the RE-1 silencing transcription factor (REST/NRSF) which is believed to be responsible for silencing the transcription of neural genes in non-neural cells (Schoenherr and Anderson, 1995; Yeo et al., 2005). SCP-1 appears to reduce transcription of the genes to which it is bound, perhaps by slowing down clearance of transcription factors from the promoter (Thompson et al., 2006).

The existing information clearly indicates that golli proteins can bind to known proteins in transcriptional complexes, but the mechanisms underlying its transport to the nucleus and the specific gene activities it might be involved in regulating still remain unknown.

Golli proteins DECREASE Ca^{++} entry into T-cells upon activation of the cells

The nuclear-cytoplasmic shift of golli observed in neural cells suggested that golli proteins might be involved in some aspect of intracellular signaling, since many signaling molecules have been shown to shuttle between the cytoplasm or plasma membrane and the nucleus. However, definite proof of a signaling role for golli-MBPs came from an examination of the activation of T cells. These cells have been well studied because they are easily stimulated and a good deal is now known about downstream signaling events after activation of the T-cell receptor. For example, synthesis of IL2 is a hallmark of T cell activation and the pathway from T cell receptor (TCR) engagement to the activation of the IL2 promoter via AP-1, NFḰB and NFAT has been elucidated. Furthermore, TCR activation also initiates a cascade of tyrosine kinase events that recruit PKCθ to the TCR and activates phospholipase Cγ. Downstream of these events, the release of IP_3 from phosphatidyl inositol by PLC causes a release of Ca^{++} from internal stores, which, in turn, triggers an influx of Ca^{++} through calcium-release activated channels (CRAC) in the plasma membrane. Both IL2 gene transcription and CRAC activation can be triggered by activation of PKC with the phorbol ester, PMA.

Feng and coworkers examined the role of golli proteins in T-cells extensively (Feng et al., 2004; 2006). They found that BG21 is the major golli isoform expressed in T cells and that when Jurkat T cells were transfected with BG21-GFP the protein behaved like an intracellular signaling molecule. For example, immunofluorescence studies showed that upon activation of the cells with PMA, the BG21-GFP fusion protein translocated from the cytoplasm to the plasma membrane in a fashion identical to PKCθ. This was confirmed by Western blots of membrane fractions, which also showed movement of BG21 from cytosolic fractions to lipid raft fractions in sucrose density gradients after activation of the cells.

In vitro transfection of golli into Jurkat T cells inhibited IL-2 reporter gene transcription upon TCR engagement (Feng et al., 2004). Although golli proteins possess several PKC phosphorylation sites, Ser 136, 141, and 188, within the MBP domain common to both BG21 and J37, the inhibitory function of golli was independent of its PKC phosphorylation (Feng et al., 2004) and resided in the golli domain alone (133aa). Subsequent studies showed that golli acted negatively on T-cell receptor signaling by inhibiting store-depletion-induced Ca^{++} entry into the T-cell through CRAC channels (Feng et al., 2006). Although phosphorylation of these sites does not appear to be necessary for golli to modulate T-cell activation, they may play another role unrelated to calcium uptake in these cells, possibly in the nucleus.

Golli proteins INCREASE Ca^{++} entry into oligodendrocytes upon activation of the cells

Golli proteins also have been shown to modulate Ca^{++} entry into oligodendrocytes, although in an opposite fashion (Jacobs et al, 2005; Paez et al., 2007). In T-cells golli inhibits Ca^{++} uptake upon stimulation of the T-cell receptor, but in oligodendrocytes it enhances Ca^{++} entry via voltage-gated channels under depolarizing conditions. These in vitro data were obtained from a comparison of Ca^{++} changes in primary cultures of normal OLs vs. OLs in which the golli products of the gene were selectively ablated. In these studies, the absence of the golli proteins decreased Ca^{++} uptake when the cells were exposed to agents known to induce Ca^{++} influx in OLs, e.g. high K^+, AMPA, PMA and caffeine (Jacobs et al, 2005). These results indicated that under normal circumstances golli proteins enhance Ca^{++} uptake into OLs.

Modulation of intracellular Ca^{++} levels is important in a number of OL activities, such as cell-cell communication (Simpson et al 1997), process extension (Yoo et al, 1999), migration (Simpson & Armstrong 1999) and oligodendrocyte differentiation and myelination (Soliven, 2001). It has also been proposed that Ca^{++} surges may be involved in the signal for myelination in a remyelinating animal model (Mateo Paz Soldan et al., 2003). Direct effects of golli proteins on Ca^{++} mediated process extension (Paez et al., 2007) and migration (P. Paez and A. Campagnoni, unpublished results) have now been shown.

Unique phenotypes of the golli KO and golli overexpressing mice

Elucidation of the function(s) of the golli proteins has been aided substantially by the generation of mice in which the golli proteins were selectively ablated in all cells and tissues (i.e., golli KO mice; Jacobs et al., 2005); and mice in which overexpression of the golli J37 isoform was targeted specifically to OLs using the classic MBP promoter (Martin et al., 2007).

Selective ablation of the golli products of the MBP gene in golli KO mice

The phenotype of a golli "knock out" (KO) mouse, which was generated through the deletion of exon 2 of the MBP gene, has been reported (Feng et al., 2006; Jacobs et al., 2005; Voskuhl et al., 2003). In this mouse, there appears to be no effect on expression of the classic MBPs, but there is complete ablation of expression of the golli MBPs in both the nervous system and the immune system (Voskuhl et al., 2003; Feng et al., 2006).

Phenotypic characteristics of the nervous system in golli KO mice

Several examples of both spontaneously occurring- and targeted-ablation of myelin proteins have been reported. Ablation of the classic MBPs causes a severe neurological phenotype in the shiverer (shi) mouse (Wolf and Billings-Gagliardi, 1984). In the shi mouse there is substantial hypomyelination associated with tremors and seizures, resulting in the premature death of some homozygous animals. The targeted ablation of the PLP and MAG (myelin-associated glycoprotein) genes has resulted in KO mice with no obvious neurological or dysmyelinating phenotype (Klugmann et al., 1997). However, neuronal abnormalities develop in both of these mice at later postnatal ages, and in the MAG KO mouse, myelin degeneration is observed (Fruttiger et al., 1995).

The golli KO mouse exhibits a phenotype unlike any of the other myelin protein KO animals (Jacobs, 2005; Jacobs et al., 2005). Like the PLP and MAG KOs, there is no overt dysmyelinating neurological phenotype, such as tremors or seizures, but the golli KO exhibits delayed expression of myelin proteins; and cortical OLs isolated from KO brains elaborate smaller and less extensive membrane sheets in culture. While hypomyelination is observed in KO brains, it does not appear to be global, but rather it is confined to selected areas of the brain (e.g., visual cortex, sensory cortex and optic nerve) as determined by Northern blots, Western blots, and immunohistochemical analysis with myelin protein markers. Ultrastructural analysis reveals abnormalities in myelin structure and in some OLs. These results suggest that OL heterogeneity with respect to golli expression and/or regulation might exist; and they are consistent with many reports describing OL heterogeneity in the cerebral cortex (for review, see Noble et al., 2003). Hypomyelination is transient in some areas, but in others, such as the visual cortex, persists into adulthood. Abnormal visual-evoked potentials indicate that the hypomyelination in the visual cortex has functional consequences in the KO brain.

Abnormalities in Ca^{++} uptake have been reported in KO OLs (Jacobs et al., 2005), but unlike the KO T-cells in which loss of golli result in increased Ca^{++} uptake (Feng et al., 2006), loss of golli in OLs resulted in depressed Ca^{++} uptake, indicating that golli normally stimulates, rather than inhibits, Ca^{++} uptake in OLs. For example, in WT OLs, depolarization of the cells with 20mM K^{+} caused an average increase in intracellular Ca^{++} levels of 47%, while in KO OL cultures, the Ca^{++} increases were only 20%. In addition, the magnitude of Ca^{++} transients induced by AMPA and PMA were also significantly reduced in the KO OLs. Importantly, these differences in the Ca^{++} response between WT and KO OLs were lost when the cells were in a medium with zero Ca^{++}. The data indicated that OLs from the golli KO mice were less responsive to stimulation of Ca^{++} influx.

Phenotypic characteristics of the immune system in golli KO mice

As early as the late 1980s, reports appeared in the literature of immunoreactivity against MBP and PLP in macrophage preparations from mouse peritoneum, bone marrow and spleen (Alliot and Pessac, 1988) (see Chapter IV). Further work eventually revealed that the major products responsible for this immunoreactivity were probably the golli proteins after the structure and relationship of these proteins to the MBPs was elucidated in mice and humans (Campagnoni et al., 1993; Pribyl et al., 1993; 1996b). Independently, Pessac and coworkers also identified transcripts that were identical to the golli transcripts, but which they called HMBP-R for hemopoietic MBP-related (Grima et al., 1992, 1994; Zelenika et al., 1993; Kalwy et al, 1998) or, more recently HMBP (Marty et al., 2002). Thus, the expression of golli transcripts and proteins in the immune system was recognized soon after their discovery, so that it was anticipated that there would be a phenotype associated with the immune system in the golli KO.

Quantitative FACS analysis showed that golli BG21 is expressed more abundantly in T-cells than in other immune cells and that the levels of its expression changed with thymocyte differentiation (Feng et al., 2000). Furthermore, golli appears to inhibit T-cell activation and Ca^{++} uptake through CRAC (i.e. Ca^{++} Release Activated Ca^{++}) channels in the T-cell (Feng et al., 2006). T cells isolated from golli KO animals were found to be hyperproliferative when activated, which is consistent with a role for golli as an inhibitor of T-cell activation.

One of the most interesting aspects of the golli KO phenotype in the immune system is the effect of the absence of golli on EAE. In a preliminary study to examine the effects of the partial ablation of the golli proteins on MBP-induced EAE, Voskuhl et al. (2003) compared the clinical course of EAE between heterozygous ($golli^{+/-}$) and wild-type ($golli^{+/+}$) mice. While they found no difference between the two groups in incidence of disease, severity of the first episode of disease, or remission after the first episode, they did observe a significant reduction in relapses in $golli^{+/-}$ mice vs. controls. More recently, Feng et al (2006) performed a more extensive examination of the susceptibility of homozygous KO mice to MOG-induced EAE. They found that the animals were completely resistant to EAE induced by MOG as well as PLP (J.M. Feng and A. Campagnoni, unpublished). Since the absence of golli results in hyperproliferative T-cells, these results might seem counterintuitive, i.e. that it might have been expected that the golli KO animals would be more susceptible to EAE. Feng et al. (2006) postulated that, in view of the role of golli as an inhibitor of Ca^{++} uptake in T cells , its absence would lead to sustained increases in internal Ca^{++} levels, which would cause memory T cells to either die or to become anergic (i.e. no longer susceptible to further activation), thereby leading to EAE resistance. This hypothesis is currently under investigation.

Selective overexpression of the golli products of the MBP gene in oligodendrocytes

A transgenic mouse in which the J37 golli isoform is overexpressed specifically within OLs under the control of the classic MBP promoter has been engineered (Reyes et al., 2003; Martin et al., 2007; Jacobs et al., 2007). This mouse model, called JOE (for J37 golli OverExpressor), displays a phenotype that is unlike other mice in which myelin proteins have been overexpressed, such as the myelin PLP and DM20 proteins (Mastronardi et al., 1993; Anderson et al., 1998; Ikenaka and Kagawa, 1995; Readhead et al., 1994). While the homozygous JOE mice die prior to birth, the hemizygous mice are severely hypomyelinated until approximately two months of age. During this time they exhibit severe intention tremors from the onset of myelination that gradually abate by adulthood, although the tremors are evident when animals are placed under proprioceptive load. After ~P50, myelin begins to accumulate in the brain but never completely reaches control levels.

One of the most striking features of JOE hemizygous brain sections, immunostained with anti-golli peptide, is the presence of prominent, multi-processed cells scattered throughout presumptive white matter tracts and grey matter, which are never observed in wild type littermates. These cells stain positively for the early OL marker NG2 but are only weakly labeled by the more mature OL marker, PLP, and they are negative for neuronal and astrocytic markers (Jacobs et al., 2007). The morphology and the expression patterns of these large JOE cells suggest that they might be OL progenitors that fail to myelinate and/or are significantly delayed in their maturation.

Martin et al. (2007) examined the JOE mice using μMRI and visual evoked potentials (VEPs) in a longitudinal study encompassing the ages of severe tremors and their subsequent abatement. JOE transgenics and their unaffected siblings were examined from 21-75 days using in vivo VEPs and 3D T2-weighted μMRI. The μMRI data revealed global hypomyelination during the period of peak myelination (21–42 days), which was partially corrected at later ages (>60 days) in the JOE mice. The μMRI data correlated well with myelin staining of tissue sections, the transient intention tremor, and the VEPs. The study provided confirmation of the histological and immunocytochemical data using non-invasive techniques supporting a delay of CNS myelin development and persistent hypomyelination in JOE mice.

Paez et al. (2007) analyzed Ca^{++} responses in oligodendrocyte precursor cells (OPCs) isolated from golli KO mice using high extracellular K^{+} to activate Ca^{++} influx through VOCCs (Voltage Operated Calcium Channels). They also performed similar experiments in OL cell lines that were forced to overexpress golli proteins through transfection of the cells with golli constructs under the control of the CMV promoter. They found an enhanced Ca^{++} influx in golli overexpressing cells, which was the predicted result, and was opposite to the effects observed in golli KO OLs. As observed in golli KO OLs, the differences in the Ca^{++} response between control and golli overexpressing cells were lost when the cells were placed in a medium without Ca^{++} indicating that golli produced its effect through Ca^{++} influx rather than Ca^{++} release from internal stores. Confirmation of the involvement of VOCCs in the Ca^{++} influx was obtained from control experiments performed in the presence of the VOCC inhibitors Ni^{++} and Cd^{++}. Both Ni^{++} and Cd^{++} produced strong effects on the amplitude of the Ca^{++} influx in control and golli overexpressing cells, but the inhibitory effect of Cd^{++} was markedly more effective than Ni^{++}. Cd^{++} completely inhibited the golli-related increase in Ca^{++} influx resulting from high K^{+} stimulation, but a component of the golli-related enhancement of Ca^{++} uptake persisted in the presence of Ni^{++}.

Thus, the phenotype of the JOE mice is substantially different from that of the golli KO in terms of obvious clinical/behavioral disorders. Both exhibit hypomyelination, which appears to be region-specific in the case of the golli KO and more developmentally specific in the case of the JOE animals. Morphological signs of unusual OL-lineage cells are quite prominent in the JOE mice but absent in the golli KO animals. Finally, both share complementary Ca^{++} uptake abnormalities in OLs that are consistent with a positive regulatory role for golli on Ca^{++} uptake in these cells. The phenotypes also indicate that golli plays a role in regulating OL differentiation, process extension (and possibly migration) through regulation of Ca^{++} levels in OLs.

Emerging relevance of golli expression in pathology and disease

Since their original description in the mouse, many papers and EST databases have documented golli sequence orthologs, including such widely divergent species as Xenopus (NCBI, Accession # NP_001016592), bovine (Larsen et al., 1999) and chimpanzee (NCBI Accession #XP_001139339). There have been some interesting studies on the expression of the protein in human diseases. Golli proteins have been found to be upregulated in adult OPCs and microglia/macrophages around MS lesions (Filipovic et al, 2002). Filipovic and Zecevic (2005a,b) have postulated a role for golli, which is upregulated in microglia during inflammation, in the proliferation of OPCs in the brain. Moscarello et al. (2002) identified golli as a component of remyelination induced by treatment of demyelinating transgenic mice with Paclitaxel (Taxol). Recently, golli proteins were found to be components of prominent intranuclear inclusions that form in neurons of individuals with Fragile X-associated tremor/ataxia syndrome, a late onset neurodegenerative disorder caused by expansion of CGG repeats in a mutation of the Fragile X mental retardation gene (Iwahashi et al., 2006). These results confirm earlier findings localizing golli to the nuclei of some neurons and OLs (Landry et al., 1996), identification of a nuclear binding partner for golli (Fernandes et al., 2004), and suggest that golli may play an important, if as yet unknown, role in the nucleus. Other studies have uncovered up-regulation of golli expression as part of EST analysis or gene arrays to identify unique genes involved in pathological situations. For example, some years ago Glasgow et al. (2000) found that golli was one of twelve genes identified as being regulated by osmotic changes in magnocellular neurons of the supraoptic nucleus. Although puzzling to the authors at the time, the more recent findings that golli regulates Ca++ homeostasis (Feng et al., 2006) would make their results more understandable. Very recently, in a search for novel target genes related to Parkinson's disease (PD), Kim et al. (2007) compared ESTs from libraries made from human normal and PD substantia nigra. They found 21 upregulated genes to be differentially expressed in human PD tissues and/or in an MPTP-treated mouse model by quantitative real-time RT–PCR. Of these, they identified golli-MBP and MBP transcripts to be the most up-regulated of the genes and suggest that these be considered useful targets for elucidating the molecular mechanisms associated with PD. They found MBP/golli to be associated with cell death activity, which might be related to the fact that sustained elevation of Ca++ in cells can lead to their death (Jiang et al, 1994). Thus, these studies on golli in pathological conditions add to the growing evidence that suggests an important role for golli in the metabolism of many cells within the immune and nervous systems.

CONCLUSION

The preponderance of available evidence supports the conclusion that at least one of the functions of the golli proteins is to regulate Ca^{++} uptake in T-cells and in oligodendrocytes. In T-cells, the action of golli is on the store-operated, CRAC channels, and in oligodendrocytes, golli appears to act on multiple types of Ca^{++} channels including store-operated, voltage-operated and ligand-gated channels. Interestingly, golli serves as a negative modulator of Ca^{++} uptake in T-cells, and it serves as a positive modulator in oligodendrocytes. In T-cells, the consequence of this regulation by golli is to influence T-cell activation and possibly anergy. In oligodendrocytes, golli regulation of Ca^{++} homeostasis appears to influence process extension and possibly migration during development. An as yet unexplored further function of golli proteins is their potential role in the nucleus, where they have been shown capable of binding to at least one important transcription factor. Further work will undoubtedly elucidate the mechanisms by which the golli proteins act, both at the level of regulating Ca^{++} channels and subsequent signaling pathways; and possibly transcriptional events in the nucleus.

ACKNOWLEDGEMENTS

This work was supported by National Institutes of Health Grants NS23022 and NS33091; and by National Multiple Sclerosis Society Grant RG2693.

REFERENCES

Ahmed, M.A., Bamm, V.V., Harauz, G., Ladizhansky, V. (2007). The BG21 isoform of Golli myelin basic protein is intrinsically disordered with a highly flexible amino-terminal domain. *Biochemistry*, 46, 9700-12.

Alliot, F., Pessac, B. (1988). Macrophages express glial markers. *Biol Cell*, 63, 109-11.

Anderson, T.J., Schneider, A., Barrie, J.A., Klugmann, M., McCulloch, M.C., Kirkham, D., Kyriakides, E., Nave, K.A., Griffiths, I.R. (1998). Late-onset neurodegeneration in mice with increased dosage of the proteolipid protein gene. *J Comp Neurol*, 394, 506-19.

Bamm, V.V., Ahmed, M.A., Ladizhansky, V., Harauz, G. (2007). Purification and spectroscopic characterization of the recombinant BG21 isoform of murine golli myelin basic protein. *J Neurosci Res*, 85, 272-84.

Campagnoni, A.T., Skoff, R.P. (2001). The pathobiology of myelin mutants reveal novel biological functions of the MBP and PLP genes. *Brain Pathol*, 11, 74-91.

Campagnoni, A.T., Pribyl, T.M., Campagnoni, C.W., Kampf, K., Amur-Umarjee, S., Landry, C.F., Handley, V.W., Newman, S.L., Garbay, B., Kitamura, K. (1993). Structure and developmental regulation of Golli-mbp, a 105-kilobase gene that encompasses the myelin basic protein gene and is expressed in cells in the oligodendrocyte lineage in the brain. *J Biol Chem*, 268, 4930-8.

Feng, J.M., Hu, Y.K., Xie, L.H., Colwell, C.S., Shao, X.M., Sun, X.P., Chen, B., Tang, H., Campagnoni, A.T. (2006). Golli protein negatively regulates store depletion-induced calcium influx in T cells. *Immunity*, 24, 717-27.

Feng, J.M., Fernandes, A.O., Campagnoni, C.W., Hu, Y.H., Campagnoni, A.T. (2004). The golli-myelin basic protein negatively regulates signal transduction in T lymphocytes. *J Neuroimmunol*, 152, 57-66.

Feng, J.M., Givogri, I.M., Bongarzone, E.R., Campagnoni, C., Jacobs, E., Handley, V.W., Schonmann, V., Campagnoni, A.T. (2000). Thymocytes express the golli products of the myelin basic protein gene and levels of expression are stage dependent. *J Immunol,* 165, 5443-50.

Fernandes, A.O., Campagnoni, C.W., Kampf, K., Feng, J.M., Handley, V.W., Schonmann, V., Bongarzone, E.R., Reyes, S., Campagnoni, A.T. (2004). Identification of a protein that interacts with the golli-myelin basic protein and with nuclear LIM interactor in the nervous system. *J Neurosci Res*, 75, 461-71.

Filipović, R., Rakic, S., Zecevic, N. (2002). Expression of Golli proteins in adult human brain and multiple sclerosis lesions. *J Neuroimmunol*, 127, 1-12.

Filipović, R., Zecević, N. (2005a). Interaction between microglia and oligodendrocyte cell progenitors involves Golli proteins. *Ann N Y Acad Sci*, 1048, 166-74.

Filipović, R., Zecevic, N. (2005b). Lipopolysaccharide affects Golli expression and promotes proliferation of oligodendrocyte progenitors. *Glia*, 49, 457-66.

Fritz, R.B., and Kalvakolanu, I. (1995). Thymic expression of the golli-myelin basic protein gene in the SJL/J mouse. *J Neuroimmunol*, 57, 93-99.

Fruttiger, M., Montag, D., Schachner, M., Martini, R. (1995). Crucial role for the myelin-associated glycoprotein in the maintenance of axon-myelin integrity. Eur *J Neurosci*, 7, 511-5.

Glasgow, E., Murase, T., Zhang, B., Verbalis, J.G., Gainer, H. (2000). Gene expression in the rat supraoptic nucleus induced by chronic hyperosmolality versus hyposmolality. Am *J Physiol Regul Integr Comp Physiol*, 279, R1239-50.

Grima, B., Zelenika, D., Javoy-Agid, F., Pessac, B. (1994). Identification of new human myelin basic protein transcripts in the immune and central nervous systems. *Neurobiol Dis*, 1, 61-6.

Grima, B., Zelenika, Pessac, B. (1992). A novel transcript overlapping the myelin basic protein gene. *J. Neurochem*, 59, 2318–2323.

Huseby, E.S., Goverman, J. (2000). Tolerating the nervous system: a delicate balance. *J Exp Med*, 191, 757-60.

Ikenaka, K., Kagawa, T. (1995). Transgenic systems in studying myelin gene expression. *Dev Neurosci*, 17, 127-36.

Iwahashi, C.K., Yasui, D.H., An, H.J., Greco, C.M., Tassone, F., Nannen, K., Babineau, B., Lebrilla, C.B., Hagerman, R.J., Hagerman, P.J. (2006). Protein composition of the intranuclear inclusions of FXTAS. *Brain*, 129, 256-71.

Jacobs, E.C., Pribyl, T.M., Feng, J.M., Kampf, K., Spreur, V., Campagnoni, C., Colwell, C.S., Reyes, S.D., Martin, M., Handley, V., Hiltner, T.D. (2005). Region-specific myelin pathology in mice lacking the golli products of the myelin basic protein gene. *J Neurosci*, 25, 7004-13.

Jacobs, E. (2005). Genetic alterations in the mouse myelin basic proteins result in a range of dysmyelinating disorders. *J Neurol Sci,* 228, 195-197.

Jacobs, E., Reyes, S., Campagnoni, C., Givogri, I., Kampf, K., Handley, V., Spreuer, S., Fisher, R., Campagnoni, A. (2007). Comparison of two transgenic lines overexpressing the J37 golli protein in oligodendrocytes. Program No. 458.26 Abstract Viewer/Itinerary Planner. San Diego, CA, Society for Neuroscience, Online.

Kalwy, S., Marty, M.C., Bausero, P., Pessac, B. (1998). Myelin basic protein-related proteins in mouse brain and immune tissues. *J Neurochem*, 70, 435-8. Erratum in, *J Neurochem,* 1998 70, 1777.

Kaur, J., Libich, D.S., Campagnoni, C.W., Wood, D.D., Moscarello, M.A., Campagnoni, A.T., Harauz, G. (2003). Expression and properties of the recombinant murine Golli-myelin basic protein isoform J37. *J Neurosci Res*, 71, 777-84.

Kim, J.M., Lee, K.H., Jeon, Y.J., Oh, J.H., Jeong, S.Y., Song, I.S., Kim, J.M., Lee, D.S., Kim, N.S. (2007). Identification of genes related to Parkinson's disease using expressed sequence tags. *DNA Res*, 13, 275-86.

Kitamura, K., Newman, S.L., Campagnoni, C.W., Verdi, J.M., Mohandas, T., Handley, V.W., Campagnoni, A.T. (1990). Expression of a novel transcript of the myelin basic protein gene. *J Neurochem,* 54, 2032-41.

Klugmann, M., Schwab, M.H., Pühlhofer, A., Schneider, A., Zimmermann, F., Griffiths, I.R., Nave, K.A. (1997). Assembly of CNS myelin in the absence of proteolipid protein. *Neuron*, 18, 59-70.

Landry, C.F., Ellison, J., Skinner, E., Campagnoni, A.T. (1997). Golli-MBP proteins mark the earliest stages of fiber extension and terminal arboration in the mouse peripheral nervous system. *J Neurosci Res*, 50, 265-71.

Landry, C.F., Ellison, J.A., Pribyl, T.M., Campagnoni, C., Kampf, K., Campagnoni, A.T. (1996). Myelin basic protein gene expression in neurons: developmental and regional changes in protein targeting within neuronal nuclei, cell bodies, and processes. *J Neurosci*, 16, 2452-62.

Larsen, N.J., Helen Hayes, H., Bishop, M., Davis, S.K., Taylor, J.F., Kirkpatrick, B.W. (1999). A comparative linkage and physical map of bovine chromosome 24 with human chromosome 18. *Mammalian Genome*, 10, 482-487.

Liu, H., MacKenzie-Graham, A.J., Palaszynski, K., Liva, S., Voskuhl, R.R. (2001). "Classic" myelin basic proteins are expressed in lymphoid tissue macrophages. *J Neuroimmunol*, 116, 83-93.

Martin, M., Reyes, S.D., Hiltner, T.D., Givogri, M.I., Tyszka, J.M., Fisher, R., Campagnoni, A.T., Fraser, S.E., Jacobs, R.E., Readhead, C. (2007). T(2)-weighted microMRI and evoked potential of the visual system measurements during the development of hypomyelinated transgenic mice. *Neurochem Res*, 32, 159-65.

Marty, M.C., Alliot, F., Rutin, J., Fritz, R., Trisler, D., Pessac, B. (2002). The myelin basic protein gene is expressed in differentiated blood cell lineages and in hemopoietic progenitors. *Proc Natl Acad Sci*, USA, 99, 8856-61.

Mastronardi, F.G., Ackerley, C.A., Arsenault, L., Roots, B.I., Moscarello, M.A. (1993). Demyelination in a transgenic mouse: a model for multiple sclerosis. J *Neurosci Res,* 36, 315-24

Mateo Paz Soldan, M., Warrington, A.E., Bieber, A.J., Bogoljub, C., Van Keulen, V., Pease L,R., Rodriguez, M. (2003). Remyelination-promoting antibodies activate distinct Ca^{2+} influx pathways in astrocytes and oligodendrocytes: relationship to the mechanism of myelin repair. *Mol Cell Neurosci*, 22, 14-24.

Maverakis, E., Mendoza, R., Southwood, S., Raja-Gabaglia, C., Abromson-Leeman, S., Campagnoni, A. T., Sette, A., and Sercarz, E. E. (2000). Immunogenicity of self antigens is unrelated to MHC-binding affinity T-cell determinant structure of golli-MBP in the BALB/c mouse. *J Autoimmun, 15,* 315-22.

Moscarello, M.A., Mak, B., Nguyen, T.A., Wood, D.D., Mastronardi, F., Ludwin, S.K. (2002). Paclitaxel (Taxol) attenuates clinical disease in a spontaneously demyelinating transgenic mouse and induces remyelination. *Mult Scler*, 8, 130-8.

Newman, S., Kitamura, K., Campagnoni, A.T. (1987). Identification of a cDNA coding for a fifth form of myelin basic protein in mouse. *Proc Natl Acad Sci*, USA, 84, 886-890.

Noble, M., Arhin, A., Gass, D., Mayer-Pröschel, M. (2003). The cortical ancestry of oligodendrocytes: common principles and novel features. *Dev Neurosci*, 25, 217-33.

Paez, P.M., Spreuer, V., Handley, V., Feng, J.M., Campagnoni, C., Campagnoni, A.T. (2007). Increased expression of golli myelin basic proteins enhances calcium influx into oligodendroglial cells. *J Neurosci,* 27, 12690-9.

Pedraza, L., Fidler, L., Staugaitis, S.M., Colman, D.R. (1997). The active transport of myelin basic protein into the nucleus suggests a regulatory role in myelination. *Neuron*, 18, 579-89.

Polverini, E., Boggs, J.M., Bates, I.R., Harauz, G., and Cavatorta, P. (2004). Electron paramagnetic resonance spectroscopy and molecular modeling of the interaction of myelin basic protein (MBP) with calmodulin (CaM)--diversity and conformational adaptability of MBP CaM-targets. *J Struct Biol,* 148, 353-369

Pribyl, T.M., Campagnoni, C.W., Kampf, K., Ellison, J.A., Landry, C.F., Kashima, T., McMahon, J., Campagnoni, A.T. (1996a). Expression of the myelin basic protein gene locus in neurons and oligodendrocytes in the human fetal central nervous system. *J Comp Neurol,* 374, 342-53.

Pribyl, T.M., Campagnoni, C.W., Kampf, K., Handley, V.W., Campagnoni, A.T. (1996b). The major myelin protein genes are expressed in the human thymus. *J Neurosci Res*, 45, 812-9.

Pribyl, T.M., Campagnoni, C.W., Kampf, K., Kashima, T., Handley, V.W., McMahon, J., Campagnoni, A.T. (1993). The human myelin basic protein gene is included within a 179-kilobase transcription unit: expression in the immune and central nervous systems. *Proc Natl Acad Sci*, USA, 90, 10695-9.

Readhead, C., Schneider, A., Griffiths, I., Nave, K.A. (1994). Premature arrest of myelin formation in transgenic mice with increased proteolipid protein gene dosage. *Neuron*, 12, 583-95.

Reyes, S.D., Campagnoni, A.T. (2002). Two separate domains in the golli myelin basic proteins are responsible for nuclear targeting and process extension in transfected cells. *J Neurosci Res*, 69, 587-96.

Reyes, S.D., Givogri, M.I., Campagnoni, C.W., Handley, V., Schonmann, V., Fisher, R., Campagnoni, A.T. (2003). Overexpression of the golli J37 isoform transgenic mice results in CNS hypomyelination. Program No. 141.17. Abstract Viewer/Itinerary Planner. Washington, DC, Society for Neuroscience, Online.

Roth, H.J., Kronquist, K.E., Kerlero de Rosbo, N., Crandall, B.F., Campagnoni, A.T. (1987). Evidence for the expression of four myelin basic protein variants in the developing human spinal cord through cDNA cloning. *J Neurosci Res*, 17, 321-8.

Schoenherr, C.J., Anderson, D.J. (1995). The neuron-restrictive silencer factor (NRSF): a coordinate repressor of multiple neuron-specific genes. *Science*, 267, 1360-3.

Simpson, P.B., Armstrong, R.C. (1999). Intracellular signals and cytoskeletal elements involved in oligodendrocyte progenitor migration. *Glia*, 26, 22-35.

Simpson, P.B., Mehotra, S., Lange, G.D., Russell, J.T. (1997). High density distribution of endoplasmic reticulum proteins and mitochondria at specialized Ca2+ release sites in oligodendrocyte processes. *J Biol Chem*, 272, 22654-61.

Soliven, B. (2001). Calcium signalling in cells of oligodendroglial lineage. *Microsc Res Tech*, 52, 672-9.

Thompson, J., Lepikhova, T., Teixido-Travesa, N., Whitehead, M.A., Palvimo, J.J., Jänne, O.A. (2006). Small carboxyl-terminal domain phosphatase 2 attenuates androgen-dependent transcription. *EMBO J,* 25, 2757-67.

Tosic, M., Rakic, S., Matthieu, J.M., Zecevic, N. (2002). Identification of Golli and myelin basic proteins in human brain during early development. *Glia,* 37, 219-28.

Voskuhl, R.R., Pribyl, T.M., Kampf, K., Handley, V., Liu, H.B., Feng, J., Campagnoni, C.W., Soldan, S.S., Messing, A., Campagnoni, A.T. (2003). Experimental autoimmune encephalomyelitis relapses are reduced in heterozygous golli MBP knockout mice. *J Neuroimmunol*, 139, 44-50.

Wolf, M.K., Billings-Gagliardi, S. (1984). CNS hypomyelinated mutant mice (jimpy, shiverer, quaking): in vitro evidence for primary oligodendrocyte defects. *Adv Exp Med Biol*, 181, 115-33.

Yeo, M., Lee, S.K., Lee, B., Ruiz, E.C., Pfaff, S.L., Gill, G.N. (2005). Small CTD phosphatases function in silencing neuronal gene expression. *Science*, 307, 596-600.

Yeo, M., Lin, P.S., Dahmus, M.E., Gill, G.N. (2003). A novel RNA polymerase II C-terminal domain phosphatase that preferentially dephosphorylates serine 5. *J Biol Chem*, 278, 26078-85.

Yoo, A.S., Krieger, C., Kim, S.U. (1999). Process extension and intracellular Ca2+ in cultured murine oligodendrocytes. *Brain Res*, 827, 19-27.

Zelenika, D., Grima, B., Pessac, B. (1993). A new family of transcripts of the myelin basic protein gene: expression in brain and in immune system. *J Neurochem*, 60, 1574-7.

In: Myelin Basic Protein
Editor: Joan M. Boggs
ISBN: 978-1-60456-699-4

Chapter II

POSTTRANSLATIONAL MODIFICATIONS OF MYELIN BASIC PROTEINS

Robert Zand*

ABSTRACT

Proteins express an extensive degree of versatility not derived solely from their primary structures. The recognition that the enzymatic introduction of prosthetic groups such as phosphate, carbohydrate, sulfate, and methyl groups, and internal backbone changes such as conversion of arginine residues to citrulline, aspartate to isoaspartate, and glutamine to glutamic acid can cause a myriad of changes involving backbone charge, conformational changes and changes in the tertiary structure of the protein that results in conferrance of a specific biological activity to the protein. Myelin basic protein (MBP) from the central nervous system undergoes an extensive number of posttranslational modifications, the biological functions of which have not been completely identified. These modifications lead to the production of what has been termed charge isomers. In the mammalian systems that have been examined to date there are typically eight charge isomers. In contrast, in MBP obtained from submammalian species, there are typically four charge isomers. The introduction of a phosphate group typically results in the addition of one negative charge per phosphate. Conversely, the conversion of an arginine residue to a citrulline residue removes a single positive charge for each such modification. It is apparent that the introduction of such charges on the backbone of the native protein can result in conformational and tertiary structural changes to the parent molecule. Technological advances in the protein separation and purification as well as the high degree of analytical identification of protein modifications using mass spectrometric methods has resulted in the identification of many *in vivo* posttranslational

* Department of Biophysics and Department of Biological Chemistry, University of Michigan, 930 N. University Ave., Ann Arbor, MI 48109-1055. Telephone: 1-734-764-5138; Fax: 1-734-764-3323; E-mail: rzand@umich.edu

changes in MBP. The present chapter is a summary of the modifications that have been identified in the MBP from various species.

Introduction

The ability of a protein to extend its biological and biochemical functions is accomplished in nature by the use of posttranslational modifications. The protein as encoded by its gene and expressed by the machinery of protein synthesis lacks any covalent modification to the synthesized protein and its amino acid components that can influence its biological and biochemical function. In many proteins the lack of such modifications makes the protein biologically and biochemically nonfunctional. The introduction of covalently attached groups such as acetyl, phosphate, methyl, carbohydrate, sulfate, or modifications such as deamidation and the conversion of the amino acid arginine to the amino acid citrulline (deimination) allows the protein to assume a function or functions that the unmodified form does not possess. This diversity of function is exhibited by many different proteins. Myelin basic protein (MBP) enjoys a unique status among proteins that are posttranslationally modified. This is the result of the diversity of posttranslational modifications and the number of each type of modification. MBP is known to undergo a number of posttranslational modifications whose biochemical and biological roles are not known or poorly understood. Some protein posttranslational modifications are reversible while others are not. The classic example is phosphorylation-dephosphorylation which controls many essential metabolic processes. The regulation of glycogen metabolism by the enzyme glycogen phosphorylase was elucidated by the studies of Fisher and Krebs (1955), Cori and Green (1943a,b), Cori and Cori (1940) and Sutherland and Cori (1951). This was the first example identified of the regulation of enzyme activity by phosphorylation-dephosphorylation.

Five posttranslational modifications have been identified in MBP, and these are: acetylation, methylation, phosphorylation, deimination (citrullination), and deamidation. Some posttranslational modifications have the potential to modify the charge present on the parent protein. The covalent attachment of a phosphate group to the parent protein adds a negative charge to the molecule. The conversion of an arginine residue to a citrulline residue removes a positive charge from the protein. In MBP, the introduction of covalent phosphate or citrullination results in the formation of "charge isomers" in which the electrophoretic pattern of the parent protein can reveal the presence of multiple bands. These bands reflect the differences in the net charge of the protein and not differences in the primary structure. Consequently, they are referred to as charge isomers. They can be separated by chromatography on a CM52 column into the components C1, C2, C3, C4, C5 and C8 in order of decreasing net positive charge.

When the MBP protein is isolated it contains each of the five modifications cited above. In addition, the MBP molecule contains receptor sites for binding GTP (Chan et.al., 1988), N-acetylgalactose (Hagopian et al., 1971), and ADP-ribose (Boulias and Moscarello, 1994). None of the latter three modifications for which receptor sites are present have been identified as actually occurring in the isolated CNS protein.

The objective of this review is to provide the reader with a list of posttranslational modifications that have been identified in the MBP's from various mammalian and submammalian animals. However, it is not intended to assign specific biochemical or physiological functions to these modifications. It is proposed that many of these posttranslational modifications are involved in as yet unassigned signal transduction pathways.

ACETYLATION

Every MBP protein that has been isolated and sequenced, to date, has its N-terminal amino acid acetylated. In addition, Moscarello et. al. (1992) have reported that human MBP also has acyl groups of butyl, hexyl and octyl attached to the N-terminal residue of the 18.5 kDa isoform. However, such diversity of acyl modification has not been reported for other MBP's.

The attachment of an acetyl group to the amino terminus is catalyzed by the enzyme amino-terminal acetyltransferase. This enzyme transfers the acetyl group from acetyl-CoA to the amino group of the N-terminal amino acid of the protein. The most common N-terminal amino acid that is acetylated is either serine or alanine. The N-terminal sequences of the human, bovine, chicken and dogfish MBP's that have been analyzed by mass spectrometry are all acetylated, as are the MBP's for rabbit, guinea pig, porcine, mouse, rat and other mammalian and submammalian species that were sequenced by more traditional methods. Acetylated MBP N-terminal sequences all have the amino acid sequence of ASQKRPSQR for human and AAQRPSR for bovine MBP. The sequence for the N-terminal chicken MBP is ASQKRSSFR. To date, throughout the MBP family of proteins from mammalian to sub-mammalian, the N-terminal residue has been found to be acetylated.

In contrast to histones, no acetylated epsilion-amino residues in MBP lysine residues have been found. The biological function for the acetylation of the N-terminal residue has not been established. It has been suggested that the acetylated amino acid that occurs in many proteins stabilizes the molecule and may also enhance the biological activity of the molecule, whatever that activity may be.

The acetylation of N terminal serine or alanine residues in MBP can occur after the cotranslational hydrolytic removal of the N-terminal methionine in concert with the action of the N-acetyltransferase enzyme (Polevoda and Sherman, 2000, 2003). The sequence of reactions for N-terminal acetylation can be viewed as:

H_3N^+Methionine + Methionine aminopeptidase $\longrightarrow$ Met + H_3N^+-amino acid-protein

N-acetyltransferase + acetyl-CoA + protein $\longrightarrow$ N-acetyl-amino acid-protein + CoA.

METHYLATION

The methylation of amino acids in proteins provides a route for the diversification of functions from the same protein backbone. The covalent attachment of methyl groups to an arginine residue in a protein can result in the formation of monomethyl-, symmetric dimethyl- and /or asymmetric dimethyl-arginine residues. The covalent attachment of methyl groups to the epsilion-amino function of a lysine residue is also known to occur in many proteins such as histones and cytochrome c, but has not been detected in MBP. For reviews of MBP methylation the following articles should be consulted (Kim et al., 1997; Paik and Kim, 1990). Methylation of the arginine residue in MBP is catalyzed by the enzyme protein methyl-arginine N-methylase I that utilizes S-adenosyl-L-methionine (SAM, or AdoMet) as the methyl donor (EC 2.1.123) (Scheme I). Demethylated AdoMet is also called S-adenosyl-L-homocysteine. The enzyme may have a number of subtypes, but not all have been isolated and characterized. The enzyme that methylates MBP is not the same enzyme that methylates histones. The enzyme that is specific for the methylation of histones has a molecular weight of 275 kDa. This enzyme consists of subunits with molecular weights of 110 and 75 kDa, whereas the molecular weight of the MBP-specific enzyme is 500 kDa with subunits of approximately 100 and 75 kDa .

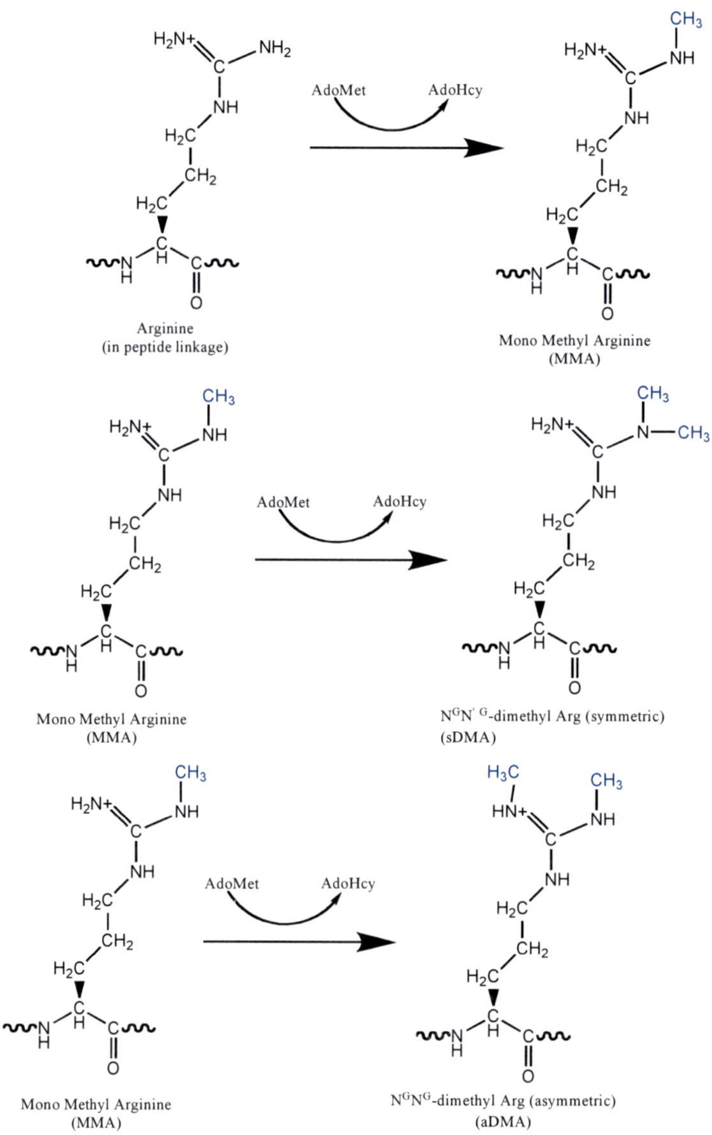

Scheme I. The methylation of the human and dimethyl arginine $N^G N'^G$.

The presence of asymmetric dimethylarginine N^GN^G in bovine MBP was reported by Brostoff and Eylar (1971). The R^{107} of bovine 18.54 kDa MBP yields primarily monomethyl arginine N^G and symmetric dimethyl arginine, $N^G N^G$ residues (Baldwin and Carnegie, 1971). Attempts to confirm the presence of the asymmetric dimethyl arginine isomer have not been successful. The *in vivo* methylation of chicken MBP was investigated by Small and Carnegie (1982), who were unable to detect the presence of unsymmetric dimethylation in chicken MBP (Young and Grynspan, 1987). These investigators reported that for bovine MBP the amount of monomethylated Arg was 0.36 mole/100mol and the symmetric dimethylarginine was 0.18 mole/100 moles of amino acid residues, with no traces of the unsymmetric dimethylated amino acid. Eylar et al. (1971) had reported that traces of the unsymmetric isomer were present. This result has not found support in the later literature (Brostoff et al., 1972)

No extensive *in vivo* studies have been reported for the amounts of methylation in mammalian-derived MBP, the ratio of methylated to unmethylated arginine has been reported to vary from species to species (Diebler and Martenson, 1973; Dunkley and Carnegie, 1974). The extent of methylation has been suggested to be under genetic control within each species. This does not appear to be a supportable conclusion since the function(s) of methylation would appear to be the same in the MBP of all animals. It seems unlikely that genetic control of an enzyme-initiated reaction would operate for a reaction whose substrate (MBP) and product (methylated arginine residues) appear to be virtually identical for all vertebrates and the product of that reaction is important for the biological function of the membrane that facilitates the nerve impulse.

Most studies of posttranslational methylation of MBP prior to 2000 were accomplished using HPLC and conventional amino acid analysis. More recent studies have utilized capillary electrophoresis, special HPLC columns and packings, and various forms of mass spectrometry that permit greater sensitivity and resolution of the modified amino acid residues. Enzymes have been isolated that demethylate lysines in histones. No enzymes are known to demethylate arginines in MBP (Holbert and Marmorstein, 2005)

The methylation of MBP charge isomers from normal human brain and from MS brain has been reported (Kim et al., 2003). Differences in the number of sites that were methylated were found. For MBP charge isomers from normal human brain, the following ratios of methylated arginine/nonmethylated arginine residues were found:

Normal MBP, mono-methyl Arg:	C1 1.6, C2 1.31, C3 0.98, C4 1.23, C5 0.85, C8 1.14
Dimethyl Arg:	C1 1.8, C2 2.40, C3 1.68, C4 2.44, C5 2.14, C8 1.90

For MBP charge isomers obtained from MS brain, the ratios of methylated arginine/nonmethylated arginine residues are listed below:

MS MBP, mono-methyl Arg:	C1 1.1, C2 0.67, C3 0.69, C4 2.47, C5 2.49, C8 1.26
Dimethyl Arg:	C1 1.1, C2 0.97, C3 1.28, C4 4.20, C5 4.29, C8 1.94

PHOSPHORYLATION

The posttranslational modification that covalently attaches phosphate groups to serine and threonine residues in MBP is probably linked to the action of more than one kinase enzyme. A review of MBP phosphorylation *in vivo* and *in vitro* by Ulmer (1988) appeared in 1988. The bovine MBP charge isomer C1 is not phosphorylated. For the other charge isomers, C3 is phosphorylated at Thr 97 and Ser 164; isomer C4 is phosphorylated at Ser 54,Thr 97, and Ser160; C5 is phosphorylated at Ser 7, Ser 54, Thr 97, and Ser 164; C6 is phosphorylated on Ser 7, Ser 54, Thr 97, Ser 160 and Ser 164 (Zand et al., 1998) (Table 1). In human MBP, the sites of phosphorylation parallel the sites found in bovine MBP (Kim et al., 2003).

In chicken MBP (Kim et al., 2008), only four charge isomer peaks were found on CMC chromatography. The number of charge isomers is reduced on going from mammalian to non-mammalian species. In mammals the number of charge isomers is eight while the number of charge isomers is reduced to four in submammalian species. In chicken MBP, the C1 isomer had no phosphorylated sites (Table 2). The C2 isomer had ten phosphorylated sites and the C3 isomer had eight phosphorylated sites. Chicken isomers C2 and C3 do not share any common sites of phosphorylation with dogfish MBP (Zand et al., 2001); however, they share common serine and threonine sites with mammalian MBP. Phosphorylation of MBP by kinases *in vivo* was reported by Miyamoto and Kakiuchi (1974) and Martenson et al. (1983).

Deamidation of glutamine at residues 103 and 147

Non-enzymatic deamidation of MBP occurs on the glutamine residues and asparagine residues of many proteins. The loss of the amide group depends on pH, ionic strength, solvent and temperature. In a paper by Robinson and Robinson (2004) it was suggested that deamidation of these residues is linked to the rates of turnover of the proteins and function as molecular clocks. For the MBP proteins, the factors that facilitate deamidation apparently are not optimal so that no asparagine residues are deamidated but some fraction of the glutamine residues are deamidated. In human C1and C2, Gln 103, 121, and 147 and in C3, Gln 81, 103, 121, and 147 are deamidated. Gln 8. 81, 103, 121 and 147 are deamidated in C4 and C5 (Kim et al., 2003). A comprehensive treatment of deamidation and of glutamine and aspaginyl residues in proteins and peptides can be found in the monograph by Robinson and Robinson (2004). Robinson and coworkers (Robinson and Rudd, 1974; Robinson, 1974; Robinson et al., 1970) have suggested that the deamidation of proteins and peptides function as control mechanisms for the rates of protein turnover, aging and development.

Deimination of arginine residues (citrullination)

The amino acid citrulline is rarely found in the sequence of amino acids that comprise the primary structure of a protein or peptide. It is not coded for among the genetic code used to generate the primary structure of a protein. How then is it incorporated into that primary structure and what is its role? The finding of Citrulline in MBP was first reported by Finch et al. (1971) and subsequently located in the C-8 fraction of MBP by Wood and Moscarello (1989) (see Chapter III). The deimination of the free amino acid arginine, by the enzyme NO synthase yields NO and the amino acid citrulline. In contrast, the deimination of arginine residues in a protein or peptide by the enzyme peptidylarginine deiminase yields citrulline residues and ammonia. This modification results in a decrease of one positive charge for each such residue modified in the protein and an increase in the molecular weight by 0.98402 Da (monoisotopic). A number of pathological processes have been correlated with the conversion of arginine to citrulline. Among such correlated illnesses are rheumatoid arthritis, Alzheimer's disease, multiple sclerosis, MS regulation of gene expression and epithelial terminal differentiation. In MS, the substitution of citrulline for arginine was reported to increase the severity of the pathology. The 18.5kDa charge isomer that has six citrulline residues is reported to be elevated in demyelinating pathologies (Moscarello et al., 1994).

The number of citrullinated residues in a protein is dependent on the methylation of the arginine residues in that protein. Under normal circumstances it appears that six citrullinated residues are present in human and bovine MBP. In chicken MBP, two citrullinated amino acids were found in MBP charge isomer C3 (Kim et al., 2008). Charge isomer C2 contained a single citrulline residue at R165 and the charge isomer C1 did not contain any citrulline. These results correspond to the charge isomer distribution in chicken MBP, where C1 is the most positively charged isomer. The mass spectral analysis of C3 confirmed that the conversion of R41 to citrulline occurred in the sequence 32-47 and R165 in the sequence 155-172. Only peptide 155-165 was found to contain citrulline in C2, and R165 was considered to be citrullinated instead of arginine 161. No evidence was found to support the demination of arginine 161 to citrulline.

CONCLUSION

The residues modified in different MBP charge isomers isolated from cow, human, chicken, and dogfish are shown in Tables 1 and 2. The specific reason for the diversity of posttranslational modifications associated with MBP is not known. Perhaps the introduction of posttranslational additions to the parent MBP may facilitate the creation of autoantigens resulting in autoimmune pathology as a secondary component to the main biological function. Another possibility is that these posttranslational modifications play a role in the signal pathway(s) of the neuronal biochemistry in the brain (see Chapters VIII and IX). These actions would be in addition to the role long postulated as being the major function of MBP in maintaining the tight wrapping of the concentric myelin membrane around the nerve axon.

The number of charge isomers of MBP undergoes an abrupt change in going from mammalian animals to submammalian animals. The number of charge isomers in human, bovine and other mammalian species has been reported to include eight forms. In chicken, the number was found to be four (Kim et al., 2008). Similarly the dogfish had only four and the alligator revealed only four charge isomers. It appears that as one descends the evolutionary scale the number of charge isomers is reduced from eight to four. This reduction in the number of charge isomers may be reflected in the reduction in the antigenicity of the MBP

Table 1. Posttranslational Changes in Human and Bovine MBP Determined by Mass Spectrometry

Residues phosphorylated in human MBP charge isomers (Kim et al., 2003).
(0 = none; + indicates at least partially phosphorylated)

Residue	C1	C2	C3	C4	C5	C8
Ser 7	0	0	0	+	+	+
Ser 12	0	0	0	+	+	+
Ser 19	+	+	+	+	+	0
Ser 56	+	+	0	+	+	0
Thr 95	+	0	0	0	0	0
Thr 98	0	0	+	+	+	+
Ser 115	0	0	+	+	+	+
Ser 151	+	+	+	+	+	0
Ser163/165	0	+	+	+	+	+

Modifications detected in bovine MBP charge isomers (Zand et al., 1998)
C1 unmodified
C2 deamidated Gln 146
C3 phosphorylated Thr 97, Ser 164
C4 phosphorylated Ser 54, Thr 97, Ser 160
C5 phosphorylated Ser 7, Ser 54, Thr 97, Ser 164
C6 phosphorylated Ser 7, Ser 54, Thr 97, Ser 160, Ser 164

Table 2. Residues Phosphorylated in Chicken and Dogfish MBP Determined by Mass Spectrometry

Residues phosphorylated in Chicken MBP charge isomers (Kim et al., 2008).
(0 = none; + indicates at least partially phosphorylated)

	C1	C2	C3
Ser 7	0	+	+
Ser 18	0	+	0
Ser 33	0	+	+
Ser 64	0	+	+
Ser 73	0	+	0
Thr 96	0	+	+
Ser 113	0	+	+
Ser 141	0	+	+
Ser 164	0	+	+
Ser 168 0	+	+	

Residues phosphorylated in Dogfish (lamprey) MBP charge isomers (Zand et al., 2001)
C1 - phosphorylated Ser 72, Ser 83, Ser 120 or 121
C2 - phosphorylated Ser 72, Ser 83; Ser 134,138 and 139 contain two or three phosphate groups
C3 - phosphorylated Ser 72, Ser 83, Ser 120 or 121

from submammalian animals and its ability to induce experimental allergic encephalomyelitis (EAE) in the usual experimental animals used for this purpose. MBP from these submammalian animals lacks the triprolyl sequence and the phosphorylated threonine residue that precedes this sequence. This domain may be the antigenic site that contributes to the induction of EAE. The MBP from shark and carp was reported by Martenson et al. (1972) not to induce EAE in guinea pigs and Lewis rats. Agrawal et. al. (1982) reported that that dogfish MBP was ineffective in inducing EAE in guinea pigs. Mixtures of whole brain and spinal cord from turtles, snakes and frogs were reported by Patterson (1957) to be ineffective in inducing EAE.

REFERENCES

Agrawal, H.C., O'Connell, K., Randle, C.L., and Agrawal, D. (1982). Phosphorylation of four basic proteins of rat brain myelin. *Biochem. J*, 201, 39-47.

Baldwin, G.S., and Carnegie, E.R. (1971). Isolation and partial characterization of methylated arginines from the encephalitogenic basic protein of myelin. *Biochem. J,* 123, 69

Baldwin, G.S., and Carnegie, E.R. (1971). Specific enzymic methylation of an arginine in the experimental allergic encephalomyelitis protein from human myelin. *Science*, 171, 579-581.

Bannister, A.J., and Kouzarides, T. (2005). Reversing histone methylation. *Nature*, 439, 1103-1106.

Boulias, C., and Moscarello, M.A. (1994). ADP-ribosylation of human myelin basic protein. *J. Neurochem*, 20, 1269-1277.

Boulias, C., Mastronardi, F.G., and Moscarello, M.A. (1995). ADP-ribosyltranferase activity in myelin membranes isolated from human brain. *Neurochem. Res*, 63, 351-359.

Brostoff, S.W., and Eylar, E.H. (1971). Localization of the methylated arginine in the A-1 protein from myelin. *Proc. Nat. Acad. Sci.* (USA), 68, 765-769.

Brostoff, S.W., and Eylar, E.H. (1972). The proposed amino acid sequence of the P1 protein of rabbit sciatic nerve myelin. *Arch. Biochem, Biophys*, 153, 590-598.

Brostoff, S., Rosegay, A., and Vandenheuval, J.A. (1972). Identification of NG,N'G-dimethylarginine and NG,NG-dimethylarginine in the basic A1 protein from bovine myelin. *Arch. Biochem, Biophys*, 148, 156-60.

Chan, K.F.J., Stoner, G.L., Hashim, G.A., and Huang, K.P. (1986). Phosphorylation at Thr 98 and Ser 165. *Biochem, Biophys. Res. Commun*, 1343, 1388-1394.

Chan, C.K., Ramwani, J., and Moscarello, M.A. (1988). Myelin basic protein binds GTP at a single site in the N-terminus, *Biochem, Biophys. Res. Commun*, 152,1468-1473.

Cori, G.T., and Cori, C.F. (1940). The kinetics of the enzymatic synthesis of glycogen from glucose-1-phosphate. *J Biol Chem*, 135(2), 733-56.

Cori, G., and Green, A. (1943). Crystalline muscle phorylase II. Prosthetic group. *J. Biol. Chem.* 151, 31-38.

Cori, C., Cori, G., and Green, A. (1943). Crystalline muscle phosphorylase phosphorylase III. Kinetics. *J. Biol. Chem*, 151, 39-55.

Deibler, G.E., Martenson, R.E., Kramer, A.J., and Kies, M.W. (1975). The contribution of phosphorylation and loss of COOH-terminal arginine to the microheterogeneity of myelin basic protein. *J. Biol. Chem*, 250, 7931-8.

Dunkley, P.R., Carnegie, P.R. (1974). Amino acid sequence of the smaller basic protein from rat brain myelin. *Biochem. J*, 141, 243-55.

Eylar, E.H., Brostoff, S., Hashim, G., Caccam, J., and Burnett, P. (1971). Basic A1 protein of the myelin membrane. The complete amino acid sequence. *J. Biol. Chem*, 246, 5770-5784.

Finch, P.R., Wood, D.D., and Moscarello, M.A. (1971). The presence of citrulline in a myelin protein fraction. FEBS Lett. 14, 145-148.

Fisher, E.H., and Krebs, E.G. (1955). Conversion of phosphorylase b to phosphorylase a in muscle extracts. *J. Biol. Chem*, 216, 121-132.

Hagopian, A., Westfall, F.C., Whitehead, J.S., and Eylar, E.H. (1971). Glycosylation of the A1 protein from myelin by a polypeptide N-acetylgalactosaminyltransferase, *J. Biol. Chem*, 246, 2519-2523.

Holbert, M.A., and Marmorstein, R. (2005). Structure and activity of enzymes that remove histone modifications. *Curr. Opin. Struct. Biol*, 15, 673-680.

Kim, S., Lim, I.K., Park, G.H., and Paik, W.K. (1997). Biological methylation of myelin basic potein: enzymology and biological significance. *Int. J. Biochem. Cell. Biol*, 29,743-751.

Kim, S., Chanderkar, L.K., and Gosh, S.K. (1990). In W.K. Paik and S. Kim (Eds.), *Protein Methylation* (pp. 77-95). Boca Raton, FL, CRC Press.

Kim, J.K., Mastronardi, F., Wood, D.D., Lubman, D.M., Zand, R., and Moscarello, M.A. (2003). Multiple Sclerosis, an important role for post-translational modificatuions of myelin basic protein in pathogenesis. *Molecular and Cellular Proteomics*, 2, 453-461.

Kim, J.K., Zhang, R., Eric, F., Strittmatter, E.F., Smith, R.D., and Zand, R. (2008). Characterization of posttranslational modifications in chicken myelin basic protein charge isomers. Phosphorylation, Methylation, deimination and deamidation. Private Communication.

Martenson, R.E., Kiess, G.E., Levine, S., and Alvord, E.C. (1972). Myelin basic protein of mammalian and submammalian vertebrates, encephalitogenic activities in guinea pigs and rats. *J. Immunol*, 109, 262-270.

Martenson, R.E., Law, M.J., and Diebler, G.E. (1983). Identification of multiple in vivo phosphorylation sites in rabbit myelin basic protein. *J. Biol. Chem*, 258, 930-937.

Miyamoto, E., and Kakiuchi, S., (1974). *In vitro* and *in vivo* phosphorylation of myelin basic protein by exogenous and endogenous adenosine 3':5'-monophosphate dependent protein kinases in brain, *J. Biol. Chem*, 249, 2769-2777.

Moscarello, M.A., Pang, H., Pace-Asciak, C.R., and Wood., D.D. (1992). The N terminus of human myelin basic protein consists of C2, C4, C6, and C8 alkyl carboxylic acids. *J. Biol. Chem*, 45, 9779-9782.

Moscarello, M.A., Wood, D.D., Ackerly. C., and Boulias, C. (1994). Myelin in Multiple Sclerosis is developmentally immature. *J. Clin. Investigation*, 94, 146-154.

Paik, W.K., and Kim, S. (1990). *Protein methylation*. Boca Raton, FL, CRC Press.

Polevoda, B., and Sherman, F. (2000). N-alpha-terminal acetylation of eukaryotic proteins. *J. Biol. Chem*, 275, 36479-36482.

Polevoda, B., and Sherman, F. (2003). N-terminal acetyl transferases and sequence requirements for N-terminal acetylation of eukaryotic proteins. *J. Mol. Biol*, 325, 595-622.

Paterson, P.Y., (1957). Study of experimental encephalomyelitis employing mammalian and nonmammalian nervous tissues. *J. Immunol*, 78, 472-

Robinson, N.E., and Robinson, A.B. (2004). *Molecular Clocks*. Cave Junction, OR, Althouse Press.

Robinson, A.B., and Rudd, C. (1974). Deamidation of glutaminyl and asparaginyl residues in peptides and proteins. *Curr. Top. Cell Regul*, 8,247-295.

Robinson, A.B., McKerrow, J.H., and Cary, P. (1970). Controlled deamidation of peptides and proteins: An experimental hazard and a possible biological timer. *Proc. Natl. Acad. Sci*, USA, 66, 753-757.

Robinson, A.B. (1974). Evolution and the distribution of glutaminyl and asparaginyl residues in proteins. *Proc. Natl. Acad. Sci*, USA,71, 885-888.

Small, D.H., and Carnegie, P.R. (1982). *In vivo* methylation of an arginine in chicken myelin basic protein. *J. Neurochem*, 38, 184-190.

Sulakke, P.V., Petrali, E.H., Davis, E.R., and Thiessen, B. (1980). Calcium stimulated endogenous protein kinase catalyzed phosphorylation of basic proteins in subfractions and myelin like membrane fraction from rat brain. *Biochemistry*, 19, 5363-5372.

Sutherland, E.W., and Cori, C.F. (1951). Effect of hyperglycemic-glycogenolytic factor and epinephrine on liver phosphorylase. *J. Biol. Chem*, 188, 531-43.

Turner, R.S., Kemp, B.E., Su, H., and Kuo, J.F. (1985). Substrate specificity of fragments of the bovine myelin basic protein. *J. Biol. Chem*, 260, 11503-11507.

Ulmer, J.B. (1988). The phosphorylation of myelin proteins. *Progress in Neurobiology*, 31, 241-259.

Wood, D.D., and Moscarello, M.A. (1989). Isolation and characterization and lipid aggregating properties of a citrulline-containing myelin basic protein. *J. Biol. Chem*, 264, 5121-5127.

Young, P.R, and Grynspan, F. (1987). Analysis of methylated amino acids by high-performance liquid chromatography: Methylation of myelin basic protein. *J. Chromatogr*, 421, 130-135.

Zand, R., Jin, X., Kim, J., Wall, D.B., Gould, R., and Lubman, D.M. (2001). Studies of posttranslational modifications in Spiny Dogfish Myelin Basic Protein. *Neurochem. Res*, 26,539-547.

Zand, R., Li, M.X., Jin, X., and Lubman, D. (1998). Determination of the sites of posttranslational modifications in the charge isomers of bovine myelin basic protein by capillary electrophoresis-mass spectrometry. *Biochemistry*, 37, 2441-2449.

In: Myelin Basic Protein
Editor: Joan M. Boggs

ISBN: 978-1-60456-699-4

Chapter III

DEIMINATION OF MYELIN BASIC PROTEIN BY PAD ENZYMES, AND THEIR ROLE IN MULTIPLE SCLEROSIS

Fabrizio G. Mastronardi* and Mario A. Moscarello*

ABSTRACT

Myelin basic protein (MBP) is a major central nervous system myelin protein. One of its functions in the multilamellar structure of myelin is to maintain the tight compaction of the cytoplasmic faces of the myelin membrane. This function of MBP is associated with the presence of a high number of positively charged arginines. In normal individuals, this function is maintained; however in the demyelinating disease multiple sclerosis (MS) the myelin becomes unstable. The loss of myelin stability is associated with the loss of positive charge on MBP. The loss of positive charge is due to the post-translational modification of arginine to the neutral amino acid citrulline. The different charged isomers of MBP can be isolated using cation exchange chromatography. Typically normal MBP contains up to 20% of the deiminated isomer called C-8. MS individuals have reduced levels of the most cationic isomer (C-1) and elevated levels of C-8, typically 35-45% of the MBP. In acute MS cases, up to 90% of the MBP is in the C-8 form. Citrullination (deimination) is mediated by a family of enzymes, the peptidylarginine deiminases (PAD), of which five exist in the human and mouse genomes. PAD2 levels and activity in oligodendrocytes are central in the pathogenesis of the MS lesion. The hypomethylation of the PAD2 promoter results in increased expression of PAD2 in myelin that leads to increased MBP citrullination and subsequently unstable myelin. The formation of the citrullinated MBP at the myelin periaxonal structures may result in an early cascade of events resulting in

* Department of Molecular Structure and Function, The Hospital for Sick Children, 555 University Avenue, Toronto, ON, Canada M5G 1X8. Telephone: 1-416-813-6850, 1-416-813-5920; Fax: 416-813-5022; E-mail: fabrizio@sickkids.ca, mam@sickkids.ca

oligodendrocyte apoptosis. Myelin instability also results in the release of immunogenic epitopes. These neoepitopes include MBP peptides capable of stimulating the immune response. The immune response feeds forward with further demyelination and more damage. Extrinsic signals, such as TNF-alpha, mobilize PAD4 enzyme into the nucleus where it deiminates histones. The mobilization of PAD4 into the nucleus may be an additional mechanism that contributes to oligodendrocyte stress.

INTRODUCTION

Multiple sclerosis (MS) is the most common primary demyelinating disease in humans and is a leading cause of serious neurological disease in young adults in North America and Western Europe (McFarlin and McFarland, 1982a, b). The hallmark feature of the MS lesion is the selective destruction of the myelin sheath leaving the underlying axons relatively intact, and it is therefore classified as a primary demyelinating disease (Barnett and Sutton, 2006). Along with myelin loss, there is also axonal degeneration (Bjartmar and Trapp, 2001, 2003; Trapp et al., 1998; Trapp et al., 1999) and a propensity for reduced axon diameter in demyelinated fibers compared with myelinated fibers in the normal brain (Prineas, 1975; Prineas and Connell, 1978). Inhibition of oligodendrocyte maturation into myelinating cells may be caused by both the dystrophic denuded axons or by an imbalance of growth factors.

The MS plaque is characterized by focal areas of myelin destruction associated with astroglial scar formation. The lesions are scattered throughout the CNS with a predilection for optic nerves, brain-stem, spinal cord and periventricular white matter (Adams, 1977). In addition to the myelinolysis (Prineas, 1975; Prineas and Connell, 1978), lymphocytic infiltration in perivascular regions along with macrophages including microglial cells and a large proliferation of astrocytes is a common pathological feature in demyelinated areas. Oligodendrocytes, the myelin-producing cells in the brain, are absent in plaques, although they are present at edges of lesions, where there is evidence of attempts at remyelination (Prineas and Connell, 1979; Prineas et al., 1984; Prineas and Wright, 1978). Although the cause of MS is not known, it is believed to be a complex disease involving genetic, environmental and immunological factors. The role of each of these factors is under intense investigation at this time.

Our thesis is that the initial process of myelin degradation results from a failure to form compact adult myelin as a result of chemical changes in MBP such as deimination of arginyl residues by the enzyme peptidylarginine deiminase (PAD). We have referred to this as "developmental" or arrested maturity (Mastronardi and Moscarello, 2005). The release or exposure of antigenic sites triggers the autoimmune phase, which propagates the disease (Moscarello et al., 2007). The primary event of myelin degradation is summarized in our scheme (Figure 1). In order to understand the role of these chemical changes to MBP in maintaining myelin, an understanding of the properties of MBP and its charge isomers is essential.

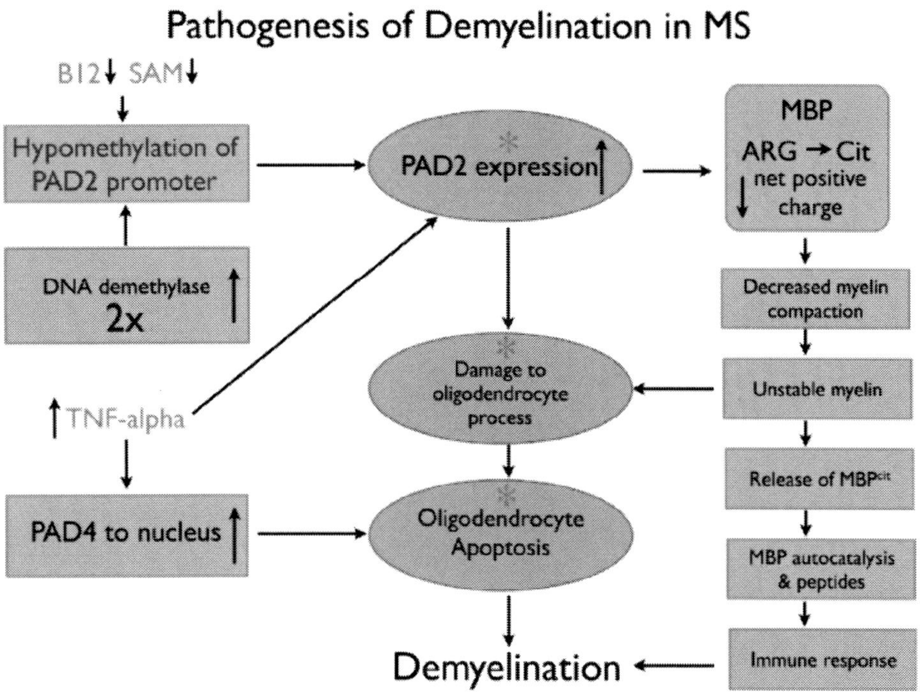

Figure 1. Schematic of the hypothesis that PAD2 levels and activity in oligodendrocytes are central in the pathogenesis of the MS lesion. The hypomethylation of genomic DNA was found to be associated with increased DNA demethylase activity, which was found to be 2x higher in MS white matter (Mastronardi et al., 2007). Hypomethylation of the PAD2 promoter results in increased expression of PAD2 in myelin that leads to increased MBP citrullination and subsequently unstable myelin. The formation of the citrullinated MBP at the myelin periaxonal structures may result in an early cascade of events resulting in oligodendrocyte apoptosis. Myelin instability also results in the release of immunogenic epitopes. These epitopes include MBP peptides capable of stimulating the immune response. The immune response feeds forward with further demyelination and more damage. Extrinsic signals, such as TNFalpha, mobilize PAD4 enzyme into the nucleus where it deiminates histones (Mastronardi et al., 2006). The mobilization of PAD4 into the nucleus may be an additional mechanism that contributes to oligodendrocyte apoptosis.

MBP is the most extensively studied protein of myelin and constitutes approximately 35% of the protein content of central nervous system myelin (Kies et al., 1972). Interest in MBP stems from the demonstration of its encephalitogenic activity in susceptible animals (Alvord and Kies, 1959; Alvord et al., 1959; Kies et al., 1965; Laatsch et al., 1962). The myelin basic proteins were first extracted from white matter by selective solubilization in dilute mineral acids (Deibler et al., 1972). Myelin basic proteins from mammalian species are separated on SDS polyacrylamide gels by electrophoresis into separate isoforms of which an 18.5 kDa protein is the major form in the adult human and bovine species (Figure 2) (Martenson et al., 1972).

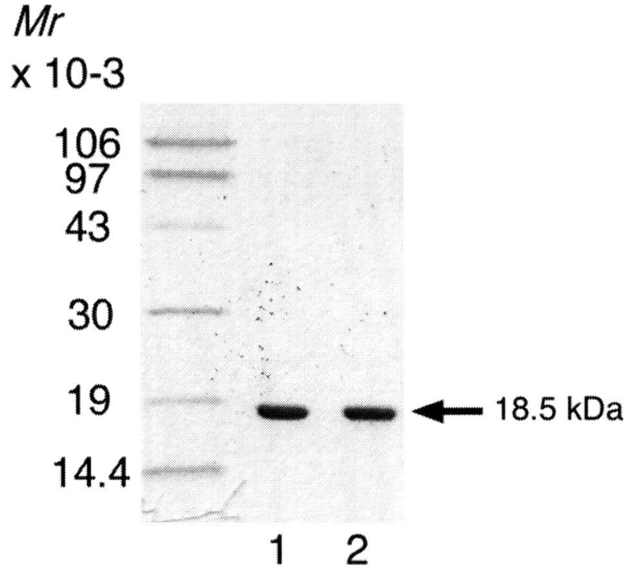

Figure 2. SDS-polyacrylamide gel of 18.5 kDa MBP from adult mammals. Human MBP (lane 1) and bovine MBP (lane 2) migrate with an apparent molecular mass of 18.5 kDa.

MBP Charge Isomers

The charge isomers add further complexity to understanding the role of the various MBPs during myelinogenesis. So far only the charge isomers of the 18.5 kDa isoform from mammalian MBP have been studied extensively, although microheterogeneity has been described in the four isoforms of the normal mouse (Fannon and Moscarello, 1991). By cation exchange chromatography, fractionation of the 18.5 kDa isoform of mammalian myelin MBP into up to 8 charge isomers (components) was achieved (Chou et al., 1976) (Figure 3). These isomers (not isomers in the mass spectrometric sense since the various modifications result in mass changes) arose from post translational modifications of MBP, including phosphorylation, deamidation, C-terminal arginine loss, and methionine oxidation (Brostoff and Eylar, 1971; Chou et al., 1976; Deber et al., 1983; Deibler et al., 1990; Wood and Moscarello, 1989) (see Chapter II). The protein fraction that does not bind to the cation exchanger, but washes through in the unbound fraction before the start of the sodium chloride gradient, is referred to as component 8 (C-8). The fraction that elutes under conditions of high salt (300 mM NaCl) is the most tightly bound fraction and is termed component 1 (C-1) Charge isomers with intermediate mobilities have been termed C-2, C-3, C-4 and C-5 respectively. These components or charge isomers are the result of the loss of positive charge or the increase in negative charge (Chou et al., 1976).

Typical CM52 profiles of MBP from normal and MS white matter are shown in Figure 3. The C-1 and C-8 peaks are indicated beneath the fraction numbers. In 1972, Finch et al. had identified a citrulline-containing myelin protein fraction. Wood and Moscarello (1989) showed that some MBP from normal brain, which contained citrulline, was found in the void volume of the CM-52 column. This MBP fraction contained citrulline in place of arginine at residues 25, 31, 122, 130, 159 and 170 preferentially located in the N and C-terminal portions of normal MBP (Wood and Moscarello 1989). The conversion of the guanidino group of Arg to the ureido group of citrulline in the C-8 protein results in a net loss of 6 positive charges on the protein. Antibodies that bind to the ureido group of citrulline on MBP showed a preferential reactivity with C-8 compared with C-1 (McLaurin et al., 1992). Immunogold localization, using this antibody, of the C-8 isomer in human brain white matter from autopsy material revealed a predominant localization to the intraperiod line of myelin, whereas a monoclonal antibody reactive to residues 130-137 of MBP was mainly localized to the major dense line of myelin, which is formed by the close apposition of the oligodendrocyte cytoplasmic surfaces of the plasma membrane (McLaurin et al., 1993). These studies indicated that the citrullinated form of MBP was localized to a specific site in myelin; however its functional significance is as yet not known.

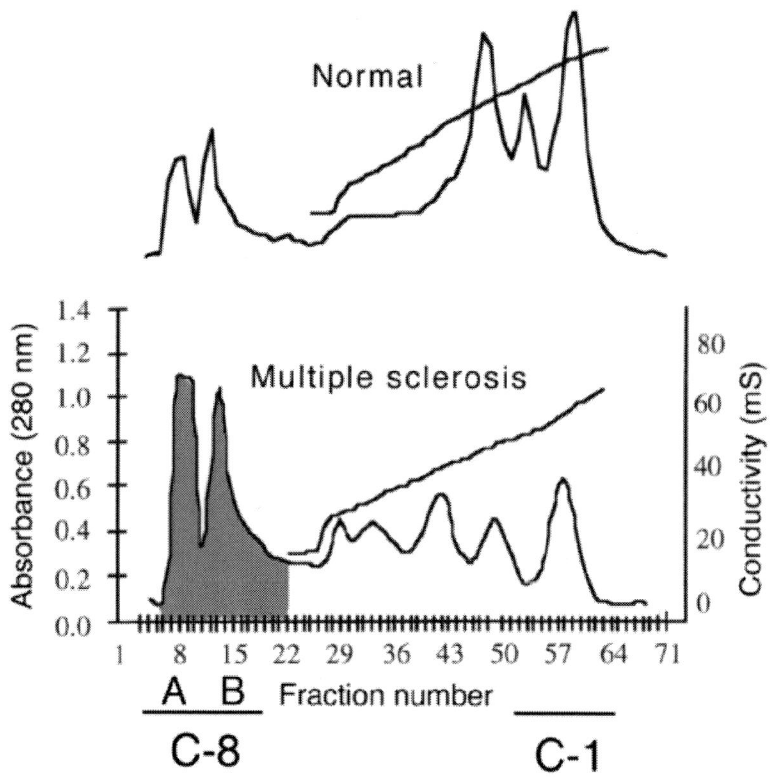

Figure 3. CM52 chromatograms of MBP isolated from normal and MS white matter. The C-8 component contains citrullinated MBP in addition to other proteins (Liu et al., 2005; Mastronardi et al., 2003). The C-8 A and B basic protein components elute before the start of the salt gradient and are found in the void volume. The C-8 B component peak (shaded) is in greater abundance in MS. The C-1 MBP component elutes last under high salt conditions due to its positive charge.

The C-8 component from MS brain is shaded in Figure 3. Typically the C-8 component is significantly increased in MS brain as compared with normal adults (Moscarello et al., 1994). The binding of an anti-citrulline antibody to MBP (McLaurin et al., 1992) from MS brain is more prominent than in normal MBP. This is demonstrated by Western immunoblot of MBP with anti-citrulline antibodies (Figure 4A) and with a commercially available citrulline detection kit (Mastronardi et al., 2007). Although the immunoreactivity of anti-citrulline antibodies with MBP from MS individuals was variable, it was significantly higher than with MBP from normal individuals. These charge isomers can also be separated by urea alkaline tube gel electrophoresis at pH 10.6. A representative urea alkaline tube gel electrophoretic separation of human MBP from normal and MS white matter is shown in Figure 4B. In this gel system the most cationic charge isomer migrates furthest (C-1) and the least cationic component (C-8) has the slowest mobility (Figure 4B).

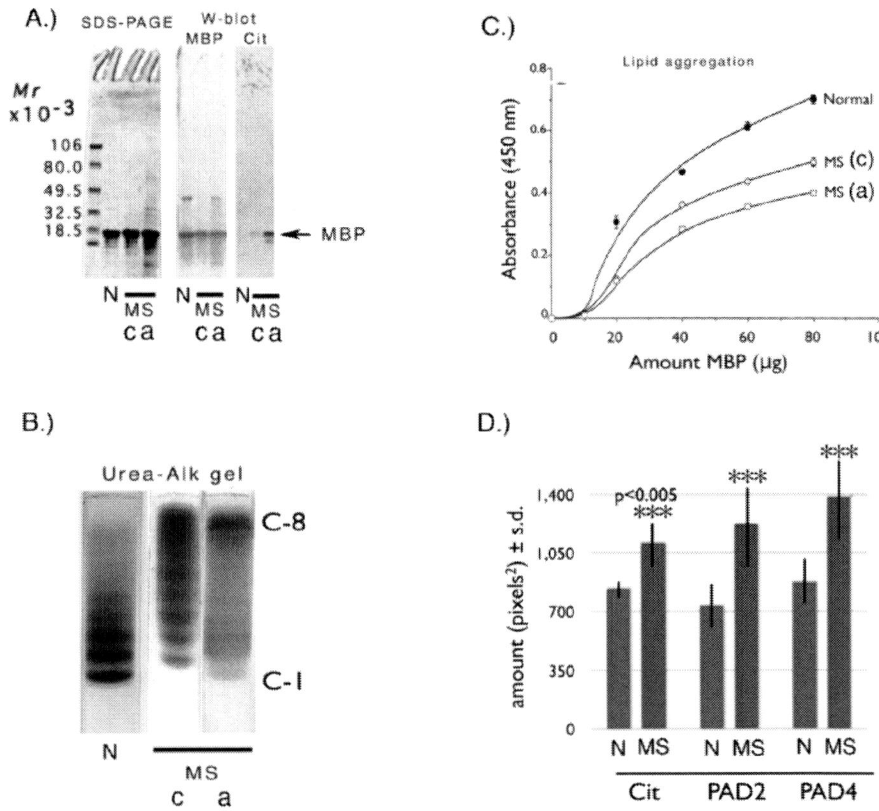

Figure 4. (A) Coomasie-blue stained gels of MBP isolated from normal (N) and MS white matter (left panel); MBP Western blot of MBP from normal (N) and MS individuals (middle panel); Citrulline Western blot of MBP from normal (N) and MS individuals (right panel). The MBP samples from MS individuals were isolated from a chronic MS case (c) and and acute MS case (a). (B) Urea-alkaline gel electrophoresis separation of MBP charge isomers of normal (N) and MS MBP. The amido black stained bands representing the C-1 (non-citrulline) and C-8 (citrulline containing) MBP isomers are indicated. The MS individuals were from a chronic MS case (c) and from an acute MS case (a). The acute MS sample contain less C-1 but much more C-8. C1 is the lowest band in each case but its position is different on the 3 different gels. (C) Lipid aggregation induced by normal and MS MBP in vitro. The acute MS case (a) has more citrullinated MBP than the Chronic MS case (c). This is reflected in the reduced ability of the MBP from the acute MS individual in aggregating lipid vesicles. (D) Quantitation of citrullinated proteins in normal (N) and MS myelin by immunoslot blot assays using anti-citrulline antibody revealed elevated levels of citrullinated protein in MS individuals (n=12) ($p<0.005$). In addition the levels of the citrullinating enzymes, PAD2 and PAD4, were both significantly elevated in MS myelin.

The charge isomers of MBP are further modified by the presence of heterogeneous fatty acylated N-termini (Moscarello et al., 1992). The heterogeneous fatty acylation of MBP residues 1-21, prepared by cyanogen bromide cleavage of the 18.5 kDa isoform at its N-terminus, affected its secondary structure in solution and its ability to be modified by phosphorylation (Costentino et al., 1994) or ADP-ribosylation (Boulias and Moscarello, 1994).

Our thesis is that the MS lesion develops as a result of increased MBP citrullination. Figure 1 illustrates the sequence of events in the pathogenesis of MS as a result of MBP citrullination. Because of the decreased positive charge of citrullinated MBP it is less able to form strong interactions with the negatively charged phospholipids and as a result compact myelin structure cannot be maintained. As part of our hypothesis, the loss of this MBP function results in demyelination and if kept unchecked results in lesion development.

Consequences of Increased Citrullination of MBP

A. Proteolysis

The consequences of citrullination of MBP involve changes in secondary structure, imparting a more open conformation (see Chapter VI) which renders the protein more susceptible to enzymatic degradation, through which immunogenic peptides are formed (Fig. 1). Citrullinated MBP was more readily digested by proteases such as cathepsin D (Cao et al., 1999), a myelin-associated protease, and stromelysin than non-citrullinated MBP. Cathepsin D and stromelysin, both released immunodominant peptides. Whereas the immunodominant peptide released by cathepsin D is part of a large peptide (residues 40-89), that released by stromelysin (MMP-3) is 17 residues, which can be accommodated by MHC class II molecules (D'Souza et al., 2002; D'Souza and Moscarello, 2006). More recently catalytic serum derived antibodies, termed abzymes, isolated from MS patients were shown to have specificity for MBP in MS patients (Belogurov et al., 2008). These antibodies generated MBP peptides containing residues 81-103 and 91-114, similar to those of MMP-3. These authors also characterized isolated serum antibodies from MS, normal and other neurological diseases and mapped their epitope specificities along the MBP molecule. High affinity antibodies were detected in three regions along MBP. These included peptides 43-68, 53-81, 123-140 and 146-170. Interestingly the region encompassing residues 81-103 had weak binding present only in the relapsing-remitting MS cohort. The primary and secondary-progressive MS cohorts contained strong antibody binding to the MBP peptide 123-140 as compared with the antibodies from healthy donors. No significant differences were found in serum antibodies isolated from other neurological diseases and relapsing-remitting MS patients along the 123-140 string of residues. These results are indicative that neo-epitope generation may be involved in disease progression.

B. MBP autocatalysis and neoepitopes

Not only is the citrullinated MBP more readily degraded by proteases such as cathepsin D (Cao et al., 1999), it also possesses a unique autocatalytic property which may contribute to disease development. Citrullinated MBP was autocatalytic i.e., it degraded itself in the absence of proteases at a 200 times faster rate than non-citrullinated MBP. Since the MBP was isolated in 0.2N sulphuric acid, contaminating proteases could not be responsible for this activity (D'Souza et al., 2005). The autocatalytic activity has important implications for the autoimmune theory of MS. In order to explain the sensitization of T-cells prior to demyelination, the theory of molecular mimicry has received considerable attention. According to this theory, a virus containing a protein with some sequence homology to MBP sensitizes T-cells in the periphery, which travel to the brain and attack myelin. We have shown the MBP from MS white matter undergoes autocleavage at slightly alkaline pH values, whereas very little autocleavage was observed with the MBP from normal tissues (D'Souza et al., 2005). Cleavage generated the immunodominant epitope (84-89) in a larger peptide. After labeling MBP from MS tissue with ^{3}H-diisopropylfluorophosphate, a labeled peptide consisting of residues 140-152 which contained a single seryl residue, the only labeled seryl residue of 18 seryl residues in MBP was isolated. This active seryl residue was analogous to active serines in seryl proteases.

The importance of this observation is that it is suggestive of a non-enzymatic auto-proteolytic activity imparted by citrullination of MBP and provides a mechanism for the release of MBP peptides directly from myelin thereby sensitizing T-cells in the periphery to activate the immune response which may result in the generation of neoepitopes and disease progression.

The Role of Myelin Basic Protein in Myelin Compaction

Studies presented by Dupouey et al., (1979), Privat et al., (1979) and Matthieu et al., (1980) (Dupouey et al., 1979; Matthieu et al., 1981; Privat et al., 1979) in a genetic mutant mouse lacking MBP, the shiverer mouse, have revealed much less myelin than normal and a much wider cytoplasmic space in the myelin suggesting that MBP may be required for myelin compaction. The rescue of the shiverer mutation via increasing dosage of the normal mouse 14 kDa MBP by transgenic technology led to, a normal phenotype when 25 to 100% MBP was expressed, while the shivering phenotype was observed in mice with 0 to 12% MBP expression. The rescue of shiverer was correlated with membrane spacing of 156-160 Å, normal for CNS myelin, and the broad reflections were indicative of compact myelin (Blaurock, 1981; Kirschner and Ganser, 1980). These studies supported a requirement of MBP for the proper compaction of CNS myelin.

Other studies have demonstrated the localization of MBP at the cytoplasmic face, by electron microscopic immunolocalization of MBP in brain sections (Herndon et al., 1973; McLaurin et al., 1993; Moscarello et al., 1994; Omlin et al., 1982). X-ray diffraction studies have revealed that the protein had a unique, compact (but not globular) structure in a lipid environment (Haas et al., 2004). The tertiary structure of the protein is thought to facilitate these interactions. MBP in aqueous solution has been shown to exist as a random coil

(Anthony and Moscarello, 1971; Gow and Smith, 1989; Krigbaum and Hsu, 1975), facilitating electrostatic interactions with lipid headgroups, which bridges the bilayers (Rispoli et al., 2007). However, some hydrophobic interactions allowing for the partial membrane penetration of MBP side chains in its N-terminal half, including an amphipathic helix in the middle of the molecule at residues Val83-Thr92 (murine sequence numbering) has been suggested (Bates et al., 2004; Boggs et al., 1999; Boggs et al., 1981; Demel et al., 1973; Gow et al., 1990; London and Vossenberg, 1973; Marsh, 1990; Wood et al., 1977). For a detailed discussion of structural models of MBP the reader is referred to Chapters VI, IX, and X.

Bilayer Structure in Normal Appearing White Matter (NAWM) in MS Brain Is Not "Normal"

Studies in our lab, on the structural analyses of myelin and its protein components, mainly MBP, have shown that MS occurs in a structurally immature myelin membrane (Moscarello et al., 1994). Studies on the transition temperature of myelin, a measure of bilayer order, during normal development by wide angle X-ray diffraction revealed a decreased order (lower transition temperature) in infants from 2 months suggestive of a less compact myelin. The bilayer order steadily increased to the adult level by age 17. MS myelin and myelin lipids revealed a lower transition temperature. This transition temperature was similar to that of young children aged 4-5, consistent with our concept that myelin in MS was developmentally immature (Brady et al., 1981a; Chia et al., 1984; Moscarello et al., 1985; Moscarello et al., 1994). Other studies have demonstrated a reduced ability of MBP from MS brain to organize single lamellar lipids into multilamellar vesicles in vitro (Brady et al., 1981a; Brady et al., 1981b). A typical lipid aggregation experiment in which the optical density is a measure of lipid aggregation is shown in Figure 4C. Increasing amounts of MBP from normal white matter was able to efficiently aggregate lipids (higher optical density), whereas MBP from MS white matter had a reduced ability. This reduced ability of lipid aggregation is associated with the reduced positive charge of the protein from MS brain. These experimental results were suggestive that the biophysical properties of myelin and MBP from MS brain were inherently different from normal individuals.

MBP Microheterogeneity in MS White Matter

Since the lipid composition in MS myelin was found to be similar to that of normal (Wood and Moscarello, 1984), changes in the bilayer ordering were attributed to protein differences (Moscarello et al., 1994) such as the elevation of C8 MBP referred to above. A more complex picture of microheterogeneity of MBP in MS was reported by (Kim et al., 2003). They used capillary electrophoresis and tandem mass spectrometry to isolate and sequence MBP isomers from normal and MS white matter. This study revealed unequivocal differences in the MBP post-translational modifications between normal and MS individuals. Two important changes that affect the net charge of MBP included phosphorylation of Serine (Ser)/Threonine (Thr) residues and deimination of Arginines. MS MBP had significantly less phosphorylated Thr98 (Yon et al., 1996) as well as reduced phosphorylation at other Ser/Thr sites throughout the molecule. The reduced phosphorylation of MBP was also found in the primary progressive demyelinating transgenic ND4 mouse prior to the onset of demyelination (Kim et al., 2003). Furthermore, the glycogen synthase kinase, GSK3 which phosphorylates MBP at Thr98 was also reduced in the ND4 mouse indicating that reduced kinase activity may be an important factor for myelin stability.

PEPTIDYL ARGININE DEIMINASES (PADS)

Protein deimination (arginyl to citrullinyl conversion) is mediated by a family of enzymes, the Ca^{2+}-dependent peptidylarginine deiminases (PAD) (EC 3.5.3.15) that display both tissue and substrate specificities. Ca^{2+} binding is essential for the formation of the active site. The deimination reaction mediated by PAD enzymes is illustrated in Figure 5. The PAD genes are clustered in a distinct genomic region in humans (1p36.1) and syntenic locations of rodent genomes (Vossenaar et al., 2003). The PAD isozymes 1-4 are expressed in adult tissue. PAD6 expression is embryonic.

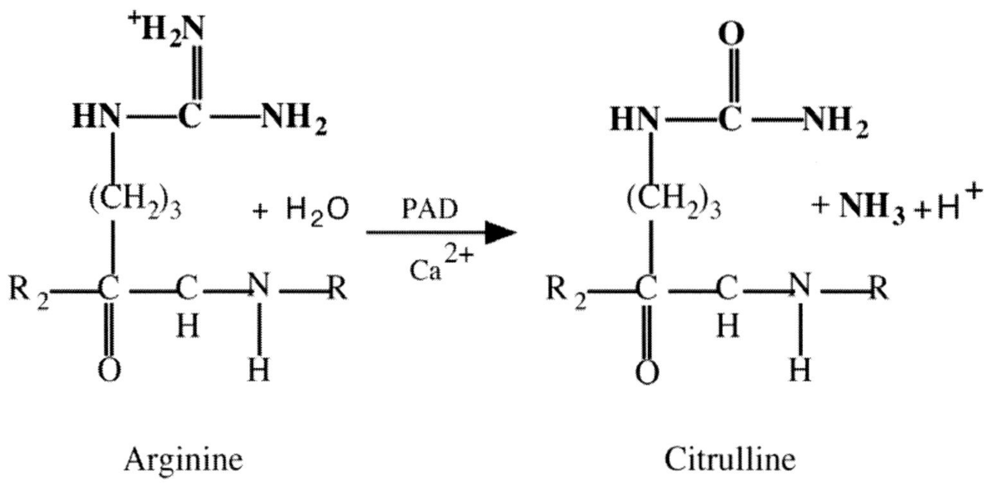

Figure 5. The deimination reaction mediated by the enzyme peptidylarginine deiminase results in the removal of a positive charge on arginine in peptides by a hydrolytic reaction resulting in a neutral citrullinyl residue in proteins such as myelin basic protein. The mass change as a result of converting an arginine to a citrulline is one Dalton.

PAD2 has been well characterized in brain. It is found in white matter and oligodendrocytes where it deiminates myelin basic protein (Finch et al., 1971; Lamensa and Moscarello, 1993). PAD4, unlike PADs 1-3, can be translocated into the nucleus (Vossenaar et al., 2003), where it deiminates histones (Cuthbert et al., 2004; Hagiwara et al., 2002; Nakashima et al., 2002; Sarmento et al., 2004; Wang et al., 2004). More recently we showed that PAD4 was consistently elevated in MS NAWM nuclear extracts (Mastronardi et al., 2006). All PADs share 50-55% sequence identity. The effect of increased nuclear PAD4 was associated with an increased histone H3 citrullination in MS NAWM. The mechanism of targeting PAD4 (which contains a nuclear localization signal) to the nucleus was associated with elevated TNF-alpha. Exposure of oligodendrocytes to TNF-alpha resulted in translocation of PAD4 into the nucleus and the appearance of citrullinated histone-3. This increase was observable within the first hour following induction and continued to increase up to 24 hours (Mastronardi et al., 2006) with the formation of activated caspase-3, which was increased at 24h. Oligodendrocyte death occurs and precludes synthesis of myelin components so that remyelination cannot take place. Eventually plaque formation occurs.

PAD isolated from bovine brain white matter was separated by SDS PAGE and migrated with an apparent Mr of 80,000 and a minor band at 50,000. Protein substrates for PAD included polyarginine and MBP. Histone was not a good substrate. With the C-1 MBP isomer, 17 of 19 Arg residues were converted to Cit, suggesting a constraint for the enzyme's specificity in vivo, possibly from lipid interactions (Lamensa and Moscarello, 1993).

Citrullinated proteins were found to be significantly elevated in MS myelin. This was consistent with elevated PAD2 and PAD4 protein in MS myelin (Figure 4D). Both PAD2 and PAD4 enzymes have been localized in myelin in intact mouse optic nerve (Wood et al., 2008). Whereas PAD2 was found in clusters in the myelin sheath, periaxonal regions of myelin, the axon and oligodendrocyte cytoplasm, PAD4 was found as single particles in myelin, the axon and oligodendrocyte cytoplasm and nucleus. Considerable labeling of the oligodendrocyte nucleus was found, consistent with the nuclear localization of this enzyme. In vitro deimination of MBP C-1 carried out by both recombinant PAD2 and PAD4 revealed some differences in the number of arginyl residues deiminated. Recombinant PAD2 deiminated 18 of 19 arginyl residues whereas recombinant PAD4 deiminated 14 of 19 arginyl residues. Extensive deimination occurred both in aqueous solution and in lipid vesicles (Wood et al., 2008). Although the mechanism by which proteins are deiminated is not understood, the active site, determined by studies with artificial substrates, such as benzoylarginine ethyl ester, is in the C-terminal portion of the molecule involving Cys 645 in PAD4, which takes part in a nucleophilic attack on the carbon atom of the guanidino group of arginine (Arita et al., 2004)

The PAD activity in MS normal appearing white matter was shown to be elevated 2-3 fold. In our ND4 transgenic mouse model (Mastronardi et al., 1993), PAD was elevated prior to any clinical or pathological signs of disease (Moscarello et al., 2002). An important role for PAD in demyelinating disease was suggested by these data.

The PAD2 CPG Island

In searching for the mechanism by which PAD2 was up-regulated, we studied the promoter region of the PAD2 gene, since promoters are known to regulate transcriptional activity. Methylation of cytosine (C) in a CpG island turns off the promoter activity of genes (Caiafa and Zampieri, 2005; Detich et al., 2002; Pennings et al., 2005). We found that the human PAD2 promoter contained 74% CG content which is much higher than bulk DNA and constitutes a CpG island. We measured the amount of methyl cytosine in white matter genomic DNA isolated from normal individuals and various cases of MS. We found that the genomic DNA from MS white matter contained 1/3 the amount of methylated cytosines as genomic DNA from normal brain. Thymus genomic DNA from the same MS patients was normally methylated as was white matter genomic DNA from patients with Alzheimer's, Parkinson's and Huntington's diseases. These data suggested that the decreased methylation was specific for MS. We concluded that the elevated PAD2 in MS was the result of increased transcription of the enzyme due to the hypomethylated promoter in brain. Demethylation of the promoter is carried out by a DNA demethylase. We showed that this enzyme activity was 2 fold increased in MS white matter over normal suggesting that this may be the reason for hypomethylation of the promoter (Mastronardi et al., 2007).

Future Directions and Concluding Remarks

The pathogenesis of MS, a variable and heterogeneous disease of the central nervous system, is currently unknown. Although considered for many years an autoimmune disorder, it is being increasingly viewed as a neurodegenerative disease involving pathogenic changes preceding the immune involvement. An early event is considered to be apoptosis of the oligodendrocyte (Barnett and Prineas, 2004; Barnett and Sutton, 2006; Ludwin, 1981; Ludwin and Johnson, 1981). The degeneration of the oligodendrocyte is believed to begin at the distal-most end of the cell process where it contacts myelin (Ludwin, 2006).

In previous studies we have shown that myelin basic protein contains an elevated amount of citrulline, formed by the deimination of arginyl residues by the enzyme peptidylarginine deiminase. The accompanying decrease in positive charge compromises the ability of the protein to compact lipid bilayers resulting in a less compact structure. The PAD enzyme, which carries out this reaction has been localized to myelin, including the periaxonal structures in myelin, and in the oligodendrocyte cell body (Wood et al., 2008). We reason that the formation of the citrullinated MBP at the myelin periaxonal region may result in an early cascade of events resulting in oligodendrocyte apoptosis (Figure 1).

The idea that citrullinated proteins, the result of increased PAD enzyme activity, are effectors in cellular stress and programmed death is starting to come to the forefront as new peptidyl-arginine substrate targets are being uncovered (Bhattacharya et al., 2006; Hagiwara et al., 2002; Mizoguchi et al., 1998). The hallmark features of apoptotic death, including DNA laddering, nuclear changes, and apoptotic bodies were observed in both Jurkat and human promyelitic leukemia HL-60 cell lines stably transfected with PAD-4 (Liu et al., 2006). Induction of calcium influx with calcium ionophores in mouse peritoneal macrophages resulted in apoptosis associated with early deimination of vimentin (Asaga et al., 1998). The association of citrullinated proteins with acute neurodegeneration in the rat brain has also been documented (Asaga and Ishigami, 2001). The association of citrullinated proteins in the regulation of apoptosis in disease is a provocative concept. Because PAD enzymes can act on multiple peptidylarginine-containing substrate targets, this would imply that unchecked PAD activity would result in pleiotropic effects. The consequences of these effects would be the loss of cellular homeostasis and metabolic support.

These studies which demonstrate a role for citrullinated proteins in development and demyelinating disease (Mastronardi et al., 2006; Moscarello et al., 2002; Moscarello et al., 1994; Pritzker et al., 1999) and in rheumatoid arthritis (Chang et al., 2005; Lundberg et al., 2005; Nissinen et al., 2003; Schellekens et al., 2000; Suzuki et al., 2003; van Venrooij and

Pruijn, 2000; Yamada et al., 2005) suggest a central role for peptidylarginine deiminases, the enzymes which generate citrulline in proteins. Our studies involving cytosine methylation of the promoter of PAD2 showed it to be only 1/3 as methylated in MS as in normals (Mastronardi et al., 2007). This suggests that remethylation of the promoter should decrease transcription, thereby decreasing the amount of enzyme synthesized. We have obtained preliminary data revealing that several CG sites in the PAD2 promoter were more methylated in the presence of vitamin B12, the methyl donor. Since PAD activity and protein are elevated in MS, direct inhibition by blocking the active site would represent a good strategy to complement the methylation studies. To this end we are exploring active site-directed reagents.

ACKNOWLEDGEMENTS

The authors acknowledge the support of the Multiple Sclerosis Society of Canada to MAM and FGM and the CIHR for supporting our research program.

REFERENCES

Adams, C.W. (1977). Pathology of multiple sclerosis: progression of the lesion. *British Medical Bulletin 33*, 15-20.

Alvord, E.C., Jr., and Kies, M.W. (1959). Clinico-pathologic correlations in experimental allergic encephalomyelitis. II. Development of an index for quantitative assay of encephalitogenic activity of antigens. *J Neuropathol Exp Neurol 18*, 447-457.

Alvord, E.C., Jr., Magee, K.R., Kies, M.W., and Goldstein, N.P. (1959). Clinico-pathologic correlations in experimental allergic encephalomyelitis. I. Observations on the early lesion. *J Neuropathol Exp Neurol 18*, 442-446.

Anthony, J.S., and Moscarello, M.A. (1971). A conformation change induced in the basic encephalitogen by lipids. *Biochim Biophys Acta 243*, 429-433.

Arita, K., Hashimoto, H., Shimizu, T., Nakashima, K., Yamada, M., and Sato, M. (2004). Structural basis for Ca(2+)-induced activation of human PAD4. *Nat Struct Mol Biol 11*, 777-783.

Asaga, H., and Ishigami, A. (2001). Protein deimination in the rat brain after kainate administration: citrulline-containing proteins as a novel marker of neurodegeneration. *Neurosci Lett 299*, 5-8.

Asaga, H., Yamada, M., and Senshu, T. (1998). Selective deimination of vimentin in calcium ionophore-induced apoptosis of mouse peritoneal macrophages. *Biochem Biophys Res Commun 243*, 641-646.

Barnett, M.H., and Prineas, J.W. (2004). Relapsing and remitting multiple sclerosis: pathology of the newly forming lesion. *Ann Neurol 55*, 458-468.

Barnett, M.H., and Sutton, I. (2006). The pathology of multiple sclerosis: a paradigm shift. *Curr Opin Neurol 19*, 242-247.

Bates, I.R., Feix, J.B., Boggs, J.M., and Harauz, G. (2004). An immunodominant epitope of myelin basic protein is an amphipathic alpha-helix. *J Biol Chem 279*, 5757-5764.

Belogurov, A.A., Jr., Kurkova, I.N., Friboulet, A., Thomas, D., Misikov, V.K., Zakharova, M.Y., Suchkov, S.V., Kotov, S.V., Alehin, A.I., Avalle, B., *et al.* (2008). Recognition and degradation of myelin basic protein peptides by serum autoantibodies: novel biomarker for multiple sclerosis. *J Immunol 180*, 1258-1267.

Bhattacharya, S.K., Crabb, J.S., Bonilha, V.L., Gu, X., Takahara, H., and Crabb, J.W. (2006). Proteomics implicates peptidyl arginine deiminase 2 and optic nerve citrullination in glaucoma pathogenesis. *Invest Ophthalmol Vis Sci 47*, 2508-2514.

Bjartmar, C., and Trapp, B.D. (2001). Axonal and neuronal degeneration in multiple sclerosis: mechanisms and functional consequences. *Curr Opin Neurol 14*, 271-278.

Bjartmar, C., and Trapp, B.D. (2003). Axonal degeneration and progressive neurologic disability in multiple sclerosis. *Neurotoxicity research 5*, 157-164.

Blaurock, A.E. (1981). The spaces between membrane bilayers within PNS myelin as characterized by X-ray diffraction. *Brain Res 210*, 383-387.

Boggs, J.M., Rangaraj, G., and Koshy, K.M. (1999). Analysis of the membrane-interacting domains of myelin basic protein by hydrophobic photolabeling. *Biochim Biophys Acta 1417*, 254-266.

Boggs, J.M., Wood, D.D., and Moscarello, M.A. (1981). Hydrophobic and electrostatic interactions of myelin basic proteins with lipid. Participation of N-terminal and C-terminal portions. *Biochemistry 20*, 1065-1073.

Boulias, C., and Moscarello, M.A. (1994). ADP-ribosylation of human myelin basic protein. *J Neurochem 63*, 351-359.

Brady, G.W., Fein, D.B., Wood, D.D., and Moscarello, M.A. (1981a). The interaction of basic proteins from normal and multiple sclerosis myelin with phosphatidylglycerol vesicles. *FEBS Lett 125*, 159-160.

Brady, G.W., Murthy, N.S., Fein, D.B., Wood, D.D., and Moscarello, M.A. (1981b). The effect of basic myelin protein on multilayer membrane formation. *Biophys J 34*, 345-350.

Brostoff, S., and Eylar, E.H. (1971). Localization of methylated arginine in the A1 protein from myelin. *Proc Natl Acad Sci U S A 68*, 765-769.

Caiafa, P., and Zampieri, M. (2005). DNA methylation and chromatin structure: the puzzling CpG islands. *J Cell Biochem 94*, 257-265.

Cao, L., Goodin, R., Wood, D., Moscarello, M.A., and Whitaker, J.N. (1999). Rapid release and unusual stability of immunodominant peptide 45-89 from citrullinated myelin basic protein. *Biochemistry 38*, 6157-6163.

Chang, X., Yamada, R., Suzuki, A., Sawada, T., Yoshino, S., Tokuhiro, S., and Yamamoto, K. (2005). Localization of peptidylarginine deiminase 4 (PADI4) and citrullinated protein in synovial tissue of rheumatoid arthritis. *Rheumatology* (Oxford) *44*, 40-50.

Chia, L.S., Thompson, J.E., and Moscarello, M.A. (1984). Alteration of lipid-phase behavior in multiple sclerosis myelin revealed by wide-angle x-ray diffraction. *Proc Natl Acad Sci U S A 81*, 1871-1874.

Chou, F.C., Chou, C.H., Shapira, R., and Kibler, R.F. (1976). Basis of microheterogeneity of myelin basic protein. *J Biol Chem 251*, 2671-2679.

Costentino, M., Pritzker, L., Boulias, C., and Moscarello, M.A. (1994). Acylation of myelin basic protein peptide 1-21 with alkyl carboxylates 2-10 carbons long affects secondary structure and posttranslational modification. *Biochemistry 33*, 4155-4162.

Cuthbert, G.L., Daujat, S., Snowden, A.W., Erdjument-Bromage, H., Hagiwara, T., Yamada, M., Schneider, R., Gregory, P.D., Tempst, P., Bannister, A.J., *et al.* (2004). Histone deimination antagonizes arginine methylation. *Cell 118*, 545-553.

D'Souza, C.A., Mak, B., and Moscarello, M.A. (2002). The up-regulation of stromelysin-1 (MMP-3) in a spontaneously demyelinating transgenic mouse precedes onset of disease. *J Biol Chem 277*, 13589-13596.

D'Souza, C.A., and Moscarello, M.A. (2006). Differences in susceptibility of MBP charge isomers to digestion by stromelysin-1 (MMP-3) and release of an immunodominant epitope. *Neurochem Res 31*, 1045-1054.

D'Souza, C.A., Wood, D.D., She, Y.M., and Moscarello, M.A. (2005). Autocatalytic cleavage of myelin basic protein: an alternative to molecular mimicry. *Biochemistry 44*, 12905-12913.

Deber, C.M., Cheifetz, S., and Moscarello, M.A. (1983). Microheterogeneity of bovine myelin basic protein studied by nuclear magnetic resonance spectroscopy. *Biopolymers 22*, 377-380.

Deibler, G.E., Martenson, R.E., and Kies, M.W. (1972). Large scale preparation of myelin basic protein from central nervous tissue of several mammalian species. *Preparative biochemistry 2*, 139-165.

Deibler, G.E., Stone, A.L., and Kies, M.W. (1990). Role of phosphorylation in conformational adaptability of bovine myelin basic protein. *Proteins 7*, 32-40.

Demel, R.A., London, Y., Geurts van Kessel, W.S., Vossenberg, F.G., and van Deenen, L.L. (1973). The specific interaction of myelin basic protein with lipids at the air-water interface. *Biochim Biophys Acta 311*, 507-519.

Detich, N., Theberge, J., and Szyf, M. (2002). Promoter-specific activation and demethylation by MBD2/demethylase. *J Biol Chem 277*, 35791-35794.

Dupouey, P., Jacque, C., Bourre, J.M., Cesselin, F., Privat, A., and Baumann, N. (1979). Immunochemical studies of myelin basic protein in shiverer mouse devoid of major dense line of myelin. *Neurosci Lett 12*, 113-118.

Fannon, A.M., and Moscarello, M.A. (1991). Characterization of myelin basic protein charge isomers from adult mouse brain. *Neuroreport 2*, 135-138.

Finch, P.R., Wood, D.D., and Moscarello, M.A. (1971). The presence of citrulline in a myelin protein fraction. *FEBS Lett 15*, 145-148.

Gow, A., Auton, W., and Smith, R. (1990). Interactions between bovine myelin basic protein and zwitterionic lysophospholipids. *Biochemistry 29*, 1142-1147.

Gow, A., and Smith, R. (1989). The thermodynamically stable state of myelin basic protein in aqueous solution is a flexible coil. *Biochem J 257*, 535-540.

Haas, H., Oliveira, C.L., Torriani, I.L., Polverini, E., Fasano, A., Carlone, G., Cavatorta, P., and Riccio, P. (2004). Small angle x-ray scattering from lipid-bound myelin basic protein in solution. *Biophys J 86*, 455-460.

Hagiwara, T., Nakashima, K., Hirano, H., Senshu, T., and Yamada, M. (2002). Deimination of arginine residues in nucleophosmin/B23 and histones in HL-60 granulocytes. *Biochem Biophys Res Commun 290*, 979-983.

Herndon, R.M., Rauch, H.C., and Einstein, E.R. (1973). Immuno-electron microscopic localization of the encephalitogenic basic protein in myelin. *Immunological communications 2*, 163-172.

Kies, M.W., Martenson, R.E., and Deibler, G.E. (1972). Myelin basic proteins. *Adv Exp Med Biol 32*, 201-214.

Kies, M.W., Thompson, E.B., and Alvord, E.C., Jr. (1965). The Relationship of Myelin Proteins to Experimental Allergic Encephalomyelitis. *Annals of the New York Academy of Sciences 122*, 148-160.

Kim, J.K., Mastronardi, F.G., Wood, D.D., Lubman, D.M., Zand, R., and Moscarello, M.A. (2003). Multiple sclerosis: an important role for post-translational modifications of myelin basic protein in pathogenesis. *Mol Cell Proteomics 2*, 453-462.

Kirschner, D.A., and Ganser, A.L. (1980). Compact myelin exists in the absence of basic protein in the shiverer mutant mouse. *Nature 283*, 207-210.

Krigbaum, W.R., and Hsu, T.S. (1975). Molecular conformation of bovine A1 basic protein, a coiling macromolecule in aqueous solution. *Biochemistry 14*, 2542-2546.

Laatsch, R.H., Kies, M.W., Gordon, S., and Alvord, E.C., Jr. (1962). The encephalomyelitic activity of myelin isolated by ultracentrifugation. *J Exp Med 115*, 77-88.

Lamensa, J.W., and Moscarello, M.A. (1993). Deimination of human myelin basic protein by a peptidylarginine deiminase from bovine brain. *J Neurochem 61*, 987-996.

Liu, A., Stadelmann, C., Moscarello, M., Bruck, W., Sobel, A., Mastronardi, F.G., and Casaccia-Bonnefil, P. (2005). Expression of Stathmin, a Developmentally Controlled Cytoskeleton-Regulating Molecule, in Demyelinating Disorders. *J Neurosci 25*, 737-747.

Liu, G.Y., Liao, Y.F., Chang, W.H., Liu, C.C., Hsieh, M.C., Hsu, P.C., Tsay, G.J., and Hung, H.C. (2006). Overexpression of peptidylarginine deiminase IV features in apoptosis of haematopoietic cells. *Apoptosis 11*, 183-196.

London, Y., and Vossenberg, F.G. (1973). Specific interaction of central nervous system myelin basic protein with lipids. Specific regions of the protein sequence protected from the proteolytic action of trypsin. *Biochim Biophys Acta 307*, 478-490.

Ludwin, S.K. (1981). Pathology of demyelination and remyelination. *Adv Neurol 31*, 123-168.

Ludwin, S.K. (2006). The pathogenesis of multiple sclerosis: relating human pathology to experimental studies. *J Neuropathol Exp Neurol 65*, 305-318.

Ludwin, S.K., and Johnson, E.S. (1981). Evidence for a "dying-back" gliopathy in demyelinating disease. *Ann Neurol 9*, 301-305.

Lundberg, K., Nijenhuis, S., Vossenaar, E.R., Palmblad, K., van Venrooij, W.J., Klareskog, L., Zendman, A.J., and Harris, H.E. (2005). Citrullinated proteins have increased immunogenicity and arthritogenicity and their presence in arthritic joints correlates with disease severity. *Arthritis Res Ther 7*, R458-467.

Marsh, D. (1990). Lipid-protein interactions in membranes. FEBS Lett *268*, 371-375.

Martenson, R.E., Deibler, G.E., Kies, M.W., Levine, S., and Alvord, E.C., Jr. (1972). Myelin basic proteins of mammalian and submammalian vertebrates: encephalitogenic activities in guinea pigs and rats. *J Immunol 109*, 262-270.

Mastronardi, F.G., Ackerley, C.A., Arsenault, L., Roots, B.I., and Moscarello, M.A. (1993). Demyelination in a transgenic mouse: a model for multiple sclerosis. *J Neurosci Res 36*, 315-324.

Mastronardi, F.G., daCruz, L.A., Wang, H., Boggs, J., and Moscarello, M.A. (2003). The amount of sonic hedgehog in multiple sclerosis white matter is decreased and cleavage to the signaling peptide is deficient. *Mult Scler 9*, 362-371.

Mastronardi, F.G., and Moscarello, M.A. (2005). Molecules affecting myelin stability: a novel hypothesis regarding the pathogenesis of multiple sclerosis. *J Neurosci Res 80*, 301-308.

Mastronardi, F.G., Noor, A., Wood, D.D., Paton, T., and Moscarello, M.A. (2007). Peptidyl argininedeiminase 2 CpG island in multiple sclerosis white matter is hypomethylated. *J Neurosci Res 85*, 2006-2016.

Mastronardi, F.G., Wood, D.D., Mei, J., Raijmakers, R., Tseveleki, V., Dosch, H.M., Probert, L., Casaccia-Bonnefil, P., and Moscarello, M.A. (2006). Increased citrullination of histone H3 in multiple sclerosis brain and animal models of demyelination: a role for tumor necrosis factor-induced peptidylarginine deiminase 4 translocation. *J Neurosci 26*, 11387-11396.

Matthieu, J.M., Ginalski-Winkelmann, H., and Jacque, C. (1981). Similarities and dissimilarities between two myelin deficient mutant mice, Shiverer and mld. *Brain Res 214*, 219-222.

McFarlin, D.E., and McFarland, H.F. (1982a). Multiple sclerosis (first of two parts). *N Engl J Med 307*, 1183-1188.

McFarlin, D.E., and McFarland, H.F. (1982b). Multiple sclerosis (second of two parts). *N Engl J Med 307*, 1246-1251.

McLaurin, J., Ackerley, C.A., and Moscarello, M.A. (1993). Localization of basic proteins in human myelin. *J Neurosci Res 35*, 618-628.

McLaurin, J., Hashim, G., and Moscarello, M.A. (1992). An antibody specific for component 8 of myelin basic protein from normal brain reacts strongly with component 8 from multiple sclerosis brain. *J Neurochem 59*, 1414-1420.

Mizoguchi, M., Manabe, M., Kawamura, Y., Kondo, Y., Ishidoh, K., Kominami, E., Watanabe, K., Asaga, H., Senshu, T., and Ogawa, H. (1998). Deimination of 70-kD nuclear protein during epidermal apoptotic events in vitro. *J Histochem Cytochem 46*, 1303-1309.

Moscarello, M.A., Chia, L.S., Leighton, D., and Absolom, D. (1985). Size and surface charge properties of myelin vesicles from normal and diseased (multiple sclerosis) brain. *J Neurochem 45*, 415-421.

Moscarello, M.A., Mastronardi, F.G., and Wood, D.D. (2007). The role of citrullinated proteins suggests a novel mechanism in the pathogenesis of multiple sclerosis. *Neurochem Res 32*, 251-256.

Moscarello, M.A., Pang, H., Pace-Asciak, C.R., and Wood, D.D. (1992). The N terminus of human myelin basic protein consists of C2, C4, C6, and C8 alkyl carboxylic acids. *J Biol Chem 267*, 9779-9782.

Moscarello, M.A., Pritzker, L., Mastronardi, F.G., and Wood, D.D. (2002). Peptidylarginine deiminase: a candidate factor in demyelinating disease. *J Neurochem 81*, 335-343.

Moscarello, M.A., Wood, D.D., Ackerley, C., and Boulias, C. (1994). Myelin in multiple sclerosis is developmentally immature. *J Clin Invest 94*, 146-154.

Nakashima, K., Hagiwara, T., and Yamada, M. (2002). Nuclear localization of peptidylarginine deiminase V and histone deimination in granulocytes. *J Biol Chem 277*, 49562-49568.

Nissinen, R., Paimela, L., Julkunen, H., Tienari, P.J., Leirisalo-Repo, M., Palosuo, T., and Vaarala, O. (2003). Peptidylarginine deiminase, the arginine to citrulline converting enzyme, is frequently recognized by sera of patients with rheumatoid arthritis, systemic lupus erythematosus and primary Sjogren syndrome. *Scand J Rheumatol 32*, 337-342.

Omlin, F.X., Webster, H.D., Palkovits, C.G., and Cohen, S.R. (1982). Immunocytochemical localization of basic protein in major dense line regions of central and peripheral myelin. *J Cell Biol 95*, 242-248.

Pennings, S., Allan, J., and Davey, C.S. (2005). DNA methylation, nucleosome formation and positioning. *Brief Funct Genomic Proteomic 3*, 351-361.

Prineas, J. (1975). Pathology of the early lesion in multiple sclerosis. *Human pathology 6*, 531-554.

Prineas, J.W., and Connell, F. (1978). The fine structure of chronically active multiple sclerosis plaques. *Neurology 28*, 68-75.

Prineas, J.W., and Connell, F. (1979). Remyelination in multiple sclerosis. *Ann Neurol 5*, 22-31.

Prineas, J.W., Kwon, E.E., Cho, E.S., and Sharer, L.R. (1984). Continual breakdown and regeneration of myelin in progressive multiple sclerosis plaques. *Annals of the New York Academy of Sciences 436*, 11-32.

Prineas, J.W., and Wright, R.G. (1978). Macrophages, lymphocytes, and plasma cells in the perivascular compartment in chronic multiple sclerosis. *Lab Invest 38*, 409-421.

Pritzker, L.B., Nguyen, T.A., and Moscarello, M.A. (1999). The developmental expression and activity of peptidylarginine deiminase in the mouse. *Neurosci Lett 266*, 161-164.

Privat, A., Jacque, C., Bourre, J.M., Dupouey, P., and Baumann, N. (1979). Absence of the major dense line in myelin of the mutant mouse "shiverer". *Neurosci Lett 12*, 107-112.

Rispoli, P., Carzino, R., Svaldo-Lanero, T., Relini, A., Cavalleri, O., Fasano, A., Liuzzi, G.M., Carlone, G., Riccio, P., Gliozzi, A., *et al.* (2007). A thermodynamic and structural study of myelin basic protein in lipid membrane models. *Biophys J 93*, 1999-2010.

Sarmento, O.F., Digilio, L.C., Wang, Y., Perlin, J., Herr, J.C., Allis, C.D., and Coonrod, S.A. (2004). Dynamic alterations of specific histone modifications during early murine development. *J Cell Sci 117*, 4449-4459.

Schellekens, G.A., Visser, H., de Jong, B.A., van den Hoogen, F.H., Hazes, J.M., Breedveld, F.C., and van Venrooij, W.J. (2000). The diagnostic properties of rheumatoid arthritis antibodies recognizing a cyclic citrullinated peptide. *Arthritis Rheum 43*, 155-163.

Suzuki, A., Yamada, R., Chang, X., Tokuhiro, S., Sawada, T., Suzuki, M., Nagasaki, M., Nakayama-Hamada, M., Kawaida, R., Ono, M., *et al.* (2003). Functional haplotypes of PADI4, encoding citrullinating enzyme peptidylarginine deiminase 4, are associated with rheumatoid arthritis. *Nat Genet 34*, 395-402.

Trapp, B.D., Peterson, J., Ransohoff, R.M., Rudick, R., Mork, S., and Bo, L. (1998). Axonal transection in the lesions of multiple sclerosis. *N Engl J Med 338*, 278-285.

Trapp, B.D., Ransohoff, R., and Rudick, R. (1999). Axonal pathology in multiple sclerosis: relationship to neurologic disability. *Curr Opin Neurol 12*, 295-302.

van Venrooij, W.J., and Pruijn, G.J. (2000). Citrullination: a small change for a protein with great consequences for rheumatoid arthritis. *Arthritis Res 2*, 249-251.

Vossenaar, E.R., Zendman, A.J., van Venrooij, W.J., and Pruijn, G.J. (2003). PAD, a growing family of citrullinating enzymes: genes, features and involvement in disease. *Bioessays 25*, 1106-1118.

Wang, Y., Wysocka, J., Sayegh, J., Lee, Y.H., Perlin, J.R., Leonelli, L., Sonbuchner, L.S., McDonald, C.H., Cook, R.G., Dou, Y., *et al.* (2004). Human PAD4 regulates histone arginine methylation levels via demethylimination. *Science 306*, 279-283.

Wood, D.D., Ackerley, C.A., Brand, B., Zhang, L., Raijmakers, R., Mastronardi, F.G., and Moscarello, M.A. (2008). Myelin localization of peptidylarginine deiminases 2 and 4: comparison of PAD2 and PAD4 activities. *Lab Invest 88*, 354-364.

Wood, D.D., Epand, R.M., and Moscarello, M.A. (1977). Localization of the basic protein and lipophilin in the myelin membrane with a non-penetrating reagent. *Biochim Biophys Acta 467*, 120-129.

Wood, D.D., and Moscarello, M.A. (1984). Is the myelin membrane abnormal in multiple sclerosis? *The Journal of membrane biology 79*, 195-201.

Wood, D.D., and Moscarello, M.A. (1989). The isolation, characterization, and lipid-aggregating properties of a citrulline containing myelin basic protein. *J Biol Chem 264*, 5121-5127.

Yamada, R., Suzuki, A., Chang, X., and Yamamoto, K. (2005). Citrullinated proteins in rheumatoid arthritis. *Front Biosci 10*, 54-64.

Yon, M., Ackerley, C.A., Mastronardi, F.G., Groome, N., and Moscarello, M.A. (1996). Identification of a mitogen-activated protein kinase site in human myelin basic protein in situ. *J Neuroimmunol 65*, 55-59.

In: Myelin Basic Protein
Editor: Joan M. Boggs

ISBN: 978-1-60456-699-4

Chapter IV

MYELIN BASIC PROTEIN-MEDIATED IMMUNOPATHOGENESIS IN MULTIPLE SCLEROSIS AND EAE

*Qingyong Ji and Joan Goverman**

ABSTRACT

Myelin basic protein (MBP) is considered to be an important self-antigen targeted in multiple sclerosis (MS). A widely used animal model for MS is induced by stimulating MBP-specific immunity in animals. After decades of comprehensive investigation in humans and animal models, much has been learned about cellular and humoral immunity toward this protein and its potential involvement in pathogenesis of MS. In this chapter, we will discuss anti-MBP immune responses mediated by $CD4^+$ and $CD8^+$ T cells and B cells, and their disease relevance. We will also discuss the central and peripheral tolerance mechanisms that maintain tolerance toward MBP because a breakdown in immune tolerance is critical for the initiation of autoimmune disease. The studies described herein lay the basis for further efforts in manipulating MBP-specific immunity and developing new therapeutic strategies for treatment and prevention of MS.

INTRODUCTION

Multiple sclerosis (MS) is a chronic inflammatory disease of the central nervous system (CNS) that affects about 2.5 million people worldwide. The cause of MS is unknown but susceptibility to the disease is dependent on both genetic and environmental factors (Sospedra and Martin, 2005). The inflammation and demyelination that are hallmark features of MS are

* Department of Immunology, HSC Box 357650, University of Washington, 1959 NE Pacific Street, Seattle, WA 98195-7650. Telephone: 1-206-685-7604; Fax: 1-206-543-1013; E-mail: goverman@u.washington.edu

believed to be mediated by self-reactive T lymphocytes that recognize components of myelin. Myelin basic protein (MBP), a major protein component of myelin, has been studied as an auto-antigen since the autoimmune hypothesis was proposed (Paterson, 1980). Here we review the studies demonstrating a role for MBP in the pathogenesis of MS and its animal model, experimental allergic encephalomyelitis (EAE).

Expression of MBP in the CNS and the Immune System

MBP was discovered as one of the major protein components of myelin and was initially believed to be synthesized only by myelin-forming cells. However, the MBP gene is now known to encode two different MBP families due to three different transcription initiation sites (Campagnoni et al., 1993; Givogri et al., 2000). Both families generate different isoforms via alternative splicing. Only one of these families generates MBP isoforms found in myelin, which are referred to as classic MBPs. Classic MBP transcripts are initiated from the last two of three transcription start sites in the locus (Campagnoni and Skoff, 2001). There are four major isoforms of classic MBPs that vary in size from 14 kDa to 21.5 kDa. The 14 kDa and 18.5 kDa are the major isoforms found in mouse and human myelin, respectively (Boggs, 2006). In the central nervous system, classic MBPs are synthesized by oligodendrocytes and account for about 30% of the total myelin protein. Schwann cells also synthesize classic MBP in the peripheral nervous system, where MBP constitutes approximately 10% of the myelin protein (Quarles, 2005). The other MBP family is referred to as golli-MBP (golli is abbreviated from "genes of the oligodendrocyte lineage" (Campagnoni and Skoff, 2001; Givogri et al., 2000) (see Chapter I). Transcription of golli-MBP family members begins at the first transcription start-site site in the locus. Alternative splicing of these transcripts includes many exons of classic MBP, such that golli-MBPs are fusion proteins containing unique amino acid sequence juxtaposed to portions of classic MBP sequence (Campagnoni and Skoff, 2001). Golli-MBPs are not components of the myelin sheath and therefore are not considered targets in MS. However, golli-MBPs are ubiquitously expressed in the nervous and immune systems (Campagnoni and Skoff, 2001; Feng et al., 2000). In particular, golli-MBP products have been found in thymic epithelial cell, dendritic cells and macrophages in the thymus, and T and B cells in the secondary lymphoid organs (Feng et al., 2000). Expression of golli-MBPs in the immune system, as well as the presence of classic MBP in the periphery, both play important roles in mediating immune tolerance to MBP (discussed below). This process is critically important in understanding the pathogenesis of MS because immune tolerance mechanisms shape the peripheral MBP-specific T cell repertoire that is available to contribute to disease.

In addition to the transcriptional complexity of MBP expression, MBP also exists as different isomers due to post-translational modification. These modifications include phosphorylation, deamidation, C-terminal arginine loss, and citrulline substitution for arginine (Moscarello et al., 2007) (see Chapters II and III). The 18.5 kDa human MBP isoform, which is the most abundant form in humans, has at least six charge isomers (C1-5 and C8). In MBP-C8, six arginines are replaced by citrulline primarily at the amino- and carboxy-terminal portions of the protein, substantially reducing the positive charge (Moscarello et al., 2007). Interestingly, it has been shown that MBP-C8 can increase from about 18% of MBP proteins in healthy humans to 45% of total MBP in MS patients (Moscarello et al., 1994). The increased percentage of MBP-C8 could contribute to MS pathogenesis in two ways (also see chapters III and VI). First, increased abundance in MBP-C8 may result in less dense and compact myelin, which may be more susceptible to degradation (Moscarello et al., 1986; Moscarello et al., 1994; Tranquill et al., 2000). Second, since citrullination of MBP generates neo-antigens, T cells that recognize this modified form of MBP may not have been subjected to immune tolerance but can still respond to myelin protein in the CNS (Anderton, 2004; Tranquill et al., 2000).

Multiple Sclerosis and EAE

MS is a complex disease that is manifested as a wide range of clinical symptoms and varying disease course among patients. About 85% of MS patients show relapsing-remitting symptoms, which typically evolve over a period of several days, stabilize, and then improve within weeks (Keegan and Noseworthy, 2002). In secondary progressive MS, neurological symptoms steadily worsen after a period of relapsing-remitting disease, while 15% of patients exhibit primary progressive MS in which the clinical signs continuously worsen from the onset of disease (Keegan and Noseworthy, 2002). The pathology of MS is also heterogeneous, and has been classified into four patterns based on the types of immune cells present in lesions, whether oligodendrocytes are spared, whether antibody and complement deposition is observed in lesions and the pattern of myelin injury (Lucchinetti et al., 2000). Interestingly, all lesions from an individual patient belong to only one pattern, suggesting that MS may result from four distinct types of pathogenic mechanisms. This immune pathology, as well as an association of MHC class II and MHC class I genes to the susceptibility of MS, led to the paradigm that MS is mediated by T lymphocytes, with a possible role for B cells as well (Friese and Fugger, 2005; Sospedra and Martin, 2005). As described below, the widely used animal model of MS, EAE, supports the notion that the pathogenic T cells are specific for myelin antigens such as MBP.

EAE is an animal model of CNS autoimmunity that is induced by triggering T cell-mediated immunity to myelin antigens. Because of the many similarities to MS, EAE is the most widely used model for this disease. Its origin can be traced back to the 1920s when Koritschoner and Schweinburg observed that rabbits developed paralysis following immunization of human spinal cord homogenate (Koritschoner and Schweinburg, 1925). Rivers et al. later extended this observation in primates by immunization of monkeys with rabbit brain homogenate (Rivers et al., 1933). MBP was first identified as an encephalitogenic antigen by fractionation of spinal cord homogenate used to induce EAE in guinea pigs (Martenson and Gaitonde, 1969). Since then, EAE has been induced in a variety of animal species by immunization with spinal cord homogenate, myelin proteins or their peptides in complete Freund's adjuvant (Stromnes and Goverman, 2006).

Most studies today focus on EAE models in the mouse because of the availability of reagents and genetically engineered mouse strains. In most rodents, EAE is typically manifested as ascending flaccid paralysis beginning with a limp tail and progressing to fore and hind limbs. These clinical signs are referred to as "classic EAE". The pathology in EAE is characterized by infiltration of lymphocytes and macrophages primarily in the white matter as well as demyelination. Like MS, the pathology in EAE is focused on the CNS rather than PNS. The lack of involvement of peripheral myelin may reflect a requirement for the higher local concentration of MBP found in the CNS compared to the periphery to activate MBP-specific T cells that have escaped immune tolerance. In addition to inducing disease in immunized animals, EAE can be induced by adoptive transfer of activated myelin protein-specific T cells. Transfer of purified $CD4^+$ T cells isolated from myelin-primed mice is sufficient to induce the disease. Therefore, EAE has generally been considered a MHC class II-restricted $CD4^+$ T cell-mediated autoimmune disease. The effector T cells in EAE have historically been defined as T_H1 cells based on the production by encephalitogenic T cell clones of interferon-γ (IFN-γ), interleukin 2 (IL-2) and tumor necrosis factor α (TNF-α), which are characteristic of the T_H1 subset. More recently, T cells that secrete IL-17 have demonstrated strong pathogenicity in EAE, suggesting that this T cell subset also plays an important role in disease (Bettelli et al., 2007).

MBP Peptide Specificity of $CD4^+$ T Cells in MS

Because of studies with EAE and the linkage of MS susceptibility to MHC class II genes (Olerup and Hillert, 1991), MS has been regarded primarily as a myelin-specific $CD4^+$ T cell-mediated disease. Accordingly, MBP-specific $CD4^+$ T cells have been extensively characterized in MS patients. The specificities of $CD4^+$ T cells that recognize MBP have been defined by stimulating PBMC in vitro with overlapping MBP peptides (summarized in Table 1). MBP1-19, MBP80-105, MBP110-130, and MBP140-171 peptides are more often recognized by $CD4^+$ T cells than other MBP regions, and have been regarded as dominant T cell epitopes (Chou et al., 1989; Jingwu et al., 1992; Liblau et al., 1991; Martin et al., 1990; Meinl et al., 1993; Muraro et al., 1997; Ota et al., 1990; Wucherpfennig et al., 1994). MBP151-170 is also a dominant region in citrullinated MBP-C8 (Tranquill et al., 2000). These immunodominant epitopes are mainly associated with HLA DR2 and DR4 haplotypes (Jingwu et al., 1992; Martin et al., 1992; Ota et al., 1990; Wucherpfennig et al., 1994). However, it has been difficult to demonstrate differences in responses to MBP epitopes between MS patients and normal controls (Burns et al., 1983; Chou et al., 1989; Martin et al., 1990; Muraro et al., 1997; Ota et al., 1990; Pette et al., 1990). Recently, this issue has been re-examined by comparing the responses of T cells from MS patients versus healthy controls to lower concentrations of antigen in vitro (Bielekova et al., 2004). Reducing the antigen concentration in these assays will preferentially expand cells that exhibit higher avidity for the self-antigen but might have a lower precursor frequency in the population. These low antigen conditions were also considered more physiological than the high peptide concentrations used in earlier assays that would preferentially expand low avidity T cells. These experiments confirmed that MBP13-32, MBP111-129, and MBP154-170 are high avidity dominant epitopes, whereas MB83-99 appeared to be low avidity as it required high peptide concentration for expansion (Bielekova et al., 2004). However, it is not yet clear in humans which types of epitopes are more likely to contribute to MS. As discussed below for EAE models, low avidity MBP epitopes are usually more pathogenic in vivo than high avidity epitopes because T cells specific for many high avidity epitopes are more subject to immune tolerance (Harrington et al., 1998).

Table 1. Immunodominant regions of human and rodent MBP recognized by $CD4^+$ T cells

	Dominant regions	MHC restriction	References
Human	MBP1-19 MBP80-105 MBP110-130 MBP140-170	Mostly HLA-A2 and A4	Chou et al., 1989; Jingwu et al., 1992; Liblau et al., 1991; Martin et al., 1990; Meinl et al., 1993; Muraro et al., 1997; Ota et al., 1990; Wucherpfennig et al., 1994; Bielekova et al., 2004
Mouse			
B10.PL, PL/J	MBPAc1-11, MBP121-150 *	H-2^u	Zamvil et al., 1986 Huseby et al., 2001
SJL	MBP17-27, MBP84-104	H-2^s	Sakai et al., 1988b; Sospedra and Martin, 2005
Rat			
Lewis rat	MBP68-88	RT1.B (I-A)	Mannie et al., 1985

* Immunodominancy only exists in MBP-/- mice.

FREQUENCY AND PHENOTYPES OF $CD4^+$ T CELLS IN MS PATIENTS

Many studies have attempted to relate the frequency of MBP-specific $CD4^+$ T cells to MS pathogenesis. Some researchers found a higher frequency of MBP-specific T cells in cerebrospinal fluid (CSF) of MS patients (Chou et al., 1992; Soderstrom et al., 1993; Zhang et al., 1994), and an increase in the frequency of these T cells correlated with disease activity (Chou et al., 1994; Tejada-Simon et al., 2000). Differences in MBP-specific T cell frequencies were not observed in the blood of MS patients (Allegretta et al., 1990; Jingwu et al., 1992; Martin et al., 1990; Stinissen et al., 1997; Zhang et al., 1994). Interestingly, a higher immune response to the highly citrullinated form of MBP, MBP-C8, in the blood of MS patients has been reported (Tranquill et al., 2000). In contrast to these findings, a recent study using a more sensitive flow cytometric assay of antigen-specific T cells found that the distribution of myelin-specific $CD4^+$ T cells was similar between MS patients and healthy subjects, but more $CD8^+$ myelin-specific T cells were observed in MS patients versus controls, including MBP-specific T cells (Crawford et al., 2004)

In addition to data regarding T cell frequencies, phenotypic differences in MBP-specific $CD4^+$ T cells between MS patients and controls have been reported. First, MBP-reactive $CD4^+$ T cells from MS patients express the high affinity IL-2 receptor and respond to IL-2 stimulation in vitro more strongly than those from normal individuals (Chou et al., 1994; Hellings et al., 2001; Zhang et al., 1994). Second, a higher percentage of MBP-reactive T cells in MS patients express CD45RO, a memory T cell marker, suggesting that a substantial proportion of MBP-reactive T cells have been previously activated in vivo (Burns et al., 1999). These data indicate that MBP-specific T cells are enriched in the $CD45RO^+$ memory population (Ponsford et al., 2001). Consistent with this, MBP-reactive $CD4^+$ T cells from MS patients are less dependent on B7 co-stimulation for their activation, a characteristic associated with memory T cells (Lovett-Racke et al., 1998; Scholz et al., 1998). Scholz et al. found that MBP-reactive $CD4^+$ T cells from MS patients are independent of co-stimulation mediated by B7.1 or B7.2-transfected cells in vitro (Scholz et al., 1998). Furthermore, Lovett-Racke et al. demonstrated that proliferation of MBP-specific T cells from MS patients is not inhibited by anti-CD28 (Lovett-Racke et al., 1998). Interestingly, Markovic-Plese et al. identified a population of $CD4^+CD28^-$ T cells in patients with MS, which produce a high amount of IFN-γ and upregulate IL-12R in the absence of costimulation (Markovic-Plese et al., 2001).

$CD4^+$ T cells mediate many of their effects by secreted cytokines. Based on cytokines they secreted, $CD4^+$ T cells were originally classified into T_H1 and T_H2 cells. T_H1 cells produce IL-2, IFN-γ, TNF α, and are primarily involved in the eradication of intracellular pathogens and autoimmunity. T_H2 cells predominantly produce IL-4, IL-5, IL-10 and IL-13 and mediate allergy and parasite clearance (Bettelli et al., 2007; Coffman, 2006). However, MBP-specific T cells from MS patients have been reported to exhibit a heterogeneous cytokine secretion profile representing a continuum between T_H1- and T_H2-like cells rather than profiles consistent with distinct T_H1 or T_H2 subsets (Hemmer et al., 1996; Hemmer et al., 1997; Hermans et al., 1997; Rohowsky-Kochan et al., 2000). The MBP reactive clones in these studies secreted high amounts of IFN-γ, TNF α/β, IL-4 and IL-10. However, a clear consensus has not emerged from studies of MBP-specific T cells in MS patients. Other reports suggested a skewed cytokine pattern and correlated it with disease activity (Tejada-Simon et al., 2000; Vandevyver et al., 1998). For example, Vandevyver et al. found a predominantly T_H1 or T_H0-like pattern in MBP-specific T cells isolated from both MS patients and control subjects (Vandevyver et al., 1998), whereas Tejada-Simon et al. observed a marked increase in the production of T_H1 cytokines among myelin-specific T cell lines obtained during exacerbation compared with those obtained during remission (Tejada-Simon et al., 2000). Using an ELISPOT assay, Hellings et al. and Modlovan et al. found that the majority of MBP-specific T cells in MS patients exhibit an IFN-γ-secreting T_H1 phenotype (Hellings et al., 2001; Moldovan et al., 2003). In studies that analyzed $CD8^+$ myelin-specific T cells, a mixed phenotype characterized by production of both IFN-γ and IL-10 was reported (Crawford et al., 2004).

MBP Peptide Specificity of $CD4^+$ T Cells in EAE

MBP-specific $CD4^+$ T cells have been extensively studied in EAE. Susceptibility to EAE induction is dependent on MHC background as well as on genes encoded outside the MHC region (Encinas et al., 1996), thus, only strains whose MHC molecules are able to bind and present MBP epitopes are susceptible to MBP-induced EAE. Lewis and DA rats, as well as PL/J, B10.PL and SJL/J mice, are susceptible and have been extensively used for MBP-induced EAE studies (Bernard and Carnegie, 1975; Fritz et al., 1983a; Stepaniak et al., 1995; Yasuda et al., 1975). MBP-induced EAE in rats is monophasic and demyelination is rarely observed (Swanborg, 2001). MBP-induced EAE in mice can exhibit a relapsing/remitting disease course with demyelination (Fritz et al., 1983b; Zamvil et al., 1985).

Initial studies of the immune response against MBP demonstrated a single dominant epitope in H-2^u mice (B10.PL and PL/J) consisting of MBPAc1-11 (acetylated N-terminal) peptide presented by I-A^u class II molecules (Zamvil et al., 1986). MBPAc1-11-specific T cells predominantly express TCR Vβ8.2 gene paired with Vα2 in B10.PL mice or Vα4 in PL/J mice (Acha-Orbea et al., 1988; Heber-Katz and Acha-Orbea, 1989; Urban et al., 1988; Zamvil et al., 1988). In vivo administration of anti-Vβ8 monoclonal Ab completely prevented EAE induced by adoptively transferred Vβ8.2 T cell clones (Zamvil et al., 1988). The Lewis rat also responds to one MBP epitope, MBP68-88 (Mannie et al., 1985). Interestingly, T cells recognizing this epitope also exhibit limited TCR heterogeneity and predominantly express the same Vα/Vβ combination as those utilized by the B10.PL mouse (Burns et al., 1989). In SJL/J mice, I-A^s-restricted MBP17-27 and MBP84-104 peptides are dominant epitopes for $CD4^+$ T cells (Sakai et al., 1988b; Sospedra and Martin, 2005). T cells reactive to the SJ/L MBP dominant epitopes also display preferential usage of certain TCR genes. For example, MBP89-101-reactive T cells mainly utilize Vβ17a gene segment paired with Vα1.1 gene segment (Padula et al., 1991; Sakai et al., 1988a; Yamamura et al., 1994).

EAE AND MS: TH1 VS. TH17?

Many studies showed that T cells isolated from mice with EAE produce high amounts of IL-2, IFN-γ and lymphotoxin but not IL-4 when cultured in vitro with MBP (Ando et al., 1989). Flow cytometry studies demonstrated that IFN-γ^+ and IL-2^+ cells are predominant at the early stage of EAE (Merrill et al., 1992), whereas mRNA for IL-10 increases at the remission stage (Kennedy et al., 1992). Transfer of MBP-specific T_H1 clones but not T_H2 clones induces EAE recipient mice (Baron et al., 1993). Treatment with IL-4 and IL-10 ameliorates clinical symptoms of EAE (Racke et al., 1994; Rott et al., 1994). Thus, EAE has historically been considered a T_H1 cell-mediated autoimmune disease (Liblau et al., 1995; Olsson, 1995).

The theory of T_H1-mediated EAE, however, is not supported by some key experimental data. First, Lafaille et al. showed that both T_H1 and T_H2 cells from MBP Ac1-11-specific TCR transgenic mice can induce EAE in recipient mice (Lafaille et al., 1997). The second point comes from studies manipulating INF-γ activity, the hallmark cytokine of T_H1 cells. According to the T_H1 theory, injection of IFN-γ should worsen EAE and neutralizing anti-IFN-γ antibody should ameliorate EAE. The results, however, are the opposite (Billiau et al., 1988; Duong et al., 1992; Voorthuis et al., 1990). Moreover, IFN-γ-deficient mice exhibit more severe EAE induced by MBP immunization than wild-type mice (Ferber et al., 1996; Krakowski and Owens, 1996). Interestingly, Lafaille and colleagues showed that adoptive transfer of MBP-specific TCR transgenic mice induced classic EAE that predominantly targeted the spinal cord, but if the TCR transgenic T cells were isolated from IFN-γ-deficient mice, the inflammation was primarily localized in the brain instead of the spinal cord (Wensky et al., 2005). These data strongly challenge the idea that EAE or MS is mediated exclusively by IFN-γ-secreting T_H1 cells.

Accumulating evidence now suggests that T_H17 cells may be more pathogenic than T_H1 cells in driving the autoimmune destruction in EAE and possibly MS. T_H17 cells are a distinct $CD4^+$ T cell lineage from T_H1 and T_H2 cells characterized by production of IL-17, IL-17F,

IL-6, TNF-α and IL-22 (Langrish et al., 2005; Liang et al., 2006; Zheng et al., 2007). The differentiation of T_H17 cells is induced by transforming growth factor β (TGF-β) and either IL-6 or IL-21 (Korn et al., 2007; Nurieva et al., 2007; Weaver et al., 2007; Zhou et al., 2007). Differentiated T_H17 cells are expanded by IL-23, whereas IL-4 and IFN-γ inhibit IL-23-driven expansion of T_H17 cells (Weaver et al., 2007). The pathogenic role of T_H17 cells in EAE has been demonstrated using mice deficient in IL-23 or IL-17 (Komiyama et al., 2006; Langrish et al., 2005). IL-23 is a heterodimeric cytokine composed of a unique p19 subunit and a common p40 subunit shared with IL-12. IL-12, which consists of p35 and p40 subunits, is a critical cytokine for the differentiation of IFN-γ producing T_H1 cells. In contrast to p35-deficient mice, which are susceptible to EAE induction, p19-deficient mice are completely resistant (Cua et al., 2003). Meanwhile, p40-deficient mice, which do not have both IL-12 and IL-23, fail to develop EAE after MBP or myelin oligodendrocyte glycoprotein (MOG) immunization (Cua et al., 2003; Segal et al., 1998). That EAE-resistant p19-deficient mice have normal T_H1 responses but lack IL-17-producing T cells indicates that IFN-γ is not sufficient for EAE development (Langrish et al., 2005). In accordance with this finding, EAE is significantly suppressed in IL-17-deficient mice (Komiyama et al., 2006). Adoptive transfer of T_H17 cells induces more severe clinical signs of EAE in recipient mice compared to transfer of T_H1 cells (Langrish et al., 2005).

MBP-Specific CD4$^+$ TCR Transgenic Mice

TCR transgenic mice have been valuable tools for studying the fate of T cells with a particular specificity because the majority of the T cells maturing in the thymus and circulating in the periphery express the transgenic TCR. A TCR specific for MBPAc1-11 presented by I-A^u was the first to be used in a TCR transgenic model of autoimmunity (Goverman et al., 1993), launching a new direction in model systems used for the study of MS. This model demonstrated that MBP-specific TCR transgenic mice can not only provide important insight into the tolerance mechanisms that shape the MBP-specific T cell repertoire, but can also be used to investigate environmental factors contributing to CNS autoimmunity. An initial observation made from this model was that the MBPAc1-11-specific T cells do not undergo tolerance induction in the thymus or in the periphery. Indeed, the MBP-specific T cells appear to be ignorant of the MBP self-antigen. However, a percentage of the TCR transgenic mice exhibit spontaneous EAE. Importantly, the incidence of spontaneous EAE never reached 100% and varied depending on the type of environment in which the mice were housed (Goverman et al., 1993). When housed under conventional conditions, 43% of the mice exhibited spontaneous EAE, but the incidence dropped to only 15% in a specific pathogen-free facility (Brabb et al., 1997). In this study, environmental parameters such as food, water, cages and other aspects of mouse husbandry were kept the same between the conventional and specific pathogen-free facility, suggesting that a difference in microbial exposure was responsible for the different incidence in spontaneous EAE, similar to the paradigm of an infectious agent influencing susceptibility to MS. Interestingly, the onset of spontaneous EAE occurred within a defined age window of 5-12 week-old mice, such that mice that remained healthy for more than 12 weeks rarely succumbed to disease (Brabb et al., 1997). Immunization with MBPAc1-11 peptide in complete Freund's adjuvant (CFA) plus injection of pertussis toxin induced EAE efficiently in these MBPAc1-11 TCR transgenic mice and, surprisingly, injection of pertussis toxin alone induced EAE with equal efficiency (Goverman et al., 1993). It is not yet clear which of pertussis toxin's multiple activities allows it to be such a potent trigger of disease in this model. Injection of naïve TCR transgenic T cells intrathecally into naïve mice induces EAE, indicating that T cell access to the CNS is critical for EAE induction (Brabb et al., 1997). However, pertussis toxin also affects T cell lineage differentiation and antigen-presenting cell (APC) co-stimulation, and these properties may also be key in triggering EAE in the TCR transgenic mice.

Since the initial MBPAc1-11-specifc TCR transgenic model, three other TCR transgenic mice with the same specificity (MBPAc1-11/I-A^u) were later generated by different laboratories (Lafaille et al., 1994; Liu et al., 1995; Pearson et al., 1997). These lines all showed the same lack of tolerance induction and ignorant T cell phenotype in the periphery. Wraith and colleagues demonstrated in their TCR transgenic model that the escape from central tolerance was due to the low affinity of MBPAc-11 for I-A^u molecules, such that insufficient amount of the ligand was present in the thymus to induce negative selection. However, the incidence of spontaneous EAE is reported to be very low in these lines, less than 15% seen in our original TCR transgenic line housed in the specific pathogen-free facility. We investigated this difference in spontaneous EAE incidence between our transgenic model and that generated by Lafaille et al. by co-housing mice from both TCR transgenic lines in our conventional facility. Interestingly, while the incidence of disease was higher than 40% in our transgenic mice, the incidence was only 15% in the other model (Goverman, 1999). The reason for this difference in susceptibility to spontaneous disease by the different TCR transgenic mice is not known.

A critical observation regarding spontaneous EAE was made by Lafaille's group while studying their MBPAc1-11 TCR transgenic mice crossed onto a Rag-deficient background. The RAG mutation prevents the rearrangement of any endogenous TCR chains. They showed that 100% of the $RAG^{-/-}$ TCR transgenic mice exhibited spontaneous EAE while the incidence was very low on a $RAG^{+/+}$ background (Lafaille et al., 1994). Transfer of $CD4^+$ T cells from wild-type mice prevented the spontaneous disease (Olivares-Villagomez et al., 1998; Van de Keere and Tonegawa, 1998). These data suggested that a regulatory $CD4^+$ T cell population exists in the $RAG^{+/+}$ TCR transgenic mice due to the rearrangement of endogenous TCR chains that suppressed spontaneous EAE. Indeed, $CD4^+CD25^+$ T cells purified from $RAG^{+/+}$ MBPAc1-11 TCR transgenic mice prevented spontaneous EAE (Hori et al., 2002).

Our laboratory has also used MBPAc1-11 TCR transgenic mice to study regulatory T cell activity specifically within the CNS. We asked whether the lack of spontaneous EAE in TCR transgenic mice older than 12 weeks reflected an increase in regulatory T cells within the target organ. We found that mononuclear cells isolated from the CNS but not the periphery of older TCR transgenic mice were able to suppress proliferation of MBPAc1-11-specific T cells in vitro to cognate antigen, while CNS cells from non-transgenic mice had no effect (Brabb et al., 2000). We have also observed a potent effect of small numbers of CNS cells transferred from older TCR transgenic mice into young TCR transgenic mice in preventing spontaneous EAE in vivo (unpublished observations).

HUMANIZED MBP-SPECIFIC CD4$^+$ TCR TRANSGENIC MICE

In a step toward translating the mouse models to human disease, several humanized TCR transgenic models have been generated in which transgenic human TCRs and MHC molecules are expressed in mice. One of the immediate benefits of humanized TCR transgenic mice is to test the potential contribution of specific human TCRs in disease development. Fugger and colleagues established the first humanized TCR transgenic mouse model specific for MBP85-99, a dominant epitope, in the context of HLA-DR2 allele (Madsen et al., 1999). This mouse model is generated by crossbreeding three different kinds of transgenic mice. DR2 transgenic mice express transgenes encoding the human HLA genes DRA*0101 and DRB1*1501 under the control of a mouse MHC class II promoter. TCR transgenic mice express a chimeric TCR composed of the variable domains of a human MBP85-99-specific TCR joined to mouse TCR constant regions. This transgenic human TCR recognizes mouse MBP85-99 since this region is identical between mouse and human. CD4 transgenic mice express the human CD4 co-receptor. Similar to the mouse MBPAc1-11 TCR transgenic mice, T cells develop normally in the humanized TCR transgenic mice and appear to circulate in the periphery as ignorant T cells. Approximately 4% of these mice develop spontaneous EAE and, similar to mouse models, the percentage increases to 100% when the triple-transgenic mice are bred onto RAG-2$^{-/-}$ mice (Madsen et al., 1999). Immunization with MBP85-99 peptide with CFA and pertussis toxin induces classic EAE (Madsen et al., 1999). These results support the involvement of the HLA-DR2-restricted MBP85-99-specific T cells in the pathogenesis of MS.

Two other humanized mouse models specific for MBP85-99 were independently generated (Ellmerich et al., 2005; Ellmerich et al., 2004). These transgenic mice were made by crossing HLA-DR15 transgenic mice expressing DRA*0101 and DRB1*1501 with a TCR transgenic line expressing a TCR specific for MBP85-99. The two models employ the same TCR transgenic line, which expresses a Vα3/Vβ2 TCR. The models differ in the percentage of CD4$^+$ T cells expressing the transgenic TCR. In line 8, about 44% of CD4$^+$ splenocytes express Vβ2, while in line 7, over 95% of CD4$^+$ splenocytes express Vβ2 (Ellmerich et al., 2005; Ellmerich et al., 2004). Spontaneous EAE is seen in only a small percentage of line 8 transgenic mice. However, line 8 transgenic mice spontaneously show slow and lethargic gait, which may resemble clinical signs seen in MS patients with milder disability (Ellmerich et al., 2004). Immunization of line 8 mice with MBP85-99 peptide in CFA and pertussis toxin increases the disease incidence and mice develop chronic paralysis. In contrast, approximately 60% of line 7 mice develop paralysis by 6 months of age (Ellmerich et al., 2005; Ellmerich et al., 2004). Breeding line 7 mice onto a RAG-2$^{-/-}$ background increases the incidence of spontaneous paralysis, reinforcing a possible role of regulatory T cells in MS (Ellmerich et al., 2005). Interestingly, "epitope spreading", which refers to the emergence of new myelin specificities during the course of disease that are independent of the initial specificity, is observed in both lines of human TCR transgenic mice (Ellmerich et al., 2004). In line 7 mice, the immune responses not only spread within MBP to MBP38-59, but also to an αB-crystallin epitope. The establishment of these two humanized MBP85-99 TCR transgenic models supports the notion that MBP85-99 epitope can be a target in MS, and provides alternative mouse models for the study of MS.

MS2-3C8 TCR transgenic mice represent another humanized MBP TCR transgenic mouse model (Quandt et al., 2004). The MS2-3C8 T cell clone was isolated from a relapsing and remitting MS patient and is specific for MBP111-129 in the context of HLA-DR4, another epitope that has been suggested to be immunodominant in MS. TCR transgenic mice were generated by joining MS2-3C8 TCR variable regions to mouse TCR constant regions. The TCR transgenic mice were then bred to HLA-DRB1*0401-expressing HLA-DR transgenic mice (Quandt et al., 2004). MS2-3C8 T cells can respond to mouse MBP111-129 although the sequences between human and mouse are different. These TCR transgenic mice do not develop spontaneous EAE, however, disease is induced by adoptively transferring MS2-3C8 transgenic T cells into irradiated HLA-DRB1*0401- transgenic mice that lack expression of mouse MHC class II molecules. While typical ascending paralysis is observed in recipient HLA-DR transgenic mice, some mice also develop atypical symptoms such as circulating movement or head-tilt (Quandt et al., 2004). These unique clinical signs suggest some heterogeneity in the types of immune responses generated in this model, which may result from the difference in the MBP peptide/HLA-DR allele complex in these mice compared to other humanized models.

Humanized MBP-specific TCR transgenic mice have also provided a means to study the association of certain HLA class II molecules with susceptibility to MS. The human MHC DR alleles DR2b and DR2a exhibit exceptionally strong linkage disequilibrium. Two human T cell clones expressing different TCRs specific for MBP85-99 were characterized and TCR transgenic models generated. The Hy.2E11 TCR recognizes the MBP epitope associated with DR2b molecules but also recognizes non-myelin-related epitopes presented by DR2a molecules. The Ob.1A12 TCR recognizes the same MBP85-99 without cross-reactivity to DR2a. By crossing the two different TCR transgenic mice to either HLA DR2b transgenic mice or HLA DR2b/ DR2a double transgenic mice, the effect of the expressing single or double MHC molecules transgenes on susceptibility to spontaneous EAE was studied (Gregersen et al., 2006). The investigators found that Hy.2E11$^+$-DR2b$^+$ mice spontaneous developed severe spontaneous EAE, however, both the disease incidence and severity significantly decreased with the addition of the DR2a transgene. In contrast, spontaneous EAE was not affected by the addition of DR2a transgene in TCR transgenic mice expressing the Ob.1A12 TCR and DR2b (Gregersen et al., 2006). These data suggested that the DR2b allele is associated with susceptibility to MS, and that co-expression of the DR2a allele decreases the potency of the DR2b allele effect on susceptibility.

MBP-Specific CD8$^+$ T Cells Involved in MS

Largely because of the widespread use of EAE as an animal model of MS, few studies addressed the pathogenic role of CD8$^+$ T cells in animal models of MS. EAE is induced by immunization with antigen emulsified in CFA, which is an effective means of priming CD4$^+$ MHC class II-restricted T cells. However, this immunization protocol is inefficient in priming CD8$^+$ MHC class I-restricted T cells, which recognize antigens that are synthesized intracellularly and then presented on the cell surface in association with MHC class I molecules. Thus, MS has been considered a CD4$^+$ T cell-mediated disease because the EAE model preferentially primes this T cell subset.

Despite the emphasis on $CD4^+$ T cells, a substantial amount of data exist implicating a pathogenic role for $CD8^+$ T cells in MS. In the 1980s, two studies found that $CD8^+$ T cells outnumber $CD4^+$ T cells almost ten-fold in MS brain tissue (Booss et al., 1983; Hauser et al., 1986). Subsequent studies confirmed a predominance of $CD8^+$ T cells in the CNS of MS patients (Cabarrocas et al., 2003; Gay et al., 1997). By analyzing the TCR Vβ gene usage, several groups found evidence of oligoclonal expansion in the $CD8^+$ but not $CD4^+$ T cell subset in the blood, CSF and brain in MS patients, suggesting recognition of specific antigen(s) in the CNS by $CD8^+$ T cells (Babbe et al., 2000; Jacobsen et al., 2002; Skulina et al., 2004). Employing novel approaches, two laboratories recently demonstrated an increase in MBP-specific $CD8^+$ T cells in MS patients compared to controls (Crawford et al., 2004; Zang et al., 2004). Using the sensitive flow cytometric assay for T cell proliferative responses, Crawford et al. showed that the percentage of MBP-reactive $CD8^+$ T cells increases in relapsing-remitting MS and primary progressive MS patients compared with healthy donors (Crawford et al., 2004). Using fluorescently labeled MBP peptide/HLA tetramer complexes, Zang et al. detected a higher frequency of $CD8^+$ T cells specific for MBP111-119 and MBP87-95 peptides in PBMCs of patients with MS compared to healthy individuals (Zang et al., 2004). MBP-specific $CD8^+$ T cells in MS patients are of $CD45RA^-$ $CD45RO^+$ memory T cells and produce TNF-α and IFN-γ (Tsuchida et al., 1994; Zang et al., 2004). $CD8^+$ T cells from secondary progressive MS patients secrete more lymphotoxin than those from normal donors when stimulated with anti-CD3 and anti-CD28 antibodies in vitro (Buckle et al., 2003). Using intravital microscopy, $CD8^+$ T cells in MS patients were observed to exhibit more adhesiveness to brain endothelium, a property dependent on expression of P-selectin glycoprotein ligand-1 (Battistini et al., 2003). In vitro, MBP-specific $CD8^+$ T cells are capable of killing human oligodendrocytes or peptide-pulsed HLA-A2-transfected Hmy2.C1R cells (Jurewicz et al., 1998).

The data described above suggest that MBP-specific $CD8^+$ T cell may play an important role in the pathogenesis of MS. The mechanisms by which $CD8^+$ T cells mediate CNS damage are not yet known. The activity of $CD8^+$ T cells may differ substantially from $CD4^+$ T cells in MS because $CD8^+$ and $CD4^+$ T cells utilize different effector functions and can recognize a different range of antigen-presenting cells (APCs) presenting MBP. $CD4^+$ T cells mediate tissue destruction via secretion of toxic soluble factors, while $CD8^+$ T cells can lyse cells directly via FAS-, perforin- or granzyme-mediated pathways, in addition to producing soluble mediators of inflammation. MHC class II molecules that present antigens to $CD4^+$ T cells are only expressed by "professional" APCs, such as dendritic cells, macrophages (or microglia cells), and B cells. In contrast, MHC class I molecules are ubiquitously expressed on almost all cell-types. In the CNS, MHC class I molecules can be expressed by astrocytes, oligodendrocytes, neurons and axons, with the highest expression on astrocytes (Hoftberger et al., 2004; Neumann et al., 2002). Inflammatory conditions that generate cytokines such as IFN-γ further increase MHC class I and II expression (Hamo et al., 2007; Lassmann et al., 1991). Thus, synthesis of MBP by oligodendrocytes and Schwann cells makes them potential targets for $CD8^+$ T cells, in addition to any cell that phagocytoses MBP and expresses class I on the surface in an inflammatory milieu.

A few studies have investigated the peptide specificity of human MBP-specific $CD8^+$ T cells (summarized in Table 2). A common strategy employed to address this question is to identify MBP peptides with highest potential binding affinity to a specific HLA class I molecule using a computer algorithm. Predicted epitopes are confirmed by generating $CD8^+$

T cell lines with the selected peptides. Using this method, Tsuchida et al. demonstrated that MBP110-118 is a dominant peptide of HLA-A2-restricted $CD8^+$ T cells (Tsuchida et al., 1994). Similarly, Zang et al. identified four MBP dominant epitopes for $CD8^+$ T cells (Zang et al., 2004). Peptides MBP111-119 and MBP87-95 bind to HLA-A2 while MBP134-142 and MBP14-22 peptides are associated with HLA-A24. An alternative approach that does not depend on computational methods is to refold the denatured HLA class I molecule with a panel of overlapping MBP nonamers, and then determining which peptides generate intact pMHC complexes. Using this technique, Tanigaki et al. were able to identify six HLA-A2-restricted peptides, nine HLA-B27-associated peptides, two HLA-B35-restricted peptides and three HLA-B51-restricted peptides (Tanigaki et al., 1994). Further characterization of T cell lines specific for these peptides will help identify the dominant peptide for each HLA class I allele.

Table 2. Immunodominant regions of human and mouse MBP recognized by $CD8^+$ T cells

	Dominant regions	MHC restriction	References
Human	MBP14-22	HLA-A24	Zang et al., 2004
	MBP87-95	HLA-A2	Zang et al., 2004
	MBP110-118	HLA-A2	Tsuchida et al., 1994; Zang et al., 2004
	MBP134-142	HLA-A24	Zang et al., 2004
Mouse C3H/FeJ	MBP79-87 *	$H\text{-}2k^k$	Huseby et al., 1999

*Immunodominancy only exists in $MBP^{-/-}$ mice.

PATHOGENICITY OF MBP-SPECIFIC $CD8^+$ T CELLS IS DEMONSTRATED IN NEW EAE MODELS

Because immunization with MBP and other myelin proteins in adjuvant typically induces $CD4^+$ T cell-mediated EAE, an animal model of MS based on $CD8^+$ T cell activity remained elusive for a long time. To address this problem, our group employed a novel strategy in which a MBP cDNA was cloned into both vaccinia virus and adenovirus vectors. The use of recombinant viruses results in intracellular synthesis of MBP and efficient presentation of MBP/MHC class I complexes on the cell surface. We also employed MBP-deficient ($MBP^{-/-}$) shiverer mice, which have a deletion in classic MBP exons 7-11 so that they lack the majority of MBP isoforms (Kimura et al., 1985; Molineaux et al., 1986; Roach et al., 1985). Both $MBP^{-/-}$ and wild-type C3HeB/Fej mice were infected with vaccinia virus encoding MBP. T cells from these primed mice were then re-stimulated in vitro with cells infected with adenovirus encoding MBP in order to detect and expand $CD8^+$ MBP-specific rather than virus-specific T cells (Huseby et al., 1999). $CD8^+$ T cell lines specific for MBP were generated readily in $MBP^{-/-}$ mice and the epitope specificity was mapped with overlapping MBP peptides. MBP79-87 was identified as a dominant epitope restricted to $H\text{-}2K^k$ (Huseby et al., 1999). Once this epitope was mapped, it was possible to immunize wild-type C3HeB/Fej mice directly with the MBP79-87 peptide and establish T cell clones. Interestingly, transfer of MBP79-87-specific T cell clones induced severe neurological symptoms and significant weight loss in recipient mice (Huseby et al., 2001a). The symptoms resemble many features of MS not seen in $CD4^+$ T cell-mediated EAE models, such as ataxia, spasticity, hyperreflexiveness and loss of coordinated movements. In contrast to $CD4^+$ T cell-mediated models of EAE in which the spinal cord is the predominant target, inflammatory infiltrates were localized in the brain but not the spinal cord. Extensive demyelination was observed in the brain as well. Neutralizing IFN-γ activity ameliorated the $CD8^+$ T cell-mediated disease, but neutralizing TNF-α or preventing FAS/FASL interactions did not affect disease induction (Huseby et al., 2001a).

To further characterize the MBP-specific $CD8^+$ T cell-mediated form of EAE in C3HeB/Fej mice, we generated TCR transgenic mice specific for MBP79-87. Two TCR transgenic lines were established that express either a Vα8/Vβ6 TCR (referred to as 8.6) or a Vα8/Vβ8 TCR (referred to 8.8). MBP-specific T cells exhibited a high degree of tolerance in the 8.6 transgenic line (discussed below) but, surprisingly, T cells in the 8.8 line developed normally and circulated in the periphery as apparently ignorant T cells (Perchellet et al., 2004). In contrast to MBP-specific $CD4^+$ TCR transgenic models, this $CD8^+$ TCR transgenic model exhibited no spontaneous EAE, even when crossed onto a $RAG^{-/-}$ background (Perchellet et al., 2004). 8.8 T cells are encephalitogenic because adoptive transfer of in vitro activated T cells isolated from the 8.8 mice caused recipient mice to lose weight and display neurological symptoms.

We also investigated EAE induction in 8.8 mice by viral infections, a long-term suspected environmental factor in triggering MS (Granieri et al., 2001; Meinl, 1999). Surprisingly, we found that wild-type vaccinia virus induced disease as efficiently as recombinant MBP-encoding vaccinia virus (Ji and Goverman, 2007) in the 8.8 TCR transgenic mice. The disease does not appear to be caused by poor viral clearance in this TCR transgenic mouse because 8.8 mice cleared vaccinia virus similarly to wild-type mice. We are currently exploring the mechanisms by which viruses trigger disease in this model.

B Cells Involved in MS and EAE

MBP-specific B cells could contribute to MS in two ways: either as APCs presenting MBP epitopes to MBP-specific T cells and/or by producing MBP-specific antibodies. Most studies have focused on the role of MBP-specific antibodies. For MBP-specific antibody secretion to occur, naïve B cells expressing an Ig receptor specific for a MBP epitope must bind to that epitope and at the same time receive "help" from $CD4^+$ T cells. For $CD4^+$ T cells to provide help, they must recognize a peptide/MHC class II complex on the B cell surface whose Ig receptor has been triggered. Typically, the peptide/MHC class II complex recognized by T cells is derived from the same antigen that bound to the B cell Ig receptor. T cells then provide growth factors for B cells and stimulate important B cell/T cell interactions that stimulate naive B cell differentiation. Upon receiving this help from T cells, naive B cells differentiate into antibody-secreting plasma cells, which produce large amounts of antigen-specific antibody. The T cell help provided during B cell activation also facilitates somatic hypermutation, which allows selection of B cells secreting antibodies with the same specificity but higher affinity for antigen. T cell help also facilitates class-switching, allowing antibodies with the same specificity but differing effector functions to be generated.

Anti-MBP antibody responses have been investigated in MS for over 20 years, however, two important questions remain to be resolved. The first question is whether anti-MBP antibodies are associated with MS. The results so far are conflicting, depending on the techniques used to detect MBP-specific antibodies. Several studies reported the detection of antibodies against MBP in either serum (Berger et al., 2003; Reindl et al., 1999) or CSF (Cruz et al., 1987; Warren and Catz, 1987; Warren and Catz, 1999; Warren et al., 1994), while others failed to detect them (Brokstad et al., 1994; Chou et al., 1983; Colombo et al., 1997). Using a solution-phase radioimmunoassay (RIA), O'Connor et al. recently failed to detect high affinity antibodies against MBP (O'Connor et al., 2003), but the possibility that anti-MBP antibodies are synthesized locally in the CNS and then bound to myelin is not excluded (Gerritse et al., 1994; Warren and Catz, 1993). Indeed, when the rearranged Ig V genes cloned from B cells isolated from the CSF of three MS patients were re-engineered and tested for specificity, 9/10 Fab fragments were reactive to MBP (Lambracht-Washington et al., 2007).

The epitope specificity of anti-MBP antibodies has also been mapped. Since MBP is an unstructured and flexible polypeptide chain, it was suggested that the immunodominant epitopes may be linear peptide segments rather than conformational epitopes (Wucherpfennig et al., 1997). Using polyclonal human anti-MBP antibodies isolated from CSF of MS patients, Whitaker et al. identified a core MBP epitope consisting of residues 80-89 (Whitaker et al., 1986). Consistent with this finding, Warren et al. used synthetic peptides to neutralize binding of antibodies purified from the CNS of MS patients and found that the most likely epitope recognized by most anti-MBP antibodies is localized between residues 84 and 95 (Warren and Catz, 1993). Another study found that B cells secreting MBP70-89-specific antibodies were detected at a higher frequency in MS patients (Martino et al., 1991). Collectively, these data indicate that MBP70-100 may constitute an immunodominant epitope for anti-MBP antibodies. Further studies suggest that the dominant MBP-specific Ig epitope is located between residue 85 and 96, a sequence recognized by MBP-specific $CD4^+$ and $CD8^+$ T cells as well (Warren et al., 1995; Wucherpfennig et al., 1997).

The other major question regarding anti-MBP antibodies is whether they are pathogenic or whether their presence merely reflects the outcome of myelin breakdown. Unlike MOG, which is an integral membrane protein, MBP is an intracellular protein, bound to the cytosolic surface of the oligodendrocytes (Antel and Bar-Or, 2003). Thus, it seems very unlikely that anti-MBP antibodies could access MBP within the myelin sheath, recruit complement components, or activate innate immune effector cells such as NK cells and macrophages. However, Ponomarenko and colleagues recently provided evidence that anti-MBP antibodies possess catalytic activity (Ponomarenko et al., 2006a; Ponomarenko et al., 2006b). They demonstrated that anti-MBP antibodies purified from human MS patients and mice with EAE can degrade purified MBP in vitro. This proteolytic activity is specific for MBP, and can be eliminated by absorption with anti-human or anti-mouse Ig. This work raises the interesting question of whether anti-MBP antibodies could penetrate the thick myelin sheath or cell membrane of oligodendrocytes, as has been observed for other cell-penetrating antibodies identified in other autoimmune diseases such as SLE (Lim and Zouali, 2006; Putterman, 2004).

The contribution of anti-MBP antibodies to the pathogenesis of EAE has been investigated. The incidence of MBP-induced EAE was decreased in Lewis rats when B cells were depleted by injecting anti-IgM serum (Gausas et al., 1982; Willenborg and Prowse, 1983). Transfer of serum containing anti-MBP antibodies to the B cell-depleted rats restored susceptibility to EAE (Willenborg et al., 1986). However, anti-MBP antibodies may only exacerbate disease in these models as EAE-resistant rats such as Wistar, Fischer, and Brown-Norway rats generated similar specificities and higher titer of anti-MBP antibodies after MBP immunization without developing clinical disease (Figueiredo et al., 1999; Rivero et al., 1999). In contrast to these studies in rats, Morris-Downes and colleagues showed that anti-MBP antibodies reduced clinical symptoms of EAE in mice (Morris-Downes et al., 2002).

In addition to their ability to generate specific antibodies, B cells can also influence the immune response by functioning as APCs. Because B cells are not very phagocytic, they are generally considered to be efficient APCs only for antigens that they have internalized via their specific Ig receptor. Once activated, B cells increase the capacity of antigen presentation, up-regulate the expression of co-stimulatory molecules necessary for activation of naïve T cells, and produce inflammatory cytokines such as lymphotoxin and TNF-α (Duddy and Bar-Or, 2006; Duddy et al., 2007). While this scenario could allow MBP-specific B cells to contribute to MBP-specific T cell activation, the frequency of B cells with any particular antigen specificity should not be very high in the normal repertoire.

There is currently little data regarding a role for MBP-specific B cells influencing MS via their APC function. Several studies have reported phenotypic and functional changes of B cells in MS. For example, two groups found that MS patients have more $B7.1^+$ B cells but fewer $B7.2^+$ B cells in the CSF than patients with non-inflammatory neurological disease (Sellebjerg et al., 1998; Svenningsson et al., 1997). Genc et al. showed that $B7.1^+$ B cells are significantly increased during MS exacerbation compared to stable periods (Genc et al., 1997). Recently, elevated lymphotoxin and TNF-α and decreased IL-10 have been detected in B cells from MS patients (Duddy et al., 2007).

A role for MBP-specific B cells as APCs in EAE has been investigated but remains controversial. In one study, SJL/J mice that had been depleted of B cells by administration of anti-IgM serum from birth were resistant to MBP-induced EAE but reconstitution with B cells prior to immunization rendered mice again susceptible to disease (Myers et al., 1992). Anti-MBP antibodies enhanced EAE induced by adoptively transferred T cells, leading the authors to conclude that B cells serve as the major APCs in EAE induction and MBP-specific antibodies enhanced MBP presentation by APCs (Myers et al., 1992). A different study using MBPAc1-11 immunization in B10.PL wild-type and $\mu mT^{-/-}$ mice found no difference in disease onset or severity but less recovery in the B-cell deficient mice, suggesting that B cells are not required for T cell activation and EAE induction, but may play a role in regulating disease after onset (Dittel et al., 2000; Wolf et al., 1996).

Recently, our laboratory reported the surprising finding that naïve, resting B cells in mice constitutively present endogenous MBP (Seamons et al., 2006). We used TCR transgenic mice specific for MBP121-140 to detect the presence of this epitope directly ex vivo on spleen and lymph node cells. The B cells acquired the MBP via an undefined BCR-independent mechanism, and B cells in lymph nodes present much more endogenous MBP than B cells in the spleen. Both naïve and activated/memory T cells proliferated in response to resting B cells presenting endogenous MBP; however, cytokines were not produced and the T cells became more refractory to subsequent stimulation. These observations suggest that presentation of endogenous MBP by resting B cells may have a tolerogenic effect on activated MBP-specific T cells. Interestingly, addition of MBP peptide to the resting B cells was sufficient to trigger cytokine production by activated/memory T cells, suggesting that an increase in the concentration of degraded MBP in vivo may help expand pathogenic T cells (Seamons et al., 2006).

Immune Tolerance to Self-Antigens

Establishing immune tolerance to self-antigens is critical to prevent autoimmune diseases (Rose and Mackay, 2006). The fact that autoimmune diseases such as MS occur demonstrates that immune tolerance is not perfect; the combination of genetic susceptibility and environmental influences can circumvent the many mechanisms of tolerance that have evolved over time. We and others have extensively investigated mechanisms responsible for immune tolerance to MBP because a breakdown in these mechanisms is believed to contribute to the pathogenesis of MS in some patients. Thus, defining mechanisms that maintain immune tolerance to MBP may help understand the pathogenesis of MS, and may also identify immunological strategies to treat or prevent this disease. We focused on T cell tolerance because MS is regarded as a T cell-mediated disease.

T cells mature in the thymus where they are subjected to selective processes designed to eliminate self-reactive T cells and to promote the maturation of T cells that are best equipped to recognize foreign antigens. T cell selection is based on the avidity of TCRs for self-peptide/MHC complexes presented by APCs in the thymus(Anderton and Wraith, 2002; Hammerling et al., 1991; Starr et al., 2003). Immature T cells expressing newly rearranged TCRs that bind too weakly to any peptide/MHC complex are not useful and undergo apoptosis (death by neglect). Thymocytes expressing TCRs that bind a peptide/MHC complex too strongly are destructive and are also eliminated (negative selection), or undergo receptor editing in which a different Vα gene segment is rearranged in order to generate a TCR that does not bind self-peptide/MHC complexes too tightly (Hammerling et al., 1991; Nemazee and Hogquist, 2003; Starr et al., 2003). Elimination of T cells expressing self-reactive TCRs in the thymus is termed central tolerance. The result of these processes is that only T cells with an intermediate avidity for peptide/MHC complexes that is not sufficient to generate self-reactive responses should be exported to the periphery.

$CD4^+$ T Cell Immune Tolerance to MBP

The ability to induce EAE by immunization with myelin self-antigens was the first demonstration that central tolerance is incomplete. Exposing peripheral T cells to myelin antigens in the presence of a strong adjuvant overcomes the ignorant state of myelin-specific T cells that escaped central tolerance. Once these previously ignorant T cells are activated, they respond aggressively to APCs presenting self-antigens. The availability of TCR transgenic mice greatly accelerated the study of how T cell tolerance toward MBP is both induced and circumvented. The MBPAc1-11-specific TCR transgenic models illustrated the ability of these T cells to escape central tolerance and exist in the periphery in a state of ignorance until they are activated either by microbes in the environment or by experimental manipulation (Goverman et al., 1993; Lafaille et al., 1994; Liu et al., 1995). Initial hypotheses suggested that MBPAc1-11-specific T cells escaped central tolerance because MBP was synthesized only by myelin-forming cells and would not be present in the thymus to mediate negative selection. To test this hypothesis, we investigated the immune response in B10.PL $MBP^{-/-}$ mice, which lack the self-antigen to induce tolerance. We found that endogenous classic MBP exerts strong immune tolerance because a much stronger T cell response is generated in $MBP^{-/-}$ compared to wild-type mice. However, the response is primarily directed toward a region within MBP121-150, while the MBPAc1-11-specific response was not increased (Harrington et al., 1998). Thus, differential immune tolerance occurs in B10.PL mice such that MBPAc1-11-specific T cells escape tolerance but MBP121-150-specific T cells do not.

Subsequent experiments demonstrated that the basis for the differential tolerance is the different affinities that the two MBP peptides have for the MHC class II molecule ($I\text{-}A^u$) that presents them. While epitopes within MBP121-150 bind $I\text{-}A^u$ with high affinity, the MBPAc1-11/$I\text{-}A^u$ complex is very unstable and has a half-life of only about 30 minutes (Loftus et al., 1999). The reason for the poor binding of MBPAc1-11 to $I\text{-}A^u$ is that the principal MHC anchor residue in the peptide is a lysine, which is not favorable for binding in the deep hydrophobic pocket of $I\text{-}A^u$ in which the MHC anchor residue must reside (Fairchild et al., 1993; He et al., 2002; Lee et al., 1998) (see Chapter V). Other studies from our group revealed a second reason why few MBPAc1-11/$I\text{-}A^u$ complexes are present on the surface of APCs. We analyzed the MBP peptides that are bound to $I\text{-}A^u$-expressing APCs after processing intact MBP purified from myelin. Interestingly, MBPAc1-17 and MBPAc1-18 were the most abundant, naturally processed MBP peptides eluted from B10.PL APCs (Seamons et al., 2003). These naturally-processed peptides have two distinct $I\text{-}A^u$ binding registers (residues that lie within the MHC groove in a pattern defined by the anchor residues), MBPAc1-9 and MBPAc5-16. The first binding register places a lysine in the hydrophobic MHC pocket, but the second register places a tyrosine in this pocket, which greatly increases the binding affinity (Seamons et al., 2003) (Fairchild et al., 1996). When the two registers contained within the same processed peptide compete for binding to $I\text{-}A^u$, MBPAc5-16 dominates, such that few complexes with MBPAc1-9 sequences bound within the MHC groove are presented on APCs (Seamons et al., 2003). Consistent with this observation, a mutated MBPAc1-9 peptide with tyrosine at residue 4 greatly increases its binding to $I\text{-}A^u$ (Fugger et al., 1996). Administering this mutated peptide into TCR transgenic mice causes the deletion of T cells reactive to MBPAc1-11, while no deletion is observed with the original peptide (Liu et al., 1995). Interestingly, enough MBPAc1-17 and MBPAc1-18 must be bound to $I\text{-}A^u$ in the MBAc1-9 register in vivo to activate the pathogenic MBPAc1-11-specific T cells that induce EAE in B10.PL mice.

Another example by which MBP processing shapes the T cell repertoire in the periphery is the demonstration that a dominant peptide epitope (MBP92-98) in humans and SJ/L mice that lies within MBP89-101 is destroyed by proteolytic cleavage during normal MBP processing (Anderton et al., 2002; Manoury et al., 2002). Two minor epitopes are generated instead in mice that are not affected by this cleavage, MBP89–94 and MBP95–101, and T cells specific for these epitopes induce EAE. Thus, using the MBP89-101 peptide to induce tolerance will eliminate MBP92-98-specific T cells, which are not pathogenic, but does not inactivate the pathogenic T cells because the MBP89-94 and 95-101 registers are poorly presented when the MBP89-101 peptide is used as their source. This study highlights the importance of understanding how a self-antigen is processed in order to design epitope-specific strategies for inducing tolerance.

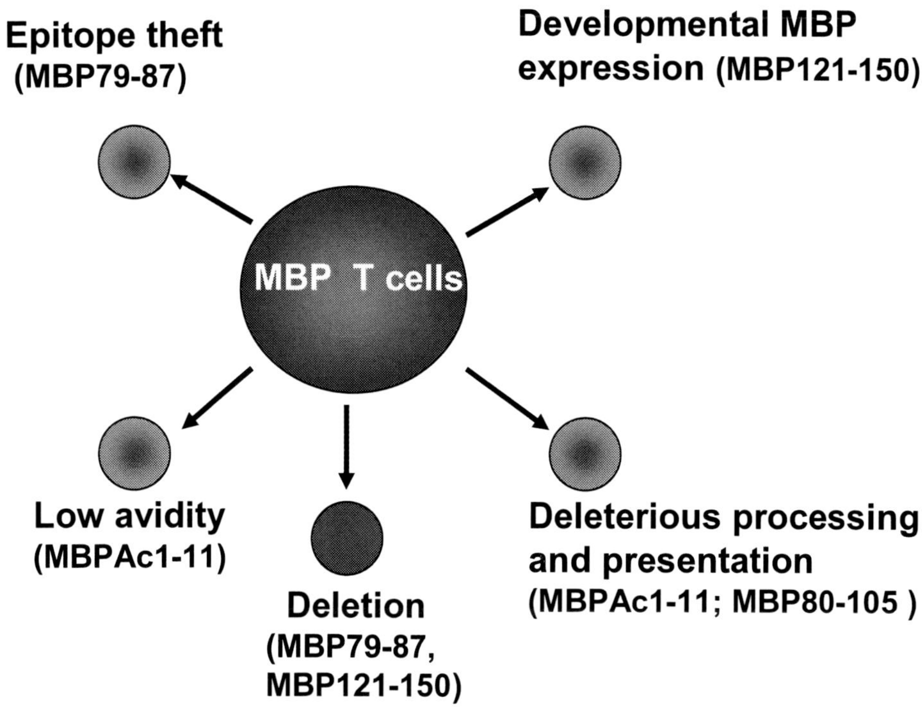

Figure 1. Escape from central tolerance by MBP-specific T cells. This graph summarizes the mechanisms by which T cells specific for different MBP epitopes escape deletion in the thymus. The red circle represents T cells that are deleted in the thymus and thus are barely detected in the periphery. Green circles represent T cells that mature in the thymus and circulate in the periphery. Epitope theft refers to a novel form of tolerance in which T cells bind MBP peptide/MHC ligands and strip these complexes from the APC cell surface without initiating T cell proliferation or triggering functional responses. This process occurs in the thymus and periphery.

To investigate mechanisms that mediate MHC class II-restricted T cell tolerance, we generated TCR transgenic mice specific for MBP121-150, the region of MBP that induces strong tolerance in wild-type but not shiverer mice. This region of MBP is not contained in any golli-MBP isoforms, thus tolerance must be induced by expression of classic MBP itself. A comparison of the maturation of MBP121-150-specific TCR transgenic thymocytes in $MBP^{+/+}$ versus $MBP^{-/-}$ mice showed that expression of endogenous MBP in wild-type mice induced strong negative selection in the thymus (Huseby et al., 2001b). We initially hypothesized that the deletion of MBP-specific thymocytes was mediated by medullar thymic epithelial cells (mTECs) expressing MBP. These cells have been shown to express peripheral tissue-specific proteins derived from the CNS, liver, eye, muscle, and placenta under the control of the transcriptional autoimmune regulator gene (aire). The consequence of this mTEC expression is that tolerance can be induced to peripheral tissue-specific antigens that would not normally be synthesized in the thymus (Anderson et al., 2002; Derbinski et al., 2001). However, we showed using bone marrow chimeras that TCR transgenic thymocyte deletion was mediated by bone-marrow derived APCs and not by epithelial cells (Huseby et al., 2001b). Further bone marrow chimeras demonstrated that APCs could acquire endogenous MBP and present it in the thymus to induce negative selection of MBP-specific T cells. This observation implied that classic MBP derived from degraded myelin is constitutively presented by APCs, which we confirmed using in vivo MBP-specific T cell proliferation assays to endogenous MBP (Huseby et al., 2001b). Mechanisms of central tolerance for MBP-specific T cells are shown in Figure 1.

Because classic MBP is developmentally expressed such that little MBP protein is detected in the CNS in the first ten days after birth (Barbarese et al., 1978; Mathisen et al., 1993), we reasoned that there would be little endogenous MBP available in young mice to mediate central tolerance. Accordingly, the MBP121-150 TCR transgenic mice exhibited age-dependent deletion of thymocytes, with little negative selection observed in the first two-three weeks of life. By four weeks of age, however, the MBP-specific thymocytes are greatly decreased, and their number diminishes further as they age. Some transgenic T cells are present in the periphery of $MBP^{+/+}$ mice, but these T cells exhibit decreased proliferative response to MBP compared to transgenic T cells isolated from $MBP^{-/-}$ mice, suggesting that the T cells have become anergic. T cell anergy is a status of unresponsiveness toward cognate antigen that can occur when the antigen is presented under non-immunogenic conditions, i.e. presentation by immature dendritic cells (Steinman and Nussenzweig, 2002), or as a result of chronic antigen stimulation. Down-regulation of the MBP-specific TCR was also observed on T cells in the periphery of older transgenic mice. Based on these observations, we investigated how MBP121-150-specific T cells that escape negative selection in young mice respond when they encounter MBP for the first in the periphery due to the increase in myelination and presentation of degraded MBP by APCs. We mimicked this situation by isolating transgenic MBP-specific T cells from shiverer mice and injecting them directly into the periphery of wild-type $MBP^{+/+}$ mice. Wild-type recipients remained healthy, despite the fact that the MBP-specific T cells proliferated vigorously in vivo in response to APCs presenting endogenous MBP. However, $RAG^{-/-}$ recipients that lack regulatory T cells succumbed to a rapid, severe autoimmune disease in which the CNS and peripheral tissues were targeted, reflecting the presentation of endogenous classic MBP by APCs in all enervated tissues (Cabbage et al., 2007). These studies further showed that regulatory T cells caused the proliferating MBP-specific T cells to differentiate into a unique, tolerized phenotype with the ability to produce IL-10 and TGF-β1 upon subsequent encounter with activated APCs presenting MBP in vivo (Cabbage et al., 2007). This tolerant response depended on continuous activity of regulatory T cells because, in their absence, the tolerized MBP-specific T cells could again induce autoimmunity. Mechanisms of peripheral tolerance for MBP-specific T cells are shown in Figure 2.

Other studies using shiverer mice have identified different MBP epitopes that induce tolerance. In C3HeB/Fej mice, T cell tolerance is induced to MBP79-87. We identified the MBP79-87 epitope as a MHC class I-restricted epitope presented by H-$2K^k$ recognized by $CD8^+$ T cells (Huseby et al., 1999). In contrast, Targoni et al. defined this same region as a

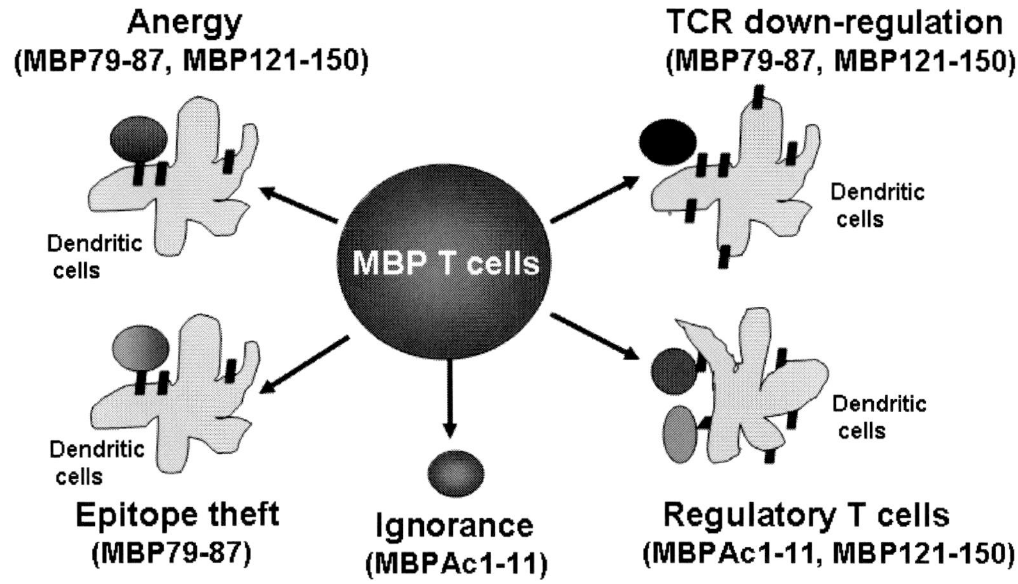

Figure 2. Regulation of MBP-specific T cells in the periphery. Mature MBP-specific T cells that escape central tolerance circulate in the periphery, where several mechanisms of peripheral tolerance prevent them from initiating autoimmunity. Different fates of T cells in the periphery are indicated by different colors

MHC class II-restricted epitope presented by I-A^k recognized by $CD4^+$ T cells (Targoni and Lehmann, 1998). It is possible that this epitope binds to both MHC molecules. Immunization of BALB/c mice (H-2^d haplotypes) on the shiverer background also revealed two regions of MBP, 59-76 and 89-101, that induce tolerance in wild-type mice. Adoptive transfer of T cell clones induced severe EAE in BALB/c recipients, which are normally resistant to EAE (Yoshizawa et al., 1998). Collectively, these studies indicate that immune tolerance eliminates a large number of MBP-specific pathogenic T cells provided that there is sufficient avidity between the T cell and the APC presenting the MBP peptide/MHC complex.

$CD8^+$ T Cell Immune Tolerance to MBP

The $CD8^+$ TCR transgenic mouse models that we described above allowed us to define several tolerance mechanisms that operate on the $CD8^+$ MBP-specific T cells, and to identify a novel form of tolerance that has not been observed in other systems. The $CD8^+$ MBP79-87-specific 8.6 transgenic line exhibited negative selection in the thymus. This result was expected because MBP79-87 is contained within isoforms of golli-MBP, which are expressed in the thymus by thymocytes, TECs, macrophages (Chignola et al., 2000; Feng et al., 2000; Marty et al., 2002; Voskuhl, 1998). The few 8.6 T cells that circulate in the periphery of 8.6 $MBP^{+/+}$ mice also exhibit anergic responses to stimulation with MBP in vitro (Perchellet et al., 2004). The 8.8 TCR transgenic line, however, did not appear to exhibit any tolerance in vivo and proliferated vigorously to MBP peptide in vitro, similar to the phenotype of ignorant

MBPAc1-11-specific T cells. We predicted that the lack of tolerance in 8.8 T cells would reflect a lower affinity of this TCR for the MBP79-87/MHC complex. Surprisingly, this did not appear to be the case, suggesting the novel hypothesis that not all high avidity MBP-specific T cells are deleted in the thymus. To investigate the tolerance mechanism in this system further, we generated mixed bone marrow chimeras by mixing 8.8 and 8.6 bone marrow cells together and adoptively transferring them into lethally irradiated wild-type mice. In these mice, the presence of the 8.8 T cells rescued the 8.6 T cells from deletion in the thymus and tolerance in the periphery (Figures 1 and 2). Additional experiments showed that 8.8 T cells appear to interact with and strip MBP79-87/H-2K^k complexes from APCs without generating an immune response of their own (Perchellet et al., 2004). While antigen-stripping by T cells has been observed before (Huang et al., 1999; Kedl et al., 2002; Taams et al., 1998), the process has always been accompanied by simultaneous T cell activation. Our finding that antigen-stripping can occur without T cell activation represents an unprecedented form of tolerance that allows high avidity CD8$^+$ T cells specific for MBP to circulate in the periphery of wild-type mice. It would be interesting to determine if a similar mechanism accounts for the presence of high avidity CD4$^+$ T cells specific for MBP13-32, MBP111-129, and MBP154-170 that have been described in patients with MS (Bielekova et al., 2004).

Conclusion

The immune response to MBP, considered to be a major antigen targeted in MS, has been extensively studied. Human T and B cell responses have been characterized in MS patients in order to define the pathogenesis of this disease. Our understanding has been greatly expanded by the use of an animal model, EAE, which is induced by stimulating MBP-specific T cell-mediated immunity. New EAE models have implicated both CD4$^+$ and CD8$^+$ T cells in the pathogenesis of CNS autoimmunity. These models have identified tolerance mechanisms and pathways that may provide therapeutic strategies for treatment of MS. TCR transgenic models specific for MBP have been particularly useful in dissecting pathogenic mechanisms. Nevertheless, many interesting questions remain unresolved. Developing and refining new models to improve correlation with human disease is needed to better understand MS and identify effective strategies for treatment and prevention.

References

Acha-Orbea, H., Mitchell, D. J., Timmermann, L., Wraith, D. C., Tausch, G. S., Waldor, M. K., Zamvil, S. S., McDevitt, H. O., and Steinman, L. (1988). Limited heterogeneity of T cell receptors from lymphocytes mediating autoimmune encephalomyelitis allows specific immune intervention. *Cell*, 54, 263-273.

Allegretta, M., Nicklas, J. A., Sriram, S., and Albertini, R. J. (1990). T cells responsive to myelin basic protein in patients with multiple sclerosis. *Science*, 247, 718-721.

Anderson, M. S., Venanzi, E. S., Klein, L., Chen, Z., Berzins, S. P., Turley, S. J., von

Boehmer, H., Bronson, R., Dierich, A., Benoist, C., and Mathis, D. (2002). Projection of an immunological self shadow within the thymus by the aire protein. *Science*, 298, 1395-1401.

Anderton, S. M. (2004). Post-translational modifications of self antigens: implications for autoimmunity. *Curr Opin Immunol*, 16, 753-758.

Anderton, S. M., Viner, N. J., Matharu, P., Lowrey, P. A., and Wraith, D. C. (2002). Influence of a dominant cryptic epitope on autoimmune T cell tolerance. *Nat Immunol*, 3, 175-181.

Anderton, S. M., and Wraith, D. C. (2002). Selection and fine-tuning of the autoimmune T-cell repertoire. *Nat Rev Immunol*, 2, 487-498.

Ando, D. G., Clayton, J., Kono, D., Urban, J. L., and Sercarz, E. E. (1989). Encephalitogenic T cells in the B10.PL model of experimental allergic encephalomyelitis (EAE) are of the Th-1 lymphokine subtype. *Cell Immunol*, 124, 132-143.

Antel, J. P., and Bar-Or, A. (2003). Do myelin-directed antibodies predict multiple sclerosis? *N Engl J Med*, 349, 107-109.

Babbe, H., Roers, A., Waisman, A., Lassmann, H., Goebels, N., Hohlfeld, R., Friese, M., Schroder, R., Deckert, M., Schmidt, S., et al. (2000). Clonal expansions of CD8(+) T cells dominate the T cell infiltrate in active multiple sclerosis lesions as shown by micromanipulation and single cell polymerase chain reaction. *J Exp Med*, 192, 393-404.

Barbarese, E., Carson, J. H., and Braun, P. E. (1978). Accumulation of the four myelin basic proteins in mouse brain during development. *J Neurochem*, 31, 779-782.

Baron, J. L., Madri, J. A., Ruddle, N. H., Hashim, G., and Janeway, C. A., Jr. (1993). Surface expression of alpha 4 integrin by CD4 T cells is required for their entry into brain parenchyma. *J Exp Med*, 177, 57-68.

Battistini, L., Piccio, L., Rossi, B., Bach, S., Galgani, S., Gasperini, C., Ottoboni, L., Ciabini, D., Caramia, M. D., Bernardi, G., et al. (2003). CD8+ T cells from patients with acute multiple sclerosis display selective increase of adhesiveness in brain venules: a critical role for P-selectin glycoprotein ligand-1. *Blood*, 101, 4775-4782.

Berger, T., Rubner, P., Schautzer, F., Egg, R., Ulmer, H., Mayringer, I., Dilitz, E., Deisenhammer, F., and Reindl, M. (2003). Antimyelin antibodies as a predictor of clinically definite multiple sclerosis after a first demyelinating event. *N Engl J Med*, 349, 139-145.

Bernard, C. C., and Carnegie, P. R. (1975). Experimental autoimmune encephalomyelitis in mice: immunologic response to mouse spinal cord and myelin basic proteins. *J Immunol*, 114, 1537-1540.

Bettelli, E., Oukka, M., and Kuchroo, V. K. (2007). T(H)-17 cells in the circle of immunity and autoimmunity. *Nat Immunol*, 8, 345-350.

Bielekova, B., Sung, M. H., Kadom, N., Simon, R., McFarland, H., and Martin, R. (2004). Expansion and functional relevance of high-avidity myelin-specific CD4+ T cells in multiple sclerosis. *J Immunol*, 172, 3893-3904.

Billiau, A., Heremans, H., Vandekerckhove, F., Dijkmans, R., Sobis, H., Meulepas, E., and Carton, H. (1988). Enhancement of experimental allergic encephalomyelitis in mice by antibodies against IFN-gamma. *J Immunol*, 140, 1506-1510.

Boggs, J. M. (2006). Myelin basic protein: a multifunctional protein. *Cell Mol Life Sci*, 63, 1945-1961.

Booss, J., Esiri, M. M., Tourtellotte, W. W., and Mason, D. Y. (1983). Immunohistological analysis of T lymphocyte subsets in the central nervous system in chronic progressive multiple sclerosis. *J Neurol Sci*, 62, 219-232.

Brabb, T., Goldrath, A. W., von Dassow, P., Paez, A., Liggitt, H. D., and Goverman, J. (1997). Triggers of autoimmune disease in a murine TCR-transgenic model for multiple sclerosis. *J Immunol*, 159, 497-507.

Brabb, T., von Dassow, P., Ordonez, N., Schnabel, B., Duke, B., and Goverman, J. (2000). In situ tolerance within the central nervous system as a mechanism for preventing autoimmunity. *J Exp Med*, 192, 871-880.

Brokstad, K. A., Page, M., Nyland, H., and Haaheim, L. R. (1994). Autoantibodies to myelin basic protein are not present in the serum and CSF of MS patients. *Acta Neurol Scand*, 89, 407-411.

Buckle, G. J., Hollsberg, P., and Hafler, D. A. (2003). Activated CD8+ T cells in secondary progressive MS secrete lymphotoxin. *Neurology*, 60, 702-705.

Burns, F. R., Li, X. B., Shen, N., Offner, H., Chou, Y. K., Vandenbark, A. A., and Heber-Katz, E. (1989). Both rat and mouse T cell receptors specific for the encephalitogenic determinant of myelin basic protein use similar V alpha and V beta chain genes even though the major histocompatibility complex and encephalitogenic determinants being recognized are different. *J Exp Med*, 169, 27-39.

Burns, J., Bartholomew, B., and Lobo, S. (1999). Isolation of myelin basic protein-specific T cells predominantly from the memory T-cell compartment in multiple sclerosis. *Ann Neurol*, 45, 33-39.

Burns, J., Rosenzweig, A., Zweiman, B., and Lisak, R. P. (1983). Isolation of myelin basic protein-reactive T-cell lines from normal human blood. Cell Immunol. 81, 435-440.

Cabarrocas, J., Bauer, J., Piaggio, E., Liblau, R., and Lassmann, H. (2003). Effective and selective immune surveillance of the brain by MHC class I-restricted cytotoxic T lymphocytes. *Eur J Immunol*, 33, 1174-1182.

Cabbage, S. E., Huseby, E. S., Sather, B. D., Brabb, T., Liggitt, D., and Goverman, J. (2007). Regulatory T cells maintain long-term tolerance to myelin basic protein by inducing a novel, dynamic state of T cell tolerance. *J Immunol*, 178, 887-896.

Campagnoni, A. T., Pribyl, T. M., Campagnoni, C. W., Kampf, K., Amur-Umarjee, S., Landry, C. F., Handley, V. W., Newman, S. L., Garbay, B., and Kitamura, K. (1993). Structure and developmental regulation of Golli-mbp, a 105-kilobase gene that encompasses the myelin basic protein gene and is expressed in cells in the oligodendrocyte lineage in the brain. *J Biol Chem*, 268, 4930-4938.

Campagnoni, A. T., and Skoff, R. P. (2001). The pathobiology of myelin mutants reveal novel biological functions of the MBP and PLP genes. *Brain Pathol*, 11, 74-91.

Chignola, R., Cestari, T., Guerriero, C., Riviera, A. P., Ferrari, S., Brendolan, A., Gobbo, M., Amato, S., Sartoris, S., Fracasso, G., et al. (2000). Expression of myelin basic protein (MBP) epitopes in human non-neural cells revealed by two anti-MBP IgM monoclonal antibodies. *Clin Exp Immunol*, 122, 429-436.

Chou, C. H., Tourtellotte, W. W., and Kibler, R. F. (1983). Failure to detect antibodies to myelin basic protein or peptic fragments of myelin basic protein in CSF of patients with MS. *Neurology*, 33, 24-28.

Chou, Y. K., Bourdette, D. N., Offner, H., Whitham, R., Wang, R. Y., Hashim, G. A., and Vandenbark, A. A. (1992). Frequency of T cells specific for myelin basic protein and myelin proteolipid protein in blood and cerebrospinal fluid in multiple sclerosis. *J Neuroimmunol*, 38, 105-113.

Chou, Y. K., Buenafe, A. C., Dedrick, R., Morrison, W. J., Bourdette, D. N., Whitham, R., Atherton, J., Lane, J., Spoor, E., Hashim, G. A., and et al. (1994). T cell receptor V beta gene usage in the recognition of myelin basic protein by cerebrospinal fluid- and blood-derived T cells from patients with multiple sclerosis. *J Neurosci Res*, 37, 169-181.

Chou, Y. K., Vainiene, M., Whitham, R., Bourdette, D., Chou, C. H., Hashim, G., Offner, H., and Vandenbark, A. A. (1989). Response of human T lymphocyte lines to myelin basic protein: association of dominant epitopes with HLA class II restriction molecules. *J Neurosci Res*, 23, 207-216.

Coffman, R. L. (2006). Origins of the T(H)1-T(H)2 model: a personal perspective. *Nat Immunol*, 7, 539-541.

Colombo, E., Banki, K., Tatum, A. H., Daucher, J., Ferrante, P., Murray, R. S., Phillips, P. E., and Perl, A. (1997). Comparative analysis of antibody and cell-mediated autoimmunity to transaldolase and myelin basic protein in patients with multiple sclerosis. *J Clin Invest*, 99, 1238-1250.

Crawford, M. P., Yan, S. X., Ortega, S. B., Mehta, R. S., Hewitt, R. E., Price, D. A., Stastny, P., Douek, D. C., Koup, R. A., Racke, M. K., and Karandikar, N. J. (2004). High prevalence of autoreactive, neuroantigen-specific CD8+ T cells in multiple sclerosis revealed by novel flow cytometric assay. *Blood*, 103, 4222-4231.

Cruz, M., Olsson, T., Ernerudh, J., Hojeberg, B., and Link, H. (1987). Immunoblot detection of oligoclonal anti-myelin basic protein IgG antibodies in cerebrospinal fluid in multiple sclerosis. *Neurology*, 37, 1515-1519.

Cua, D. J., Sherlock, J., Chen, Y., Murphy, C. A., Joyce, B., Seymour, B., Lucian, L., To, W., Kwan, S., Churakova, T., *et al.* (2003). Interleukin-23 rather than interleukin-12 is the critical cytokine for autoimmune inflammation of the brain. *Nature*, 421, 744-748.

Derbinski, J., Schulte, A., Kyewski, B., and Klein, L. (2001). Promiscuous gene expression in medullary thymic epithelial cells mirrors the peripheral self. *Nat Immunol*, 2, 1032-1039.

Dittel, B. N., Urbania, T. H., and Janeway, C. A., Jr. (2000). Relapsing and remitting experimental autoimmune encephalomyelitis in B cell deficient mice. *J Autoimmun*, 14, 311-318.

Duddy, M., and Bar-Or, A. (2006). B-cells in multiple sclerosis. *Int MS J*, 13, 84-90.

Duddy, M., Niino, M., Adatia, F., Hebert, S., Freedman, M., Atkins, H., Kim, H. J., and Bar-Or, A. (2007). Distinct effector cytokine profiles of memory and naive human B cell subsets and implication in multiple sclerosis. *J Immunol*, 178, 6092-6099.

Duong, T. T., St Louis, J., Gilbert, J. J., Finkelman, F. D., and Strejan, G. H. (1992). Effect of anti-interferon-gamma and anti-interleukin-2 monoclonal antibody treatment on the development of actively and passively induced experimental allergic encephalomyelitis in the SJL/J mouse. *J Neuroimmunol*, 36, 105-115.

Ellmerich, S., Mycko, M., Takacs, K., Waldner, H., Wahid, F. N., Boyton, R. J., King, R. H., Smith, P. A., Amor, S., Herlihy, A. H., *et al.* (2005). High incidence of spontaneous disease in an HLA-DR15 and TCR transgenic multiple sclerosis model. *J Immunol*, 174, 1938-1946.

Ellmerich, S., Takacs, K., Mycko, M., Waldner, H., Wahid, F., Boyton, R. J., Smith, P. A., Amor, S., Baker, D., Hafler, D. A., et al. (2004). Disease-related epitope spread in a humanized T cell receptor transgenic model of multiple sclerosis. *Eur J Immunol*, 34, 1839-1848.

Encinas, J. A., Lees, M. B., Sobel, R. A., Symonowicz, C., Greer, J. M., Shovlin, C. L., Weiner, H. L., Seidman, C. E., Seidman, J. G., and Kuchroo, V. K. (1996). Genetic analysis of susceptibility to experimental autoimmune encephalomyelitis in a cross between SJL/J and B10.S mice. *J Immunol*, 157, 2186-2192.

Fairchild, P. J., Pope, H., and Wraith, D. C. (1996). The nature of cryptic epitopes within the self-antigen myelin basic protein. *Int Immunol*, 8, 1035-1043.

Fairchild, P. J., Wildgoose, R., Atherton, E., Webb, S., and Wraith, D. C. (1993). An autoantigenic T cell epitope forms unstable complexes with class II MHC: a novel route for escape from tolerance induction. *Int Immunol*, 5, 1151-1158.

Feng, J. M., Givogri, I. M., Bongarzone, E. R., Campagnoni, C., Jacobs, E., Handley, V. W., Schonmann, V., and Campagnoni, A. T. (2000). Thymocytes express the golli products of the myelin basic protein gene and levels of expression are stage dependent. J *Immunol*, 165, 5443-5450.

Ferber, I. A., Brocke, S., Taylor-Edwards, C., Ridgway, W., Dinisco, C., Steinman, L., Dalton, D., and Fathman, C. G. (1996). Mice with a disrupted IFN-gamma gene are susceptible to the induction of experimental autoimmune encephalomyelitis (EAE). *J Immunol*, 156, 5-7.

Figueiredo, A. C., Cohen, I. R., and Mor, F. (1999). Diversity of the B cell repertoire to myelin basic protein in rat strains susceptible and resistant to EAE. *J Autoimmun*, 12, 13-25.

Friese, M. A., and Fugger, L. (2005). Autoreactive CD8+ T cells in multiple sclerosis: a new target for therapy? *Brain*, 128, 1747-1763.

Fritz, R. B., Chou, C. H., and McFarlin, D. E. (1983a). Induction of experimental allergic encephalomyelitis in PL/J and (SJL/J x PL/J)F1 mice by myelin basic protein and its peptides: localization of a second encephalitogenic determinant. *J Immunol*, 130, 191-194.

Fritz, R. B., Chou, C. H., and McFarlin, D. E. (1983b). Relapsing murine experimental allergic encephalomyelitis induced by myelin basic protein. *J Immunol*, 130, 1024-1026.

Fugger, L., Liang, J., Gautam, A., Rothbard, J. B., and McDevitt, H. O. (1996). Quantitative analysis of peptides from myelin basic protein binding to the MHC class II protein, I-Au, which confers susceptibility to experimental allergic encephalomyelitis. *Mol Med*, 2, 181-188.

Gausas, J., Paterson, P. Y., Day, E. D., and Dal Canto, M. C. (1982). Intact B-cell activity is essential for complete expression of experimental allergic encephalomyelitis in Lewis rats. *Cell Immunol*, 72, 360-366.

Gay, F. W., Drye, T. J., Dick, G. W., and Esiri, M. M. (1997). The application of multifactorial cluster analysis in the staging of plaques in early multiple sclerosis. Identification and characterization of the primary demyelinating lesion. *Brain*, 120 (Pt 8), 1461-1483.

Genc, K., Dona, D. L., and Reder, A. T. (1997). Increased CD80(+) B cells in active multiple sclerosis and reversal by interferon beta-1b therapy. *J Clin Invest*, 99, 2664-2671.

Gerritse, K., Deen, C., Fasbender, M., Ravid, R., Boersma, W., and Claassen, E. (1994). The involvement of specific anti myelin basic protein antibody-forming cells in multiple sclerosis immunopathology. *J Neuroimmunol*, 49, 153-159.

Givogri, M. I., Bongarzone, E. R., and Campagnoni, A. T. (2000). New insights on the biology of myelin basic protein gene: the neural-immune connection. *J Neurosci Res*, 59, 153-159.

Goverman, J. (1999). Tolerance and autoimmunity in TCR transgenic mice specific for myelin basic protein. *Immunol Rev*, 169, 147-159.

Goverman, J., Woods, A., Larson, L., Weiner, L. P., Hood, L., and Zaller, D. M. (1993). Transgenic mice that express a myelin basic protein-specific T cell receptor develop spontaneous autoimmunity. *Cell*, 72, 551-560.

Granieri, E., Casetta, I., Tola, M. R., and Ferrante, P. (2001). Multiple sclerosis: infectious hypothesis. *Neurol Sci*, 22, 179-185.

Gregersen, J. W., Kranc, K. R., Ke, X., Svendsen, P., Madsen, L. S., Thomsen, A. R., Cardon, L. R., Bell, J. I., and Fugger, L. (2006). Functional epistasis on a common MHC haplotype associated with multiple sclerosis. *Nature*, 443, 574-577.

Hammerling, G. J., Schonrich, G., Momburg, F., Auphan, N., Malissen, M., Malissen, B., Schmitt-Verhulst, A. M., and Arnold, B. (1991). Non-deletional mechanisms of peripheral and central tolerance: studies with transgenic mice with tissue-specific expression of a foreign MHC class I antigen. *Immunol Rev*, 122, 47-67.

Hamo, L., Stohlman, S. A., Otto-Duessel, M., and Bergmann, C. C. (2007). Distinct regulation of MHC molecule expression on astrocytes and microglia during viral encephalomyelitis. *Glia*, 55, 1169-1177.

Harrington, C. J., Paez, A., Hunkapiller, T., Mannikko, V., Brabb, T., Ahearn, M., Beeson, C., and Goverman, J. (1998). Differential tolerance is induced in T cells recognizing distinct epitopes of myelin basic protein. *Immunity*, 8, 571-580.

Hauser, S. L., Bhan, A. K., Gilles, F., Kemp, M., Kerr, C., and Weiner, H. L. (1986). Immunohistochemical analysis of the cellular infiltrate in multiple sclerosis lesions. *Ann Neurol*, 19, 578-587.

He, X. L., Radu, C., Sidney, J., Sette, A., Ward, E. S., and Garcia, K. C. (2002). Structural snapshot of aberrant antigen presentation linked to autoimmunity: the immunodominant epitope of MBP complexed with I-Au. *Immunity*, 17, 83-94.

Heber-Katz, E., and Acha-Orbea, H. (1989). The V-region disease hypothesis: evidence from autoimmune encephalomyelitis. *Immunol Today*, 10, 164-169.

Hellings, N., Baree, M., Verhoeven, C., D'Hooghe M, B., Medaer, R., Bernard, C. C., Raus, J., and Stinissen, P. (2001). T-cell reactivity to multiple myelin antigens in multiple sclerosis patients and healthy controls. *J Neurosci Res*, 63, 290-302.

Hemmer, B., Vergelli, M., Calabresi, P., Huang, T., McFarland, H. F., and Martin, R. (1996). Cytokine phenotype of human autoreactive T cell clones specific for the immunodominant myelin basic protein peptide (83-99). *J Neurosci Res*, 45, 852-862.

Hemmer, B., Vergelli, M., Tranquill, L., Conlon, P., Ling, N., McFarland, H. F., and Martin, R. (1997). Human T-cell response to myelin basic protein peptide (83-99), extensive heterogeneity in antigen recognition, function, and phenotype. *Neurology*, 49, 1116-1126.

Hermans, G., Stinissen, P., Hauben, L., Van den Berg-Loonen, E., Raus, J., and Zhang, J. (1997). Cytokine profile of myelin basic protein-reactive T cells in multiple sclerosis and healthy individuals. *Ann Neurol*, 42, 18-27.

Hoftberger, R., Aboul-Enein, F., Brueck, W., Lucchinetti, C., Rodriguez, M., Schmidbauer, M., Jellinger, K., and Lassmann, H. (2004). Expression of major histocompatibility complex class I molecules on the different cell types in multiple sclerosis lesions. *Brain Pathol*, 14, 43-50.

Hori, S., Haury, M., Coutinho, A., and Demengeot, J. (2002). Specificity requirements for selection and effector functions of CD25+4+ regulatory T cells in anti-myelin basic protein T cell receptor transgenic mice. *Proc Natl Acad Sci*, USA, 99, 8213-8218.

Huang, J. F., Yang, Y., Sepulveda, H., Shi, W., Hwang, I., Peterson, P. A., Jackson, M. R., Sprent, J., and Cai, Z. (1999). TCR-Mediated internalization of peptide-MHC complexes acquired by T cells. *Science*, 286, 952-954.

Huseby, E. S., Liggitt, D., Brabb, T., Schnabel, B., Ohlen, C., and Goverman, J. (2001a). A pathogenic role for myelin-specific CD8(+) T cells in a model for multiple sclerosis. *J Exp Med,* 194, 669-676.

Huseby, E. S., Ohlen, C., and Goverman, J. (1999). Cutting edge: myelin basic protein-specific cytotoxic T cell tolerance is maintained in vivo by a single dominant epitope in H-2k mice. *J Immunol*, 163, 1115-1118.

Huseby, E. S., Sather, B., Huseby, P. G., and Goverman, J. (2001b). Age-dependent T cell tolerance and autoimmunity to myelin basic protein. *Immunity,* 14, 471-481.

Jacobsen, M., Cepok, S., Quak, E., Happel, M., Gaber, R., Ziegler, A., Schock, S., Oertel, W. H., Sommer, N., and Hemmer, B. (2002). Oligoclonal expansion of memory CD8+ T cells in cerebrospinal fluid from multiple sclerosis patients. *Brain*, 125, 538-550.

Ji, Q., and Goverman, J. (2007). Experimental autoimmune encephalomyelitis mediated by CD8+ T cells. *Ann N Y Acad Sci*, 1103, 157-166.

Jingwu, Z., Medaer, R., Hashim, G. A., Chin, Y., van den Berg-Loonen, E., and Raus, J. C. (1992). Myelin basic protein-specific T lymphocytes in multiple sclerosis and controls: precursor frequency, fine specificity, and cytotoxicity. *Ann Neurol*, 32, 330-338.

Jurewicz, A., Biddison, W. E., and Antel, J. P. (1998). MHC class I-restricted lysis of human oligodendrocytes by myelin basic protein peptide-specific CD8 T lymphocytes. *J Immunol*, 160, 3056-3059.

Kedl, R. M., Schaefer, B. C., Kappler, J. W., and Marrack, P. (2002). T cells down-modulate peptide-MHC complexes on APCs in vivo. *Nat Immunol*, 3, 27-32.

Keegan, B. M., and Noseworthy, J. H. (2002). Multiple sclerosis. *Annu Rev Med*, 53, 285-302.

Kennedy, M. K., Torrance, D. S., Picha, K. S., and Mohler, K. M. (1992). Analysis of cytokine mRNA expression in the central nervous system of mice with experimental autoimmune encephalomyelitis reveals that IL-10 mRNA expression correlates with recovery. *J Immunol*, 149, 2496-2505.

Kimura, M., Inoko, H., Katsuki, M., Ando, A., Sato, T., Hirose, T., Takashima, H., Inayama, S., Okano, H., Takamatsu, K., and et al. (1985). Molecular genetic analysis of myelin-deficient mice: shiverer mutant mice show deletion in gene(s) coding for myelin basic protein. *J Neurochem*, 44, 692-696.

Komiyama, Y., Nakae, S., Matsuki, T., Nambu, A., Ishigame, H., Kakuta, S., Sudo, K., and Iwakura, Y. (2006). IL-17 plays an important role in the development of experimental autoimmune encephalomyelitis. *J Immunol*, 177, 566-573.

Koritschoner, R. S., and Schweinburg, F. (1925). Inducktion von Paralyse und Ruckenmarksentzundung durch Immunisierung von Kaninchen mit menschlichem Ruckenmarksgewebe. *Z Immunitatsf Exp Therapie*, 42, 217-283.

Korn, T., Bettelli, E., Gao, W., Awasthi, A., Jager, A., Strom, T. B., Oukka, M., and Kuchroo, V. K. (2007). IL-21 initiates an alternative pathway to induce proinflammatory T(H)17 cells. *Nature*, 448, 484-487.

Krakowski, M., and Owens, T. (1996). Interferon-gamma confers resistance to experimental allergic encephalomyelitis. *Eur J Immunol*, 26, 1641-1646.

Lafaille, J. J., Keere, F. V., Hsu, A. L., Baron, J. L., Haas, W., Raine, C. S., and Tonegawa, S. (1997). Myelin basic protein-specific T helper 2 (Th2) cells cause experimental autoimmune encephalomyelitis in immunodeficient hosts rather than protect them from the disease. *J Exp Med*, 186, 307-312.

Lafaille, J. J., Nagashima, K., Katsuki, M., and Tonegawa, S. (1994). High incidence of spontaneous autoimmune encephalomyelitis in immunodeficient anti-myelin basic protein T cell receptor transgenic mice. *Cell*, 78, 399-408.

Lambracht-Washington, D., O'Connor, K. C., Cameron, E. M., Jowdry, A., Ward, E. S., Frohman, E., Racke, M. K., and Monson, N. L. (2007). Antigen specificity of clonally expanded and receptor edited cerebrospinal fluid B cells from patients with relapsing remitting MS. *J Neuroimmunol*, 186, 164-176.

Langrish, C. L., Chen, Y., Blumenschein, W. M., Mattson, J., Basham, B., Sedgwick, J. D., McClanahan, T., Kastelein, R. A., and Cua, D. J. (2005). IL-23 drives a pathogenic T cell population that induces autoimmune inflammation. *J Exp Med*, 201, 233-240.

Lassmann, H., Zimprich, F., Rossler, K., and Vass, K. (1991). Inflammation in the nervous system. Basic mechanisms and immunological concepts. *Rev Neurol* (Paris), 147, 763-781.

Lee, C., Liang, M. N., Tate, K. M., Rabinowitz, J. D., Beeson, C., Jones, P. P., and McConnell, H. M. (1998). Evidence that the autoimmune antigen myelin basic protein (MBP) Ac1-9 binds towards one end of the major histocompatibility complex (MHC) cleft. *J Exp Med*, 187, 1505-1516.

Liang, S. C., Tan, X. Y., Luxenberg, D. P., Karim, R., Dunussi-Joannopoulos, K., Collins, M., and Fouser, L. A. (2006). Interleukin (IL)-22 and IL-17 are coexpressed by Th17 cells and cooperatively enhance expression of antimicrobial peptides. *J Exp Med*, 203, 2271-2279.

Liblau, R., Tournier-Lasserve, E., Maciazek, J., Dumas, G., Siffert, O., Hashim, G., and Bach, M. A. (1991). T cell response to myelin basic protein epitopes in multiple sclerosis patients and healthy subjects. *Eur J Immunol*, 21, 1391-1395.

Liblau, R. S., Singer, S. M., and McDevitt, H. O. (1995). Th1 and Th2 CD4+ T cells in the pathogenesis of organ-specific autoimmune diseases. *Immunol Today*, 16, 34-38.

Lim, P. L., and Zouali, M. (2006). Pathogenic autoantibodies: emerging insights into tissue injury. *Immunol Lett*, 103, 17-26.

Liu, G. Y., Fairchild, P. J., Smith, R. M., Prowle, J. R., Kioussis, D., and Wraith, D. C. (1995). Low avidity recognition of self-antigen by T cells permits escape from central tolerance. *Immunity*, 3, 407-415.

Loftus, C., Huseby, E., Gopaul, P., Beeson, C., and Goverman, J. (1999). Highly cross-reactive T cell responses to myelin basic protein epitopes reveal a nonpredictable form of TCR degeneracy. *J Immunol*, 162, 6451-6457.

Lovett-Racke, A. E., Trotter, J. L., Lauber, J., Perrin, P. J., June, C. H., and Racke, M. K. (1998). Decreased dependence of myelin basic protein-reactive T cells on CD28-

mediated costimulation in multiple sclerosis patients. A marker of activated/memory T cells. *J Clin Invest*, 101, 725-730.

Lucchinetti, C., Bruck, W., Parisi, J., Scheithauer, B., Rodriguez, M., and Lassmann, H. (2000). Heterogeneity of multiple sclerosis lesions: implications for the pathogenesis of demyelination. *Ann Neurol*, 47, 707-717.

Madsen, L. S., Andersson, E. C., Jansson, L., krogsgaard, M., Andersen, C. B., Engberg, J., Strominger, J. L., Svejgaard, A., Hjorth, J. P., Holmdahl, R., et al. (1999). A humanized model for multiple sclerosis using HLA-DR2 and a human T-cell receptor. *Nat Genet*, 23, 343-347.

Mannie, M. D., Paterson, P. Y., U'Prichard, D. C., and Flouret, G. (1985). Induction of experimental allergic encephalomyelitis in Lewis rats with purified synthetic peptides: delineation of antigenic determinants for encephalitogenicity, in vitro activation of cellular transfer, and proliferation of lymphocytes. *Proc Natl Acad Sci,* USA, 82, 5515-5519.

Manoury, B., Mazzeo, D., Fugger, L., Viner, N., Ponsford, M., Streeter, H., Mazza, G., Wraith, D. C., and Watts, C. (2002). Destructive processing by asparagine endopeptidase limits presentation of a dominant T cell epitope in MBP. *Nat Immunol*, 3, 169-174.

Markovic-Plese, S., Cortese, I., Wandinger, K. P., McFarland, H. F., and Martin, R. (2001). CD4+CD28- costimulation-independent T cells in multiple sclerosis. *J Clin Invest*, 108, 1185-1194.

Martenson, R. E., and Gaitonde, M. K. (1969). Electrophoretic analysis of the highly basic proteins of the rat brain fraction which induces experimental allergic encephalomyelitis. *J Neurochem*, 16, 333-347.

Martin, R., Jaraquemada, D., Flerlage, M., Richert, J., Whitaker, J., Long, E. O., McFarlin, D. E., and McFarland, H. F. (1990). Fine specificity and HLA restriction of myelin basic protein-specific cytotoxic T cell lines from multiple sclerosis patients and healthy individuals. *J Immunol*, 145, 540-548.

Martin, R., Utz, U., Coligan, J. E., Richert, J. R., Flerlage, M., Robinson, E., Stone, R., Biddison, W. E., McFarlin, D. E., and McFarland, H. F. (1992). Diversity in fine specificity and T cell receptor usage of the human CD4+ cytotoxic T cell response specific for the immunodominant myelin basic protein peptide 87-106. *J Immunol*, 148, 1359-1366.

Martino, G., Olsson, T., Fredrikson, S., Hojeberg, B., Kostulas, V., Grimaldi, L. M., and Link, H. (1991). Cells producing antibodies specific for myelin basic protein region 70-89 are predominant in cerebrospinal fluid from patients with multiple sclerosis. *Eur J Immunol*, 21, 2971-2976.

Marty, M. C., Alliot, F., Rutin, J., Fritz, R., Trisler, D., and Pessac, B. (2002). The myelin basic protein gene is expressed in differentiated blood cell lineages and in hemopoietic progenitors. *Proc Natl Acad Sci*, USA, 99, 8856-8861.

Mathisen, P. M., Pease, S., Garvey, J., Hood, L., and Readhead, C. (1993). Identification of an embryonic isoform of myelin basic protein that is expressed widely in the mouse embryo. *Proc Natl Acad Sci*, USA, 90, 10125-10129.

Meinl, E. (1999). Concepts of viral pathogenesis of multiple sclerosis. *Curr Opin Neurol*, 12, 303-307.

Meinl, E., Weber, F., Drexler, K., Morelle, C., Ott, M., Saruhan-Direskeneli, G., Goebels, N., Ertl, B., Jechart, G., Giegerich, G., and et al. (1993). Myelin basic protein-specific T lymphocyte repertoire in multiple sclerosis. Complexity of the response and dominance of nested epitopes due to recruitment of multiple T cell clones. *J Clin Invest*, 92, 2633-2643.

Merrill, J. E., Kono, D. H., Clayton, J., Ando, D. G., Hinton, D. R., and Hofman, F. M. (1992). Inflammatory leukocytes and cytokines in the peptide-induced disease of experimental allergic encephalomyelitis in SJL and B10.PL mice. *Proc Natl Acad Sci*, USA, 89, 574-578.

Moldovan, I. R., Rudick, R. A., Cotleur, A. C., Born, S. E., Lee, J. C., Karafa, M. T., and Pelfrey, C. M. (2003). Interferon gamma responses to myelin peptides in multiple sclerosis correlate with a new clinical measure of disease progression. *J Neuroimmunol*, 141, 132-140.

Molineaux, S. M., Engh, H., de Ferra, F., Hudson, L., and Lazzarini, R. A. (1986). Recombination within the myelin basic protein gene created the dysmyelinating shiverer mouse mutation. *Proc Natl Acad Sci*, USA, 83, 7542-7546.

Morris-Downes, M. M., Smith, P. A., Rundle, J. L., Piddlesden, S. J., Baker, D., Pham-Dinh, D., Heijmans, N., and Amor, S. (2002). Pathological and regulatory effects of anti-myelin antibodies in experimental allergic encephalomyelitis in mice. *J Neuroimmunol*, 125, 114-124.

Moscarello, M. A., Brady, G. W., Fein, D. B., Wood, D. D., and Cruz, T. F. (1986). The role of charge microheterogeneity of basic protein in the formation and maintenance of the multilayered structure of myelin: a possible role in multiple sclerosis. *J Neurosci Res*, 15, 87-99.

Moscarello, M. A., Mastronardi, F. G., and Wood, D. D. (2007). The role of citrullinated proteins suggests a novel mechanism in the pathogenesis of multiple sclerosis. *Neurochem Res*, 32, 251-256.

Moscarello, M. A., Wood, D. D., Ackerley, C., and Boulias, C. (1994). Myelin in multiple sclerosis is developmentally immature. *J Clin Inves*t, 94, 146-154.

Muraro, P. A., Vergelli, M., Kalbus, M., Banks, D. E., Nagle, J. W., Tranquill, L. R., Nepom, G. T., Biddison, W. E., McFarland, H. F., and Martin, R. (1997). Immunodominance of a low-affinity major histocompatibility complex-binding myelin basic protein epitope (residues 111-129) in HLA-DR4 (B1*0401) subjects is associated with a restricted T cell receptor repertoire. *J Clin Invest*, 100, 339-349.

Myers, K. J., Sprent, J., Dougherty, J. P., and Ron, Y. (1992). Synergy between encephalitogenic T cells and myelin basic protein-specific antibodies in the induction of experimental autoimmune encephalomyelitis. *J Neuroimmunol*, 41, 1-8.

Nemazee, D., and Hogquist, K. A. (2003). Antigen receptor selection by editing or downregulation of V(D)J recombination. *Curr Opin Immunol*, 15, 182-189.

Neumann, H., Medana, I. M., Bauer, J., and Lassmann, H. (2002). Cytotoxic T lymphocytes in autoimmune and degenerative CNS diseases. *Trends Neurosci*, 25, 313-319.

Nurieva, R., Yang, X. O., Martinez, G., Zhang, Y., Panopoulos, A. D., Ma, L., Schluns, K., Tian, Q., Watowich, S. S., Jetten, A. M., and Dong, C. (2007). Essential autocrine regulation by IL-21 in the generation of inflammatory T cells. *Nature*, 448, 480-483.

O'Connor, K. C., Chitnis, T., Griffin, D. E., Piyasirisilp, S., Bar-Or, A., Khoury, S., Wucherpfennig, K. W., and Hafler, D. A. (2003). Myelin basic protein-reactive autoantibodies in the serum and cerebrospinal fluid of multiple sclerosis patients are characterized by low-affinity interactions. *J Neuroimmunol*, 136, 140-148.

Olerup, O., and Hillert, J. (1991). HLA class II-associated genetic susceptibility in multiple sclerosis: a critical evaluation. *Tissue Antigens*, 38, 1-15.

Olivares-Villagomez, D., Wang, Y., and Lafaille, J. J. (1998). Regulatory CD4(+) T cells expressing endogenous T cell receptor chains protect myelin basic protein-specific transgenic mice from spontaneous autoimmune encephalomyelitis. *J Exp Med*, 188, 1883-1894.

Olsson, T. (1995). Cytokine-producing cells in experimental autoimmune encephalomyelitis and multiple sclerosis. *Neurology*, 45, S11-15.

Ota, K., Matsui, M., Milford, E. L., Mackin, G. A., Weiner, H. L., and Hafler, D. A. (1990). T-cell recognition of an immunodominant myelin basic protein epitope in multiple sclerosis. *Nature*, 346, 183-187.

Padula, S. J., Lingenheld, E. G., Stabach, P. R., Chou, C. H., Kono, D. H., and Clark, R. B. (1991). Identification of encephalitogenic V beta-4-bearing T cells in SJL mice. Further evidence for the V region disease hypothesis? *J Immunol*, 146, 879-883.

Paterson, P. Y. (1980). Autoimmune diseases of myelin. *Prog Clin Biol Res*, 49, 19-36.

Pearson, C. I., van Ewijk, W., and McDevitt, H. O. (1997). Induction of apoptosis and T helper 2 (Th2) responses correlates with peptide affinity for the major histocompatibility complex in self-reactive T cell receptor transgenic mice. *J Exp Med*, 185, 583-599.

Perchellet, A., Stromnes, I., Pang, J. M., and Goverman, J. (2004). CD8+ T cells maintain tolerance to myelin basic protein by 'epitope theft'. *Nat Immunol,* 5, 606-614.

Pette, M., Fujita, K., Kitze, B., Whitaker, J. N., Albert, E., Kappos, L., and Wekerle, H. (1990). Myelin basic protein-specific T lymphocyte lines from MS patients and healthy individuals. *Neurology*, 40, 1770-1776.

Ponomarenko, N. A., Durova, O. M., Vorobiev, II., Belogurov, A. A. Jr., Kurkova, I. N., Petrenko, A. G., Telegin, G. B., Suchkov, S. V., Kiselev, S. L., Lagarkova, M. A., et al. (2006a). Autoantibodies to myelin basic protein catalyze site-specific degradation of their antigen. *Proc Natl Acad Sci*, USA, 103, 281-286.

Ponomarenko, N. A., Durova, O. M., Vorobiev, II, Belogurov, A. A., Telegin, G. B., Suchkov, S. V., Misikov, V. K., Morse, H. C., 3rd, and Gabibov, A. G. (2006b). Catalytic activity of autoantibodies toward myelin basic protein correlates with the scores on the multiple sclerosis expanded disability status scale. *Immunol Lett*, 103, 45-50.

Ponsford, M., Mazza, G., Coad, J., Campbell, M. J., Zajicek, J., and Wraith, D. C. (2001). Differential responses of CD45+ve T-cell subsets to MBP in multiple sclerosis. *Clin Exp Immunol*, 124, 315-322.

Putterman, C. (2004). New approaches to the renal pathogenicity of anti-DNA antibodies in systemic lupus erythematosus. *Autoimmun Rev*, 3, 7-11.

Quandt, J. A., Baig, M., Yao, K., Kawamura, K., Huh, J., Ludwin, S. K., Bian, H. J., Bryant, M., Quigley, L., Nagy, Z. A., et al. (2004). Unique clinical and pathological features in HLA-DRB1*0401-restricted MBP 111-129-specific humanized TCR transgenic mice. *J Exp Med*, 200, 223-234.

Quarles, R. H. (2005). Comparison of CNS and PNS myelin proteins in the pathology of myelin disorders. *J Neurol Sci*, 228, 187-189.

Racke, M. K., Bonomo, A., Scott, D. E., Cannella, B., Levine, A., Raine, C. S., Shevach, E. M., and Rocken, M. (1994). Cytokine-induced immune deviation as a therapy for inflammatory autoimmune disease. *J Exp Med*, 180, 1961-1966.

Reindl, M., Linington, C., Brehm, U., Egg, R., Dilitz, E., Deisenhammer, F., Poewe, W., and Berger, T. (1999). Antibodies against the myelin oligodendrocyte glycoprotein and the myelin basic protein in multiple sclerosis and other neurological diseases: a comparative study. *Brain,* 122 (Pt 11), 2047-2056.

Rivero, V. E., Riera, C. M., and Roth, G. A. (1999). Humoral response against myelin antigens in two strains of rats with different susceptibility to experimental allergic encephalomyelitis (EAE). *Autoimmunity*, 29, 129-137.

Rivers, T. M., Sprunt, D. H., and Berry, G. P. (1933). Observations on attempts to produce acute disseminated encephalomyelitis in monkeys. *J Exp Med*, 58, 39-53.

Roach, A., Takahashi, N., Pravtcheva, D., Ruddle, F., and Hood, L. (1985). Chromosomal mapping of mouse myelin basic protein gene and structure and transcription of the partially deleted gene in shiverer mutant mice. *Cell*, 42, 149-155.

Rohowsky-Kochan, C., Molinaro, D., and Cook, S. D. (2000). Cytokine secretion profile of myelin basic protein-specific T cells in multiple sclerosis. *Mult Scler*, 6, 69-77.

Rose, N. R., and Mackay, I. R. (2006). *The autoimmune diseases.* (4th ed.) London: Elsevier/Academic Press.

Rott, O., Fleischer, B., and Cash, E. (1994). Interleukin-10 prevents experimental allergic encephalomyelitis in rats. *Eur J Immunol*, 24, 1434-1440.

Sakai, K., Sinha, A. A., Mitchell, D. J., Zamvil, S. S., Rothbard, J. B., McDevitt, H. O., and Steinman, L. (1988a). Involvement of distinct murine T-cell receptors in the autoimmune encephalitogenic response to nested epitopes of myelin basic protein. *Proc Natl Acad Sci*, USA, 85, 8608-8612.

Sakai, K., Zamvil, S. S., Mitchell, D. J., Lim, M., Rothbard, J. B., and Steinman, L. (1988b). Characterization of a major encephalitogenic T cell epitope in SJL/J mice with synthetic oligopeptides of myelin basic protein. *J Neuroimmunol*, 19, 21-32.

Scholz, C., Patton, K. T., Anderson, D. E., Freeman, G. J., and Hafler, D. A. (1998). Expansion of autoreactive T cells in multiple sclerosis is independent of exogenous B7 costimulation. *J Immunol*, 160, 1532-1538.

Seamons, A., Perchellet, A., and Goverman, J. (2006). Endogenous myelin basic protein is presented in the periphery by both dendritic cells and resting B cells with different functional consequences. *J Immunol*, 177, 2097-2106.

Seamons, A., Sutton, J., Bai, D., Baird, E., Bonn, N., Kafsack, B. F., Shabanowitz, J., Hunt, D. F., Beeson, C., and Goverman, J. (2003). Competition between two MHC binding registers in a single peptide processed from myelin basic protein influences tolerance and susceptibility to autoimmunity. *J Exp Med*, 197, 1391-1397.

Segal, B. M., Dwyer, B. K., and Shevach, E. M. (1998). An interleukin (IL)-10/IL-12 immunoregulatory circuit controls susceptibility to autoimmune disease. *J Exp Med*, 187, 537-546.

Sellebjerg, F., Jensen, J., and Ryder, L. P. (1998). Costimulatory CD80 (B7-1) and CD86 (B7-2) on cerebrospinal fluid cells in multiple sclerosis. *J Neuroimmunol*, 84, 179-187.

Skulina, C., Schmidt, S., Dornmair, K., Babbe, H., Roers, A., Rajewsky, K., Wekerle, H., Hohlfeld, R., and Goebels, N. (2004). Multiple sclerosis: brain-infiltrating CD8+ T cells persist as clonal expansions in the cerebrospinal fluid and blood. *Proc Natl Acad Sci,* USA, 101, 2428-2433.

Soderstrom, M., Link, H., Sun, J. B., Fredrikson, S., Kostulas, V., Hojeberg, B., Li, B. L., and Olsson, T. (1993). T cells recognizing multiple peptides of myelin basic protein are found in blood and enriched in cerebrospinal fluid in optic neuritis and multiple sclerosis. *Scand J Immunol*, 37, 355-368.

Sospedra, M., and Martin, R. (2005). Immunology of multiple sclerosis. *Annu Rev Immunol*, 23, 683-747.

Starr, T. K., Jameson, S. C., and Hogquist, K. A. (2003). Positive and negative selection of T cells. *Annu Rev Immunol*, 21, 139-176.

Steinman, R. M., and Nussenzweig, M. C. (2002). Avoiding horror autotoxicus: the importance of dendritic cells in peripheral T cell tolerance. *Proc Natl Acad Sci*, USA, 99, 351-358.

Stepaniak, J. A., Gould, K. E., Sun, D., and Swanborg, R. H. (1995). A comparative study of experimental autoimmune encephalomyelitis in Lewis and DA rats. *J Immunol*, 155, 2762-2769.

Stinissen, P., Raus, J., and Zhang, J. (1997). Autoimmune pathogenesis of multiple sclerosis: role of autoreactive T lymphocytes and new immunotherapeutic strategies. *Crit Rev Immunol*, 17, 33-75.

Stromnes, I. M., and Goverman, J. M. (2006). Active induction of experimental allergic encephalomyelitis. *Nat Protocols*, 1, 1810-1819.

Svenningsson, A., Dotevall, L., Stemme, S., and Andersen, O. (1997). Increased expression of B7-1 costimulatory molecule on cerebrospinal fluid cells of patients with multiple sclerosis and infectious central nervous system disease. *J Neuroimmunol*, 75, 59-68.

Swanborg, R. H. (2001). Experimental autoimmune encephalomyelitis in the rat: lessons in T-cell immunology and autoreactivity. *Immunol Rev*, 184, 129-135.

Taams, L. S., van Rensen, A. J., Poelen, M. C., van Els, C. A., Besseling, A. C., Wagenaar, J. P., van Eden, W., and Wauben, M. H. (1998). Anergic T cells actively suppress T cell responses via the antigen-presenting cell. *Eur J Immunol*, 28, 2902-2912.

Tanigaki, N., Fruci, D., Groome, N., Butler, R. H., Londei, M., and Tosi, R. (1994). Exploring myelin basic protein for HLA class I-binding sequences. *Eur J Immunol*, 24, 2196-2202.

Targoni, O. S., and Lehmann, P. V. (1998). Endogenous myelin basic protein inactivates the high avidity T cell repertoire. *J Exp Med*, 187, 2055-2063.

Tejada-Simon, M. V., Zang, Y. C., Yang, D., Hong, J., Li, S., Singh, R. A., Van den Berg-Loonen, E., Killian, J. M., Rivera, V. M., and Zhang, J. Z. (2000). Aberrant T cell responses to myelin antigens during clinical exacerbation in patients with multiple sclerosis. *Int Immunol*, 12, 1641-1650.

Tranquill, L. R., Cao, L., Ling, N. C., Kalbacher, H., Martin, R. M., and Whitaker, J. N. (2000). Enhanced T cell responsiveness to citrulline-containing myelin basic protein in multiple sclerosis patients. *Mult Scler*, 6, 220-225.

Tsuchida, T., Parker, K. C., Turner, R. V., McFarland, H. F., Coligan, J. E., and Biddison, W. E. (1994). Autoreactive CD8+ T-cell responses to human myelin protein-derived peptides. *Proc Natl Acad Sci*, USA, 91, 10859-10863.

Urban, J. L., Kumar, V., Kono, D. H., Gomez, C., Horvath, S. J., Clayton, J., Ando, D. G., Sercarz, E. E., and Hood, L. (1988). Restricted use of T cell receptor V genes in murine autoimmune encephalomyelitis raises possibilities for antibody therapy. *Cell,* 54, 577-592.

Van de Keere, F., and Tonegawa, S. (1998). CD4(+) T cells prevent spontaneous experimental autoimmune encephalomyelitis in anti-myelin basic protein T cell receptor transgenic mice. *J Exp Med*, 188, 1875-1882.

Vandevyver, C., Motmans, K., Stinissen, P., Zhang, J., and Raus, J. (1998). Cytokine mRNA profile of myelin basic protein reactive T-cell clones in patients with multiple sclerosis. *Autoimmunity*, 28, 77-89.

Voorthuis, J. A., Uitdehaag, B. M., De Groot, C. J., Goede, P. H., van der Meide, P. H., and Dijkstra, C. D. (1990). Suppression of experimental allergic encephalomyelitis by intraventricular administration of interferon-gamma in Lewis rats. *Clin Exp Immunol*, 81, 183-188.

Voskuhl, R. R. (1998). Myelin protein expression in lymphoid tissues: implications for peripheral tolerance. *Immunol Rev*, 164, 81-92.

Warren, K. G., and Catz, I. (1987). A correlation between cerebrospinal fluid myelin basic protein and anti-myelin basic protein in multiple sclerosis patients. *Ann Neurol*, 21, 183-189.

Warren, K. G., and Catz, I. (1993). Increased synthetic peptide specificity of tissue-CSF bound anti-MBP in multiple sclerosis. *J Neuroimmunol*, 43, 87-96.

Warren, K. G., and Catz, I. (1999). An extensive search for autoantibodies to myelin basic protein in cerebrospinal fluid of non-multiple-sclerosis patients: implications for the pathogenesis of multiple sclerosis. *Eur Neurol*, 42, 95-104.

Warren, K. G., Catz, I., Johnson, E., and Mielke, B. (1994). Anti-myelin basic protein and anti-proteolipid protein specific forms of multiple sclerosis. *Ann Neurol,* 35, 280-289.

Warren, K. G., Catz, I., and Steinman, L. (1995). Fine specificity of the antibody response to myelin basic protein in the central nervous system in multiple sclerosis: the minimal B-cell epitope and a model of its features. *Proc Natl Acad Sci*, USA, 92, 11061-11065.

Weaver, C. T., Hatton, R. D., Mangan, P. R., and Harrington, L. E. (2007). IL-17 family cytokines and the expanding diversity of effector T cell lineages. *Annu Rev Immunol*, 25, 821-852.

Wensky, A. K., Furtado, G. C., Marcondes, M. C., Chen, S., Manfra, D., Lira, S. A., Zagzag, D., and Lafaille, J. J. (2005). IFN-gamma determines distinct clinical outcomes in autoimmune encephalomyelitis. *J Immunol*, 174, 1416-1423.

Whitaker, J. N., Gupta, M., and Smith, O. F. (1986). Epitopes of immunoreactive myelin basic protein in human cerebrospinal fluid. *Ann Neurol*, 20, 329-336.

Willenborg, D. O., and Prowse, S. J. (1983). Immunoglobulin-deficient rats fail to develop experimental allergic encephalomyelitis. *J Neuroimmunol*, 5, 99-109.

Willenborg, D. O., Sjollema, P., and Danta, G. (1986). Immunoglobulin deficient rats as donors and recipients of effector cells of allergic encephalomyelitis. *J Neuroimmunol*, 11, 93-103.

Wolf, S. D., Dittel, B. N., Hardardottir, F., and Janeway, C. A., Jr. (1996). Experimental autoimmune encephalomyelitis induction in genetically B cell-deficient mice. *J Exp Med*, 184, 2271-2278.

Wucherpfennig, K. W., Catz, I., Hausmann, S., Strominger, J. L., Steinman, L., and Warren, K. G. (1997). Recognition of the immunodominant myelin basic protein peptide by autoantibodies and HLA-DR2-restricted T cell clones from multiple sclerosis patients. Identity of key contact residues in the B-cell and T-cell epitopes. *J Clin Invest*, 100, 1114-1122.

Wucherpfennig, K. W., Zhang, J., Witek, C., Matsui, M., Modabber, Y., Ota, K., and Hafler, D. A. (1994). Clonal expansion and persistence of human T cells specific for an immunodominant myelin basic protein peptide. *J Immunol*, 152, 5581-5592.

Yamamura, T., Kondo, T., Sakanaka, S., Kozovska, M., Geng, T. C., Takahashi, K., and Tabira, T. (1994). Analysis of T cell antigen receptors of myelin basic protein specific T cells in SJL/J mice demonstrates an alpha chain CDR3 motif associated with encephalitogenic T cells. *Int Immunol*, 6, 947-954.

Yasuda, T., Tsumita, T., Nagai, Y., Mitsuzawa, E., and Ohtani, S. (1975). Experimental allergic encephalomyelitis (EAE) in mice. I. Induction of EAE with mouse spinal cord homogenate and myelin basic protein. *Jpn J Exp Med*, 45, 423-427.

Yoshizawa, I., Bronson, R., Dorf, M. E., and Abromson-Leeman, S. (1998). T-cell responses to myelin basic protein in normal and MBP-deficient mice. *J Neuroimmunol*, 84, 131-138.

Zamvil, S., Nelson, P., Trotter, J., Mitchell, D., Knobler, R., Fritz, R., and Steinman, L. (1985). T-cell clones specific for myelin basic protein induce chronic relapsing paralysis and demyelination. *Nature*, 317, 355-358.

Zamvil, S. S., Mitchell, D. J., Lee, N. E., Moore, A. C., Waldor, M. K., Sakai, K., Rothbard, J. B., McDevitt, H. O., Steinman, L., and Acha-Orbea, H. (1988). Predominant expression of a T cell receptor V beta gene subfamily in autoimmune encephalomyelitis. *J Exp Med*, 167, 1586-1596.

Zamvil, S. S., Mitchell, D. J., Moore, A. C., Kitamura, K., Steinman, L., and Rothbard, J. B. (1986). T-cell epitope of the autoantigen myelin basic protein that induces encephalomyelitis. *Nature*, 324, 258-260.

Zang, Y. C., Li, S., Rivera, V. M., Hong, J., Robinson, R. R., Breitbach, W. T., Killian, J., and Zhang, J. Z. (2004). Increased CD8+ cytotoxic T cell responses to myelin basic protein in multiple sclerosis. *J Immunol*, 172, 5120-5127.

Zhang, J., Markovic-Plese, S., Lacet, B., Raus, J., Weiner, H. L., and Hafler, D. A. (1994). Increased frequency of interleukin 2-responsive T cells specific for myelin basic protein and proteolipid protein in peripheral blood and cerebrospinal fluid of patients with multiple sclerosis. *J Exp Med*, 179, 973-984.

Zheng, Y., Danilenko, D. M., Valdez, P., Kasman, I., Eastham-Anderson, J., Wu, J., and Ouyang, W. (2007). Interleukin-22, a T(H)17 cytokine, mediates IL-23-induced dermal inflammation and acanthosis. *Nature*, 445, 648-651.

Zhou, L., Ivanov, II, Spolski, R., Min, R., Shenderov, K., Egawa, T., Levy, D. E., Leonard, W. J., and Littman, D. R. (2007). IL-6 programs T(H)-17 cell differentiation by promoting sequential engagement of the IL-21 and IL-23 pathways. *Nat Immunol*, 8, 967-974.

In: Myelin Basic Protein
Editor: Joan M. Boggs
ISBN: 978-1-60456-699-4

Chapter V

A STRUCTURAL PERSPECTIVE OF PEPTIDES FROM MYELIN BASIC PROTEIN

Maria Katsara[1,2], Paul A. Ramsland[1], Theodore Tselios[2], John Matsoukas[2] and Vasso Apostolopoulos[1*]

ABSTRACT

Major histocompatibility complex molecules bind and present short antigenic peptide fragments on the surface of antigen presenting cells to T cell receptors. Recognition of peptide-MHC by T cells initiates a cascade of signals to T cells. Th1 cytokines secreted by $CD4^+$ T cells and recognition of self peptides are believed to be vital in the initiation of autoimmune diseases. Multiple sclerosis (MS) is an autoimmune inflammatory disease whereby $CD4^+$ T cells primarily recognize self proteins of the myelin sheath. Myelin basic protein is one of the auto-antigens involved in MS. Crystal structures of MBP peptides with MHC and/or T cell receptor, NMR, molecular modeling and docking studies have all played a pivotal role in shaping our understanding of the interactions in initiating immune responses. Furthermore, in the design of molecular vaccines for the treatment of MS, an understanding of 3-dimensional structures is important for peptide-based vaccine design, all of which are discussed herein.

[1] Burnet Institute , Immunology and Vaccine Laboratory, Studley Road, Heidelberg, VIC, 3084, Australia
[2] Department of Chemistry, Section of Organic Chemistry, Biochemistry, and Natural Products, University of Patras, Patras, 26500 Greece
[*] Burnet Institute (Austin Campus), Immunology and Vaccine Laboratory, Studley Road, Heidelberg, VIC, 3084, Australia. Telephone: +613-92870666/05; Fax: +613 92870600. E-mail: vasso@burnet.edu.au

ABBREVIATIONS

CNS, central nervous system
EAE, experimental autoimmune encephalomyelitis
MBP, myelin basic protein
MHC, major histocompatibility complex
MOG, myelin oligodendrocyte glycoprotein
MS, multiple sclerosis
PLP, proteolipid protein
TCR, T cell receptor

INTRODUCTION

Multiple sclerosis (MS) is a commonly occurring chronic, inflammatory and disabling disorder of the central nervous system (CNS). It affects 0.05-0.15% of Caucasians with the onset of the disease in young adulthood and women with the disease outnumber men two to one. It is widely considered that $CD4^+$ T helper type 1 (Th1) cells play a pivotal role in mediating an autoimmune attack against components of myelin sheath (see Chapter IV). Additional cells, such as $CD8^+$ T cells, macrophages and complement are also involved in axonal damage and neurodegeneration (Hemmer et al., 2002). Several autoantigens, such as myelin basic protein (MBP), proteolipid protein (PLP), myelin oligodendrocyte glycoprotein (MOG) have been proposed as candidate antigens in the induction of MS based on auto-T cells and auto-antibodies which are present in patients with MS. In addition, these peptides show their encephalitogenicity in experimental autoimmune encephalomyelitis (EAE), an animal model for MS (Hafler et al., 2005).

Genetic and environmental factors play an important role in the pathogenesis of MS (Steinman, 1996). Genome-wide studies have revealed that genes in the major histocompatibility complex (MHC) class II on chromosome 6p21.3, are related to the susceptibility of MS (Haines et al., 1996). In particular, the DR2 haplotype is correlated with MS. This haplotype contains three different MHC class II alleles-DRB1*1501, DRB5*0101, and DQB1*0602 which are in strong linkage disequilibrium. These alleles encode the following MHC class II molecules: HLA-DR2b (DRA, DRB1*1501), HLA-DR2a (DRA, DRB5*0101), and HLA-DQ6 (DQA1*0102, DQB1*0602) (Oksenberg et al., 1996). Importantly, two functional cell surface heterodimers, HLA-DR2b (DRA, DRB1*1501) and HLA-DR2a (DRA, DRB5*0101), are believed to confer an increased susceptibility to the development of MS (Steinman, 1996). In addition, HLA-DR1 (DRB1*0101) and HLA-DR4 (DRB1*0401) also bind peptides of MBP and have implications in the development of MS (Matsoukas et al., 2005; Sireci et al., 2003).

CRYSTAL STRUCTURE OF HLA-DR2 (DRA*0101, DRB1*1501) COMPLEXED WITH A PEPTIDE FROM HUMAN MYELIN BASIC PROTEIN MBP_{85-99}

Peptide interactions with HLA-DR2A

Many studies have attempted to provide insight into the interaction of peptides with MHC class II, and have focused primarily on human MBP (Table 1). A study revealed that the human MBP_{82-102} peptide was immunodominant for MBP specific T cells of MS patients with HLA-DR2 haplotype (Martin et al., 1990; Ota et al., 1990; Pette et al., 1990). HLA-DR2b (DRA, DRB5*1501)-restricted T cell clones from MS patients proliferated in response to MBP_{85-99} (ENPVV^{89}HFF^{92}KNIVTPR) peptide. A binding assay using MBP_{85-99} peptide and purified HLA-DR2b indicated that it bound with medium-high affinity (Valli et al., 1993; Wucherpfennig et al., 1994a). Positions Val^{89} and Phe^{92} of the epitope 85-99 were accommodated as P1 and P4 anchor residues, respectively, for the haplotype HLA-DR2b. However, Phe^{92} was identified as the P1 anchor residue when the peptide MBP_{85-99} was bound to HLA-DR2a (DRA, DRB5*0101). The binding of MBP_{85-99} to HLA-DR2a was of comparable binding affinity to HLA-DR2b (Wucherpfennig et al., 1994a).

Table 1. Structures of MBP peptides

MHC	Peptide	TCR αβ	Reference
Crystal structures of MBP-MHC			
HLA-DR2b	MBP_{85-99}		Smith et al., 1998
HLA-DR2a	MBP_{86-105}		Li et al., 2000
H2-IA^{u}	Ac[Y^{4}]MBP_{1-9}		He et al., 2002
Crystal structures of MBP-MHC-TCR			
HLA-DR2b	MBP_{85-99}	Ob.1A12	Hahn et al., 2005
HLA-DR2a	MBP_{89-101}	3A6	Li et al., 2005
H2-IA^{u}	Ac[Y^{4}]MBP_{1-9}	172.10	Maynard et al., 2005
Molecular modeling of MBP-MHC-TCR			
HLA-DR2b	MBP_{86-105}		Li et al., 2000
HLA-DR2a	MBP_{86-105}		Li et al., 2000
Molecular modeling of MBP-MHC			
H2-IA^{u}	AcMBP_{1-9}		Tate et al., 1995
H2-IA^{k}	AcMBP_{1-9}		Tate et al., 1995
H2-IA^{u}	Ac[Y^{4}]MBP_{1-9}		Lee et al., 1998
H2-IA^{u}	OVA-MBP_{1-9}		Lee et al., 1998
H2-IA^{u}	OVA-[Y^{4}]MBP_{1-9}		Lee et al., 1998
HLA-DR1301	$MBP_{152-165}$		Koehler et al., 2004
NMR and modeling of MBP peptides			
	MBP_{74-85}		Tselios et al., 1999
	cyclo(75-81)MBP_{74-85}		Tselios et al., 1999
	[A^{81}]MBP_{74-85}		Matsoukas et al., 2002
	[Arg^{91},Ala^{96}]MBP_{87-99}		Tselios et al., 2000
	[Arg^{91},Ala^{96}]MBP_{87-99}		Tselios et al., 2002
	cyclo(87-99)[Arg^{91},Ala^{96}]MBP_{87-99}		Tselios et al., 2000
	[Ala^{96}]MBP_{87-99}		Matsoukas et al., 2005
	cyclo(91-99)[Ala^{96}]MBP_{87-99}		Matsoukas et al., 2005
Docking of MBP-MHC			
H2-IA^{u}	MBP_{74-85}		Tzakos et al., 2004

The crystal structure of HLA-DR2b in complex with MBP_{85-99} was determined at 2.6 Å resolution, Table 1 (Smith et al., 1998). In this structure, residues P1, P4, P6 and P9 of the peptide occupy pockets within the groove of DR2b, Figure 1. Overall, the MBP peptide occupies the DR2 binding groove in a similar fashion when compared to other peptide/HLA-DR complexes. In particular, hydrophobic residues Val P1 and Phe P4 represent the primary anchor residues for DR2b binding of the MBP peptide, Figure 1. Positions His P2, Phe P3 and Lys P5 have been identified to be TCR contact residues for the human MBP-specific T cell clone, and in the crystal structure of MPB_{85-99} and HLA-DR2b, these residues have been shown to be solvent exposed for TCR recognition (Smith et al., 1998). Earlier studies have demonstrated that positions P2 His, P3 Phe, P5 Lys are critical residues for TCR recognition. In particular Phe P3 is the major residue for TCR recognition in human MBP-specific T cell clones. Substitution of Phe greatly inhibits T cell proliferation of human MBP_{85-99} T cell clones without affecting its binding to DR2b (Wucherpfennig et al., 1994a; Wucherpfennig et al., 1994b). The DR2b-MBP_{85-99} complex suggests that the peptide side-chains of P2 His, P3 Phe and P5 Lys are solvent exposed and available for TCR recognition, Figure 1, (Smith et al., 1998).

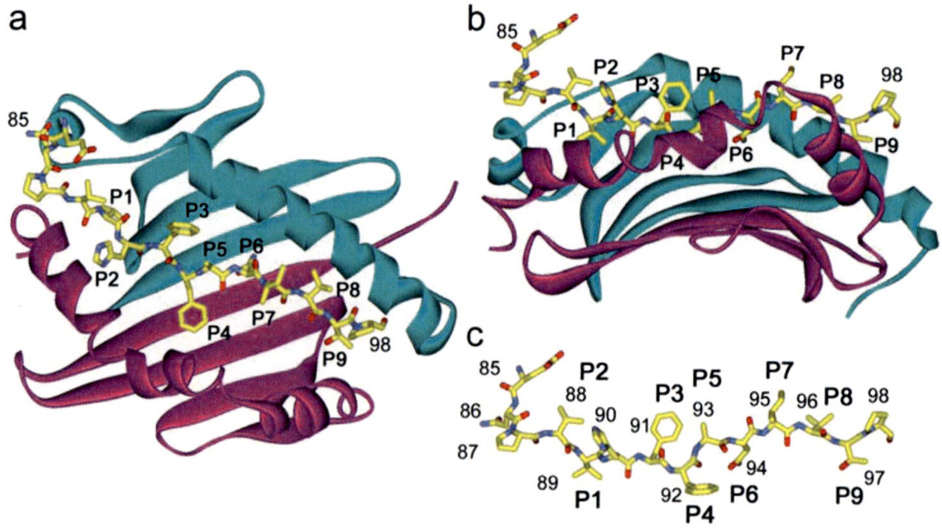

Figure 1. Views of the MBP_{85-99} peptide interaction with HLA-DR2b generated from the crystal structure of the complex, which was determined to a resolution of 2.6 Å (PDB code 1BX2; Smith et al., 1998). End-on (a) and side-views (b) of the peptide binding groove are shown with α- (cyan) and β-chains (magenta) as ribbon-style diagrams. The bound peptide (c) is also shown by itself as a side-view with the MHC contact residues pointing downwards and TCR contact residues pointing upwards. Peptide atoms are colored by atom-type (C, yellow; O, red, N, steel blue) and residues are numbered according to their position in the MBP epitope. Key MHC or TCR contact residues are designated as P1-9 (in boldface).

It is of interest that the polymorphism at the position DRβ71 of DR2 due to the presence of alanine creates available space for a bulky residue. Thus, the aromatic side-chain of phenylalanine is accommodated within the hydrophobic P4 pocket, and represents the major anchor for DR2 binding (Smith *et al.*, 1998). Alanine at position DRβ71 is only found in DR2 alleles, (such as DRB1*1501, DRB1*1506 and DRB1*1309), and not in DRB1 alleles or

DR4 alleles. Therefore, this polymorphism makes the P4 pocket of HLA-DR2 distinct from other DR molecules. Moreover, the P4 pocket of HLA-DR4 preferably accommodates negatively charged residues, some aliphatic amino acids, and has low affinity binding to aromatic residues (Hammer et al., 1994; Hammer et al., 1995); the P4 pocket in DR2 molecules nestles large, aromatic residues. In addition, the polymorphism that results in a specificity of the P4 pocket for large and aromatic residues in DR2, and negatively charged molecules for DR4, could be very useful for further development of peptide analogs for specific autoimmune diseases.

Furthermore, a T cell epitope MBP_{84-102} has been identified as immunodominant (Martin et al., 1990; Ota et al., 1990; Pette et al., 1990) and studies have demonstrated the binding of this peptide with HLA-DR2b using DR2b transgenic mice and TCR specific for MBP_{84-102} (Madsen et al., 1999). There is strong evidence that self-antigen MBP_{84-102} can be presented by both HLA-DR2a and HLA-DR2b to T cell clones from MS patients (Vergelli et al., 1997; Vogt et al., 1994)

Peptide interactions with Ob.1A.12 TCR

The crystal structure of HLA-DR2b/MBP/TCR was reported at 3.5 Å resolution, Table 1 (Hahn et al., 2005). Interestingly, MBP_{85-99} TCR-binding topology and geometry is different from antimicrobial TCRs, and these unusual properties may be responsible for allowing the autoreactive T cells to escape deletion. The MBP_{85-99} peptide was linked to the N-terminus of the TCR β chain to improve the yield and stability of the DR2b/MBP/TCR complex. The TCR clone was derived from a relapsing-remitting MS patient (Wucherpfennig et al., 1994a; Wucherpfennig et al., 1994b). The TCR (Ob.1A.12) recognizes the encephalitogenic MBP_{85-99} epitope (Wucherpfennig et al., 1994a) and when transgenic mice, expressing this human TCR and HLA-DR2, were immunized with the MBP peptide, central nervous system autoimmunity was induced (Madsen et al., 1999).

The crystal structure of the trimolecular complex (DR2b/MBP_{85-99}/TCR) has revealed an unusual interaction between the autoreactive Ob.1A.12 TCR and both the peptide and MHC class II molecule, Figure 2 (Hahn et al., 2005). In particular, earlier studies between HA1.7 TCR, a peptide from influenza hemagglutinin bound to HLA-DR1 (Hennecke et al., 2000), and the mouse D10 TCR with a relevant peptide bound to I-A^k (Reinherz et al., 1999), revealed that the CDR3 loops of the TCR created a pocket for the side-chain of the MBP peptide at position P5, and the topology of TCRs is diagonal-to-orthogonal. However, the dominant CDR3α and CDR3β loops of Ob.1A.12 TCR accommodates the side-chain of the peptide at position P2 (His), and only the N-terminus of the peptide interacts with the Ob.1A.12 TCR, which binds asymmetrically with the helices of DR2b, Figure 2 (Hahn et al., 2005). It is of interest, that the CDR3 loops (CDR3α and CDR3β) create an unusually large cavity compared to other TCRs, which accommodates both a DR2b residue (DRβ81 His) and the side-chain of a His residue (P2). The CDR3α and CDR3β directly interact with the His of the peptide, while His of the DR2 molecule is only contacted by CDR3α.

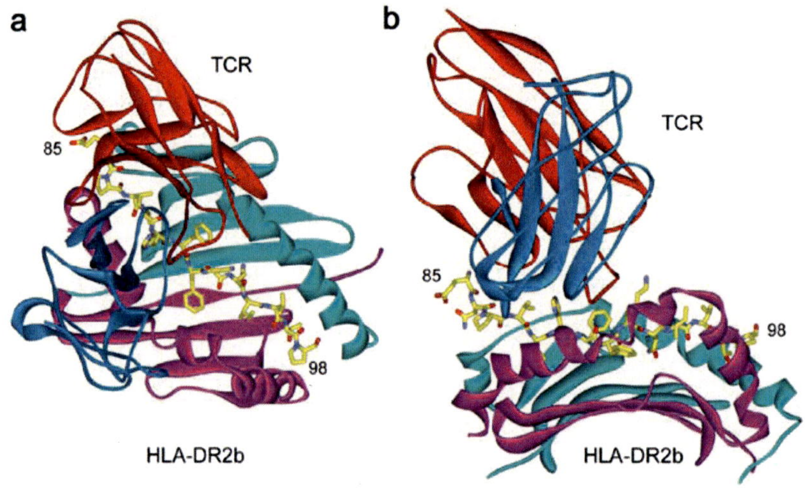

Figure 2. Crystal structure of TCR, MBP_{85-99} and HLA-DR2b complex determined to a resolution of 3.5 Å (PDB code 1YMM; Hahn et al., 2005). End-on (a) and side-views (b) of the trimolecular complex are shown with coloring as for Fig. 1, except for the TCR variable domains (Vα, blue and Vβ, red). Note that unlike most TCR/peptide/MHC complexes, the TCR binds the distal end of the MBP peptide.

Another important peptide residue, Phe P3, is interfaced between a hydrophobic area of DR2b and a large surface of the TCR CDR3β loop (Hahn et al., 2005). In addition, while only the CDR3α and CDR3β contact the peptide segment for the Ob.1A.12 TCR, four loops (CDR1 and CDR3 from both TCR chains) of HA1.7 interact with the hemagglutinin peptide. Moreover, there are many differences between the Ob.1A.12 and HA1.7 interactions with the MHC. The CDR1 and CDR2 TCR loops mostly interact with HA.1.7 in a diagonal orientation, however only hypervariable CDR3 loops interface with the MHC helices (Hahn et al., 2005). Overall, unusual TCR-binding characteristics of Ob.1A.12 compared to other TCRs, may be critical for permission of the autoreactive T cells to escape negative selection, and therefore play a role in autoimmunity.

Structure of Human MHC Class II Complexed with a Longer Epitope Peptide (MBP_{86-105}) from Human Myelin Basic Protein

Interactions with HLA-DR2a / 2b

The structure of HLA-DR2a complexed with MBP_{86-105} was determined, in complex with two different alleles (DRB5*1501 and DRB5*0101) and the binding was somewhat different, Table 1. As we previously discussed, the MBP_{85-99} binds to DR2b using Val^{89} and Phe^{92} as primary anchor residues in pockets P1 and P4, respectively (Smith et al., 1998). However, the MBP_{86-105} binds to DR2a with anchor residue Phe^{92} in the P1 pocket and Ile^{95} in the P4 pocket, thus indicating that the MBP peptide is shifted by three residues in the DR2a binding groove, Figure 3 (Li et al., 2000).

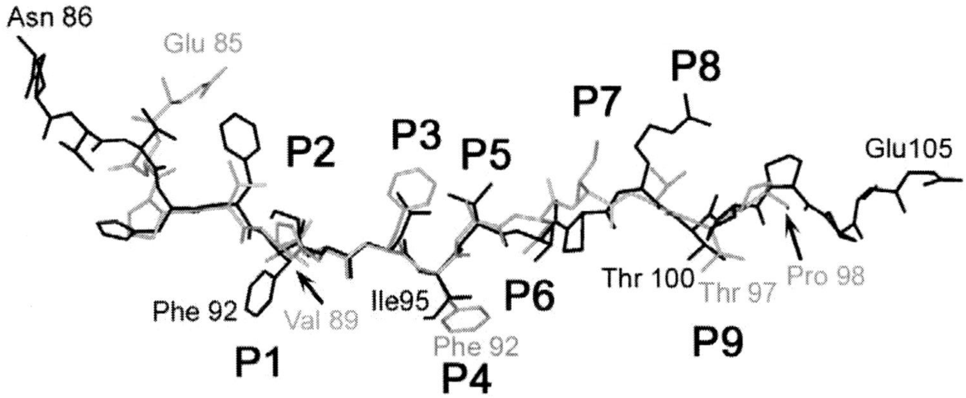

Figure 3. Structural overlay of two MBP peptides bound to different HLA-DR2 alleles. MBP86-105 (black) peptide complex with the DR2a allele was determined at a resolution of 1.9 Å (PDB code 1FV1; Li et al., 2000). The shorter MBP85-99 (grey) peptide was in complex with the DR2b allele (PDB code 1BX2; Smith et al., 1998), and although containing the same sequence, its "binding register" is shifted in the MHC binding groove. Peptides are shown as a side view with the MHC binding residues pointing downwards and the TCR recognition residues pointing upwards.

The latter peptide underwent a shift, in order to accommodate a bulky aromatic residue Phe^{92} in a large P1 pocket and a less bulky aliphatic residue Ile^{95} in a shallow P4 pocket, Figure 3 (Li et al., 2000). It was also noted that the C-terminus of the MBP peptides presented by DR2a and DR2b are positioned higher in the DR binding groove, compared to other peptides bound to DR1 (Stern et al., 1994), DR3 (Ghosh et al., 1995) and DR4 (Dessen et al., 1997). Thus, the same peptide sequences can adopt different conformations when bound in the grooves of closely related MHC molecules. Furthermore, the conformational plasticity of peptide-MHC interactions is mediated by sometimes small topological changes in the peptide-binding groove.

Interactions with TCR

Since there is no three-dimensional structure of DR2/$MBP_{86\text{-}105}$/TCR, the complex was modeled based on a TCR complex to conalbumin peptide bound to mouse MHC class II, Table 1 (Reinherz et al., 1999). The TCR contact residues were predicted to be different due to the change of the peptide alignment between the DR2a and DR2b complexes. Only Lys^{93}, Asn^{94} and Val^{96} are suggested to be the common TCR contact residues for both complexes using molecular dynamics (data not shown) (Li et al., 2000).

Structure of a human TCR complexed with a peptide from human $MBP_{89\text{-}101}$ and a MHC class II molecule

As we discussed above, the $MBP_{84\text{-}102}$ peptide represents a highly immunodominant T cell epitope (Ota et al., 1990; Steinman, 1996), and a clinical trial showed its encephalitogenic potential in a subgroup of MS patients, in which several patients developed exacerbations of the disease (Bielekova et al., 2000). A recent study elucidated the interactions between a human TCR, the epitope (89-101) and the MS-associated MHC class II molecule (HLA-DR2a) (Li et al., 2005).

Both HLA-DR2a and HLA-DR2b can present MBP_{84-102} to T cell clones from MS patients (Sospedra and Martin, 2005; Wucherpfennig et al., 1994a), but presentation by HLA-DR2a and HLA-DR2b is different in that the peptide is shifted by three amino acids (Li et al., 2000; Smith et al., 1998). Most of the studies have focused on complexes between TCR and microbial or other foreign epitopes (Garboczi et al., 1996; Hennecke et al., 2000; Kjer-Nielsen et al., 2003; Reinherz et al., 1999). Thus, there was a need to investigate whether autoreactive TCRs engage MHC complexes with MBP peptides in the same manner as anti-microbial TCR. A recently published study reported the binding of TCR 3A6 to MBP_{89-101} presented by HLA-DR2a and the crystal structure of TCR/MBP/HLA-DR2a complex.

Li et al. (Li et al., 2005) have expressed 3A6 TCR (Vα22/Vβ5.1) and HLA-DR2a in transgenic mice and the binding efficacy of TCR to HLA-DR2a bearing the MBP_{89-101} peptide was successfully tested. The 3A6 TCR/MBP/HLA-DR2a complex was crystallized with the MBP fused to the N-terminal of the TCR Vβ to enhance crystal formation, Table 1 (Li et al., 2005). The orientation angle of 3A6 TCR/MBP/HLA-DR2a is 110° (Hahn et al., 2005; Li et al., 2005), in comparison to 47° for the Ob.1A12 TCR/MBP/HLA-DR2b complex (Hahn et al., 2005).

Residues of CDR2 form numerous van der Waals interactions with residues 55-65 of HLA-DR2a, and CDR3 interacts with the HLA-DR2a β1 α-helical residues $Asp^{66\beta}$, $Phe^{67\beta}$, $Asp^{70\beta}$, and $Thr^{77\beta}$. A relatively short CDR2α loop is not able to interact with HLA-DR2a or the MBP peptide. However, two Ser residues (100 and 101) of CDR3α are contact residues with $Glu^{55\alpha}$ of DR2a. In addition, it appears that the 3A6 TCR binds weakly to MBP/DR2a, due to a lack of hydrogen bonds and salt bridges in the interaction (Li et al., 2005).

In the MBP_{89-101} peptide TCR contact residues are Phe P1, Lys P2, Val P5, and Pro P7. The CDR3α loop binds over the N-terminus of the peptide, while the middle portion and the C-terminus of the MBP peptide interact with all CDR loops of Vβ. The P1 pocket accommodates Phe^{90} (primary anchor), which is engaged by the CDR3α loop. It is of interest, that the P1 pocket of other TCRs nestles a single side-chain of the antigen peptide (Hahn et al., 2005; Rudolph et al., 2002), but that of the 3A6 TCR can accommodate both residues P1 and P2 of the MBP peptide. Both 3A6 and Ob.1A12 TCR recognize the N-terminal region of the MBP peptide (Li et al., 2005). Additionally, 3A6 primarily recognizes the N-terminal region of the MBP peptide, while antimicrobial and alloreactive TCRs predominantly interact with the central portion of the antigen peptide. Using combinational peptide libraries, it was found that a Phe to Trp mutation (at the P1 position) displayed superagonist properties with increased T cell proliferation (Hemmer et al., 2000). However, when N-terminal Val P3 was modified to Leu, the resulting peptide was not a T cell stimulator (Hemmer et al., 2000; Hemmer et al., 1998). Moreover, substitutions at positions P2, P5 and P7, which are critical for the 3A6 TCR recognition, could modulate T cell stimulation. All of the above results agreed well with interactions found by analyzing the crystal structure of the 3A6/MBP/DR2a complex. Finally, TCRs like 3A6, which primarily recognize the N-terminal region of the peptide are more likely to be cross-reactive, than other TCRs contacting the middle region of the peptide. Thus, this cross-reactivity could enhance self antigen/MHC recognition, increasing the susceptibility to autoimmunity.

STRUCTURAL INTERACTIONS BETWEEN AUTOIMMUNE ANTIGENS OF ACETYLATED N-TERMINAL 9-MER OR 11-MER PEPTIDES OF MYELIN BASIC PROTEIN AND MHC

Structural Modeling of the AcMBP$_{1-9}$-I-A^{u} complex

Two MHC class II molecules are involved in presenting the AcMBP$_{1-11}$ peptide (AcASQKRPSQRHG), the I-A^{u} or I-A^{k} (Tate et al., 1995). The acetylated N-terminal 9- or 11-mer peptides were encephalitogenic in H-2^{u} mice (PL/J or B10.PL), and could induce EAE (Gautam et al., 1992; Smilek et al., 1991; Wraith et al., 1989). However, both of these peptides (9- or 11-mer) dissociated rapidly, when complexed with MHC I-A^{u} or I-A^{k} (Mason et al., 1995).

Functional molecular studies have characterized complexes between the shorter epitope (9-mer) with the I-A^{u} or I-A^{k} class II molecule. Tate et al., suggested that changes within polymorphic MHC residues β38, β61 had effects on I-A^{u} and I-A^{k} molecules, due to neighboring polymorphism I-A β9, and influence the binding selectivity of the peptide (Tate et al., 1995). Therefore, interaction studies between MHC residue β9, with other peptides containing mutations at the position Lys^{4} with Tyr, Glu, or Ala, showed increased binding efficacy to I-A^{k} (Hβ9), and reduced to I-A^{u} (Vβ9). Thus, polymorphism β9 prefers a negatively charged E residue in the peptide for I-A^{k}, and a hydrophobic Tyr residue for I-A^{u} (Lee et al., 1998). Moreover, a molecular model of AcMBP$_{1-9}$ (AcASQ<u>K</u>RPSQR) with I-A^{u} predicts that the MBP peptide interacts only with a part of the cleft, leaving the N-terminal side-chain unoccupied, and the first amino acid (Ala) buried in pocket 3. In addition, there is a lack of MBP anchor residues in the MHC groove, and the tested peptide does not even occupy the P1 pocket. The P9 pocket contains the side-chain of Ser^{7}, but this is bound weakly. Indeed the only residue deeply buried in the hydrophobic P6 pocket is Lys^{4}, Table 1 (Lee et al., 1998).

An addition of six amino acids from the OVA$_{323-328}$ peptide to the MBP peptide, and the mutation of Lys^{4} to Tyr^{4} with a large aromatic side chain, resulted in increasing the binding stability of AcMBP$_{1-11}$ with I-A^{u}. Molecular models of Ac[Tyr^{4}]MBP$_{1-9}$, extended OVA-MBP$_{1-9}$, and OVA-[Tyr^{4}]MBP$_{1-9}$ with I-A^{u}, shows that the peptides predominantly use position 4 (for MBP peptide) to anchor into the P6 pocket. The above changes to the original Ac1-9MBP peptide led to increased binding affinity (Lee et al., 1998).

Structural analysis of the AcMBP$_{1-11}$-I-A^{u} crystal complex

In the early 1990's, an acetylated N-terminal 1-11 peptide of MBP$_{1-11}$ (AcASQ<u>K</u>RPSQRHG) and an altered peptide ligand (AcASQ<u>A</u>RPSQRHG) with a single amino acid substitution of Lys^{4} to Ala^{4}, Ac[Ala^{4}]MBP$_{1-11}$ were studied. Ac[Ala^{4}]MBP$_{1-11}$ showed increased binding to MHC class II compared to Ac1-11 and was more potent in vitro at stimulating encephalitogenic T cells (Smilek et al., 1991). Moreover, EAE was prevented when Ac[Ala^{4}]MBP$_{1-11}$ was administrated before or co-injected with Ac[Tyr^{4}]MBP$_{1-11}$, and administration of Ac[Ala^4]MBP$_{1-11}$ peptide later or near the time of disease onset also prevented EAE (Smilek et al., 1991).

Earlier studies indicated that substitution of Lys^4 with Ala^4 or Tyr^4 enhanced the stability of the I-A^u–peptide complexes, without affecting T cell recognition (Fugger et al., 1996; Lee et al., 1998). Therefore, Ac[Tyr^4]MBP$_{1-11}$ was linked to the N-terminus of the I-A^u β chain through an eight residue linker and crystallized, Table 1. The crystal structure suggested that the MBP peptide sits in an unusual shifted register in the MHC groove with only seven amino acids, and leaves P1 and P2 pockets empty, which are occupied by several ordered water molecules (He et al., 2002). It is of interest that the exposed Gln^3 and Pro^6 amino acids would be expected to fill the MHC pockets, however they appear to be TCR contact residues. The large, hydrophobic P6 pocket accommodates the aromatic ring of Tyr^4, which is stacked between Phe$^{11\beta}$ and Tyr$^{30\beta}$ in the pocket. The Ser^2 residue is partially buried in the P4 pocket, and the P9 pocket is filled with Ser^7 (He et al., 2002). All the above structural interactions for MBP/I-A^u complex are consistent with functional and modeling studies of AcMBP$_{1-11}$ peptide, previously discussed (Lee et al., 1998).

Structural analysis of the 172.10 TCR-AcMBP$_{1-11}$-I-A^u crystal complex

The very short half-life of wild-type AcMBP$_{1-11}$ bound to I-A^u may indicate a short lifetime in the thymus and an ineffective deletion of autoreactive TCRs (Fairchild et al., 1993). However, the Ac[Tyr^4]MBP$_{1-11}$ peptide has been shown to block development of EAE, thus indicating deletion of MBP-reactive T cells (Anderton et al., 2001). A crystal structure of the 172.10 TCR (Goverman et al., 1993) in complex with I-A^u-Ac[Tyr^4]MBP$_{1-11}$ was determined to address and explain the immunological findings.

The Vα TCR domain (containing CDR1α and CDR2α) binds the N-terminal region and β1 helix of MHC, and does not form any hydrogen bond with the MHC or the peptide, Table 1. However, the Vβ domain interacts with the C-terminal end of the peptide and the α1-helix of the MHC α-chain (Maynard et al., 2005). As we previously discussed, the AcMBP$_{1-11}$ peptide is shifted by two pockets allowing the entire peptide to lie toward the C-terminal end of the MHC groove (He, 2002). Thus, the peptide sits underneath the CDR3α loop, with the majority of the peptide positioned underneath the 172.10 Vβ footprint. The interactions between the 172.10 Vα and the I-A^u β1 helix are predominately van der Waals, while the interactions with the I-A^u α1 helix are van der Waals and hydrogen bonds. Ala P1, Gln P3, Arg P5 and Pro P6 have been found to be the wild-type MBP TCR contact residues (Maynard et al., 2005). These studies revealed the binding mode of AcMBP$_{1-11}$ peptide with I-A^u, and offered insights in the design of new altered peptide ligands, in particular at the P4 residue, for the immunotherapy of MS.

Furthermore, numerous peptide libraries were scanned in order to determine if a degenerate peptide repertoire could facilitate autoimmunity by forming more complementary interactions with 172.10 CDR3s. Interestingly, all peptides can activate the EAE T cell clone 172.10, although there is a clear trend that the 172.10 TCR prefers the native MBP contact residues (Maynard et al., 2005).

Conformational Studies of Linear and Cyclic MBP Peptides: NMR and Molecular Modeling

Low energy conformer of ROESY data of linear $MBP_{74\text{-}85}$ revealed a cyclic conformation by NOE connectivities, between Arg^{78}-Lys^{75}, Gln^{74}-Pro^{84} and Val^{85}-Gln^{74} amino acids (Matsoukas et al., 2001; Tselios et al., 1999). This conformation prompted the synthesis of a cyclic analog by connecting side chains of Lys^{75}-Glu^{82} which resulted in an active peptide almost equipotent to its linear counterpart (Tselios et al., 1999). Likewise, a cyclic peptide of $MBP_{74\text{-}85}$ with an Ala mutation at position 81 cyclo(74-85)[Ala^{81}]$MBP_{74\text{-}85}$, had comparable potency to the linear [Ala^{81}]$MBP_{74\text{-}85}$ peptide (Matsoukas et al., 2001). In addition, NMR and molecular dynamics of $MBP_{74\text{-}85}$ and [Ala^{81}]$MBP_{74\text{-}85}$ peptides both in water and Me_2SO solution revealed a compact conformation, and, docking studies of $MBP_{74\text{-}85}$ with I-A^u indicated that residues Lys^{75}, Arg^{78} and Asp^{81} were TCR contact residues (Tzakos et al., 2004). Molecular modeling of the linear [Arg^{91},Ala^{96}]$MBP_{87\text{-}99}$ peptide where mutations are made at important TCR contact residues 91 and 96, resulting in the ability to inhibit EAE, aided in the design and synthesis of a cyclic analog containing the central region of $MBP_{87\text{-}99}$ residues between 90-96 (Tselios et al., 2000a; Tselios et al., 2000b). In this analog Lys91, Pro96 were replaced with Arg^{91}, Ala^{96} and aminocaproic acid was added at the C-terminus as a spacer/linker resulting in a constrained cyclic moiety. This octapeptide was not able to inhibit EAE in contrast to the full length [Arg^{91},Ala^{96}]$MBP_{87\text{-}99}$ peptide. Thus, length, ring size and availability of terminal domains are important for binding to MHC.

NMR studies of linear $MBP_{87\text{-}99}$ and linear [Arg^{91},Ala^{96}]$MBP_{87\text{-}99}$ revealed a head to tail intramolecular proximity for both peptides (NOE connectivity) between Val^{87}-Arg^{97}, Lys^{91}-Arg^{97} and Phe^{89}-Pro^{99}, Val^{87}-Thr^{98}, respectively (Tselios et al., 2002). On the basis of possible cyclic conformations, cyclic peptides were designed and synthesized by connecting the side-chain amino group of Lys^{91} with the C-terminal carboxy group and between the N- and C-termini, respectively (Tselios et al., 2002). The cyclic peptides showed similar effects in animal models of EAE and on T cell proliferation of human peripheral blood mononuclear cells (PBMC). Furthermore, based on ROESY/NMR distance information and modeling of the linear [Ala^{96}]$MBP_{87\text{-}99}$ peptide, a cyclic analog cyclo(91-99)[Ala^{96}]$MBP_{87\text{-}99}$ was designed and synthesized (Matsoukas et al., 2005). Both cyclo(91-99)[Ala^{96}]$MBP_{87\text{-}99}$ and cyclo(87-99)[Arg^{91},Ala^{96}]$MBP_{87\text{-}99}$ peptides were able to suppress EAE, suppress proliferation of a human $CD4^+$ T cell clone, and induce a strong Th2/Th1 cytokine ratio from human PBMC. Furthermore, the cyclic peptides bound to HLA-DR4 and were more stable to lysosomal enzymes and Cathepsin B, D, H compared to their linear analogs (Matsoukas et al., 2005). Thus, cyclic conformation of MBP peptides predicted using NMR and computational analysis constitutes examples where rational design can lead to important peptide analogs.

Organic Compounds Can Block T Cell Presentation of Myelin Basic Protein Peptide $MBP_{152\text{-}165}$

$MBP_{152\text{-}165}$ (KIFKLGGRDSRSGS) peptide has been identified to stimulate auto-reactive T cells, which are frequently restricted by DR1301 (DRα,β1*1301) (Martin et al., 1990;

Richert et al., 1991). The predominantly MHC anchor residues are Phe^{154}, Arg^{159}, and Arg^{162} (Hastings et al., 1996), were used to model the $MBP_{152-165}$ peptide with DR1301, Table 1. The side-chain of Phe^{154} occupies a hydrophobic pocket, while both of Arg^{159} and Arg^{162} are accommodated in a negatively charged pocket in the DR1301 groove (Koehler et al., 2004).

Using the obtained HLA-DR1301 3D complex with $MBP_{152-165}$ peptide, 150,000 organic compounds were screened for binding efficacy to the peptide groove of the MHC and 1,800 compounds were selected. Of these, 39 analogs were found to be non-toxic in a high concentration. DR1301-restricted TCR-transfected cells were used to test the potency of the compounds in blocking IL-2 production. Two non-peptidic compounds were considered as promising leads and both had the ability to inhibit IL-2 secretion, when DR1301-restricted TCR transfectants were tested in the presence of different concentrations of $MBP_{152-165}$. Extensive binding affinity, activity and specificity experiments were employed to identify the most promising of these organic analogs for interfering with DR1301-restricted MBP presentation (Koehler et al., 2004).

Since, computational structure-based studies can elucidate interactions between non-peptide organic compounds and MHC molecules, such in silico approaches may be very useful for studying and developing new therapeutic organic analogs for the treatment of MS or other CNS related diseases.

Future Prospects

Increasing our basic understanding of cellular immune responses and the structural basis for peptide presentation by MHC and how T cells recognize these peptide-MHC complexes are important in peptide-based vaccine design. Where X-ray structures are not available, NMR, molecular modeling and docking studies are useful. Collectively, these structural investigations have provided important insights of peptide interactions with MHC, TCR, or how free peptides behave in solution. Structural studies, together with immunological data, are important prerequisites in drug design and development of novel peptide-based vaccines against many diseases such as MS.

Acknowledgements

MK was supported by the Ministry of Development Secretariat of Research and Technology of Greece (Grant Aus. 005) and Du Pré grant from MSIF. VA (223316) and PAR (365209) were supported by NHMRC of Australia R. Douglas Wright Fellowships. VA was supported by an NHMRC project grant 223310.

References

Anderton, S.M., Radu, C.G., Lowrey, P.A., Ward, E.S., and Wraith, D.C. (2001). Negative selection during the peripheral immune response to antigen. *The Journal of experimental medicine*, 193, 1-11.

Bielekova, B., Goodwin, B., Richert, N., Cortese, I., Kondo, T., Afshar, G., Gran, B., Eaton, J., Antel, J., Frank, J.A., et al. (2000). Encephalitogenic potential of the myelin basic protein peptide (amino acids 83-99) in multiple sclerosis: results of a phase II clinical trial with an altered peptide ligand. *Nat Med*, 6, 1167-1175.

Dessen, A., Lawrence, C.M., Cupo, S., Zaller, D.M., and Wiley, D.C. (1997). X-ray crystal structure of HLA-DR4 (DRA*0101, DRB1*0401) complexed with a peptide from human collagen II. *Immunity*, 7, 473-481.

Fairchild, P.J., Wildgoose, R., Atherton, E., Webb, S., and Wraith, D.C. (1993). An autoantigenic T cell epitope forms unstable complexes with class II MHC: a novel route for escape from tolerance induction. *International immunology*, 5, 1151-1158.

Fugger, L., Liang, J., Gautam, A., Rothbard, J.B., and mcdevitt, H.O. (1996). Quantitative analysis of peptides from myelin basic protein binding to the MHC class II protein, I-Au, which confers susceptibility to experimental allergic encephalomyelitis. *Molecular medicine* (Cambridge), Mass 2, 181-188.

Garboczi, D.N., Ghosh, P., Utz, U., Fan, Q.R., Biddison, W.E., and Wiley, D.C. (1996). Structure of the complex between human T-cell receptor, viral peptide and HLA-A2. *Nature*, 384, 134-141.

Gautam, A.M., Pearson, C.I., Smilek, D.E., Steinman, L., and mcdevitt, H.O. (1992). A polyalanine peptide with only five native myelin basic protein residues induces autoimmune encephalomyelitis. *The Journal of experimental medicine*, 176, 605-609.

Ghosh, P., Amaya, M., Mellins, E., and Wiley, D.C. (1995). The structure of an intermediate in class II MHC maturation: CLIP bound to HLA-DR3. *Nature*, 378, 457-462.

Goverman, J., Woods, A., Larson, L., Weiner, L.P., Hood, L., and Zaller, D.M. (1993). Transgenic mice that express a myelin basic protein-specific T cell receptor develop spontaneous autoimmunity. *Cell*, 72, 551-560.

Hafler, D.A., Slavik, J.M., Anderson, D.E., O'Connor, K.C., De Jager, P., and Baecher-Allan, C. (2005). Multiple sclerosis. Immunological reviews, 204, 208-231.

Hahn, M., Nicholson, M.J., Pyrdol, J., and Wucherpfennig, K.W. (2005). Unconventional topology of self peptide-major histocompatibility complex binding by a human autoimmune T cell receptor. *Nature immunology*, 6, 490-496.

Haines, J.L., Ter-Minassian, M., Bazyk, A., Gusella, J.F., Kim, D.J., Terwedow, H., Pericak-Vance, M.A., Rimmler, J.B., Haynes, C.S., Roses, A.D., et al. (1996). A complete genomic screen for multiple sclerosis underscores a role for the major histocompatability complex. The Multiple Sclerosis Genetics Group. *Nat Genet*, 13, 469-471.

Hammer, J., Bono, E., Gallazzi, F., Belunis, C., Nagy, Z., and Sinigaglia, F. (1994). Precise prediction of major histocompatibility complex class II-peptide interaction based on peptide side chain scanning. *The Journal of experimental medicine,* 180, 2353-2358.

Hammer, J., Gallazzi, F., Bono, E., Karr, R.W., Guenot, J., Valsasnini, P., Nagy, Z.A., and Sinigaglia, F. (1995). Peptide binding specificity of HLA-DR4 molecules: correlation with rheumatoid arthritis association. *The Journal of experimental medicine,* 181, 1847-1855.

Hastings, A.E., Hurley, C.K., Robinson, E.D., Salerno, K., Hernandez, E., and Richert, J.R. (1996). Molecular interactions between transfected human TCR, immunodominant myelin basic protein peptide 152-165, and HLA-DR13. *J Immunol*, 157, 3460-3471.

He, X.L., Radu, C., Sidney, J., Sette, A., Ward, E.S., and Garcia, K.C. (2002). Structural snapshot of aberrant antigen presentation linked to autoimmunity: the immunodominant epitope of MBP complexed with I-Au. *Immunity,* 17, 83-94.

Hemmer, B., Archelos, J.J., and Hartung, H.P. (2002). New concepts in the immunopathogenesis of multiple sclerosis. *Nat Rev Neurosci,* 3, 291-301.

Hemmer, B., Pinilla, C., Gran, B., Vergelli, M., Ling, N., Conlon, P., mcfarland, H.F., Houghten, R., and Martin, R. (2000). Contribution of individual amino acids within MHC molecule or antigenic peptide to TCR ligand potency. *J Immunol*, 164, 861-871.

Hemmer, B., Vergelli, M., Gran, B., Ling, N., Conlon, P., Pinilla, C., Houghten, R., mcfarland, H.F., and Martin, R. (1998). Predictable TCR antigen recognition based on peptide scans leads to the identification of agonist ligands with no sequence homology. *J Immunol,* 160, 3631-3636.

Hennecke, J., Carfi, A., and Wiley, D.C. (2000). Structure of a covalently stabilized complex of a human alphabeta T-cell receptor, influenza HA peptide and MHC class II molecule, HLA-DR1. *The EMBO journal*, 19, 5611-5624.

Kjer-Nielsen, L., Clements, C.S., Purcell, A.W., Brooks, A.G., Whisstock, J.C., Burrows, S.R., mccluskey, J., and Rossjohn, J. (2003). A structural basis for the selection of dominant alphabeta T cell receptors in antiviral immunity. *Immunity*, 18, 53-64.

Koehler, N.K., Yang, C.Y., Varady, J., Lu, Y., Wu, X.W., Liu, M., Yin, D., Bartels, M., Xu, B.Y., Roller, P.P., et al. (2004). Structure-based discovery of nonpeptidic small organic compounds to block the T cell response to myelin basic protein. *Journal of medicinal chemistry,* 47, 4989-4997.

Lee, C., Liang, M.N., Tate, K.M., Rabinowitz, J.D., Beeson, C., Jones, P.P., and mcconnell, H.M. (1998). Evidence that the autoimmune antigen myelin basic protein (MBP) Ac1-9 binds towards one end of the major histocompatibility complex (MHC) cleft. *The Journal of experimental medicine*, 187, 1505-1516.

Li, Y., Huang, Y., Lue, J., Quandt, J.A., Martin, R., and Mariuzza, R.A. (2005). Structure of a human autoimmune TCR bound to a myelin basic protein self-peptide and a multiple sclerosis-associated MHC class II molecule. *The EMBO journal*, 24, 2968-2979.

Li, Y., Li, H., Martin, R., and Mariuzza, R.A. (2000). Structural basis for the binding of an immunodominant peptide from myelin basic protein in different registers by two HLA-DR2 proteins. *Journal of molecular biology*, 304, 177-188.

Madsen, L.S., Andersson, E.C., Jansson, L., krogsgaard, M., Andersen, C.B., Engberg, J., Strominger, J.L., Svejgaard, A., Hjorth, J.P., Holmdahl, R., et al. (1999). A humanized model for multiple sclerosis using HLA-DR2 and a human T-cell receptor. *Nat Genet*, 23, 343-347.

Martin, R., Jaraquemada, D., Flerlage, M., Richert, J., Whitaker, J., Long, E.O., mcfarlin, D.E., and mcfarland, H.F. (1990). Fine specificity and HLA restriction of myelin basic protein-specific cytotoxic T cell lines from multiple sclerosis patients and healthy individuals. *J Immunol*, 145, 540-548.

Mason, K., Denney, D.W. Jr., and MCconnell, H.M. (1995). Myelin basic protein peptide complexes with the class II MHC molecules I-Au and I-Ak form and dissociate rapidly at neutral ph. *J Immunol*, 154, 5216-5227.

Matsoukas, J., Apostolopoulos, V., Kalbacher, H., Papini, A.M., Tselios, T., Chatzantoni, K., Biagioli, T., Lolli, F., Deraos, S., Papathanassopoulos, P., et al. (2005). Design and synthesis of a novel potent myelin basic protein epitope 87-99 cyclic analogue: enhanced stability and biological properties of mimics render them a potentially new class of immunomodulators. Journal of medicinal chemistry, 48, 1470-1480.

Matsoukas, J., Apostolopoulos, V., and Mavromoustakos, T. (2001). Designing peptide mimetics for the treatment of multiple sclerosis. Mini reviews in medicinal chemistry, 1, 273-282.

Maynard, J., Petersson, K., Wilson, D.H., Adams, E.J., Blondelle, S.E., Boulanger, M.J., Wilson, D.B., and Garcia, K.C. (2005). Structure of an autoimmune T cell receptor complexed with class II peptide-MHC: insights into MHC bias and antigen specificity. Immunity, 22, 81-92.

Oksenberg, J.R., Seboun, E., and Hauser, S.L. (1996). Genetics of demyelinating diseases. Brain Pathol, 6, 289-302.

Ota, K., Matsui, M., Milford, E.L., Mackin, G.A., Weiner, H.L., and Hafler, D.A. (1990). T-cell recognition of an immunodominant myelin basic protein epitope in multiple sclerosis. *Nature*, 346, 183-187.

Pette, M., Fujita, K., Wilkinson, D., Altmann, D.M., Trowsdale, J., Giegerich, G., Hinkkanen, A., Epplen, J.T., Kappos, L., and Wekerle, H. (1990). *Myelin autoreactivity in multiple sclerosis: recognition of myelin basic protein in the context of HLA-DR2 products by T lymphocytes of multiple-sclerosis patients and healthy donors.* Proceedings of the National Academy of Sciences of the United States of America, 87, 7968-7972.

Reinherz, E.L., Tan, K., Tang, L., Kern, P., Liu, J., Xiong, Y., Hussey, R.E., Smolyar, A., Hare, B., Zhang, R., et al. (1999). The crystal structure of a T cell receptor in complex with peptide and MHC class II. *Science* (New York, NY), 286, 1913-1921.

Richert, J.R., Robinson, E.D., Johnson, A.H., Bergman, C.A., Dragovic, L.J., Reinsmoen, N.L., and Hurley, C.K. (1991). Heterogeneity of the T-cell receptor beta gene rearrangements generated in myelin basic protein-specific T-cell clones isolated from a patient with multiple sclerosis. *Annals of neurology*, 29, 299-306.

Rudolph, M.G., Luz, J.G., and Wilson, I.A. (2002). Structural and thermodynamic correlates of T cell signaling. *Annual review of biophysics and biomolecular structure,* 31, 121-149.

Sireci, G., Dieli, F., Caccamo, N., Barera, A., Carta, P., Di Sano, C., Meraviglia, S., Bonanno, C.T., and Salerno, A. (2003). A human leucocyte antigen-DR1 transgene confers susceptibility to experimental allergic encephalomyelitis elicited by an epitope of myelin basic protein. *Scandinavian journal of immunology*, 58, 188-194.

Smilek, D.E., Wraith, D.C., Hodgkinson, S., Dwivedy, S., Steinman, L., and mcdevitt, H.O. (1991). *A single amino acid change in a myelin basic protein peptide confers the capacity to prevent rather than induce experimental autoimmune encephalomyelitis.* Proceedings of the National Academy of Sciences of the United States of America, 88, 9633-9637.

Smith, K.J., Pyrdol, J., Gauthier, L., Wiley, D.C., and Wucherpfennig, K.W. (1998). Crystal structure of HLA-DR2 (DRA*0101, DRB1*1501) complexed with a peptide from human myelin basic protein. The Journal of experimental medicine, 188, 1511-1520.

Sospedra, M., and Martin, R. (2005). Immunology of multiple sclerosis. *Annual review of immunology*, 23, 683-747.

Steinman, L. (1996). Multiple sclerosis: a coordinated immunological attack against myelin in the central nervous system. *Cell*, 85, 299-302.

Stern, L.J., Brown, J.H., Jardetzky, T.S., Gorga, J.C., Urban, R.G., Strominger, J.L., and Wiley, D.C. (1994). Crystal structure of the human class II MHC protein HLA-DR1 complexed with an influenza virus peptide. *Nature,* 368, 215-221.

Tate, K.M., Lee, C., Edelman, S., Carswell-Crumpton, C., Liblau, R., and Jones, P.P. (1995). Interactions among polymorphic and conserved residues in MHC class II proteins affect MHC-peptide conformation and T cell recognition. *International immunology*, 7, 747-761.

Tselios, T., Apostolopoulos, V., Daliani, I., Deraos, S., Grdadolnik, S., Mavromoustakos, T., Melachrinou, M., Thymianou, S., Probert, L., Mouzaki, A., et al. (2002). Antagonistic effects of human cyclic MBP(87-99) altered peptide ligands in experimental allergic encephalomyelitis and human T-cell proliferation. *Journal of medicinal chemistry*, 45, 275-283.

Tselios, T., Daliani, I., Deraos, S., Thymianou, S., Matsoukas, E., Troganis, A., Gerothanassis, I., Mouzaki, A., Mavromoustakos, T., Probert, L., et al. (2000a). Treatment of experimental allergic encephalomyelitis (EAE) by a rationally designed cyclic analogue of myelin basic protein (MBP) epitope 72-85. *Bioorganic and medicinal chemistry letters*, 10, 2713-2717.

Tselios, T., Daliani, I., Probert, L., Deraos, S., Matsoukas, E., Roy, S., Pires, J., Moore, G., and Matsoukas, J. (2000b). Treatment of experimental allergic encephalomyelitis (EAE) induced by guinea pig myelin basic protein epitope 72-85 with a human MBP(87-99) analogue and effects of cyclic peptides. *Bioorganic and medicinal chemistry,* 8, 1903-1909.

Tselios, T., Probert, L., Daliani, I., Matsoukas, E., Troganis, A., Gerothanassis, I.P., Mavromoustakos, T., Moore, G.J., and Matsoukas, J.M. (1999). Design and synthesis of a potent cyclic analogue of the myelin basic protein epitope MBP72-85: importance of the Ala81 carboxyl group and of a cyclic conformation for induction of experimental allergic encephalomyelitis. *Journal of medicinal chemistry*, 42, 1170-1177.

Tzakos, A.G., Fuchs, P., van Nuland, N.A., Troganis, A., Tselios, T., Deraos, S., Matsoukas, J., Gerothanassis, I.P., and Bonvin, A.M. (2004). NMR and molecular dynamics studies of an autoimmune myelin basic protein peptide and its antagonist: structural implications for the MHC II (I-Au)-peptide complex from docking calculations. *European journal of biochemistry / FEBS,* 271, 3399-3413.

Valli, A., Sette, A., Kappos, L., Oseroff, C., Sidney, J., Miescher, G., Hochberger, M., Albert, E.D., and Adorini, L. (1993). Binding of myelin basic protein peptides to human histocompatibility leukocyte antigen class II molecules and their recognition by T cells from multiple sclerosis patients. *The Journal of clinical investigation*, 91, 616-628.

Vergelli, M., Kalbus, M., Rojo, S.C., Hemmer, B., Kalbacher, H., Tranquill, L., Beck, H., mcfarland, H.F., De Mars, R., Long, E.O., et al. (1997). T cell response to myelin basic protein in the context of the multiple sclerosis-associated HLA-DR15 haplotype: peptide binding, immunodominance and effector functions of T cells. *Journal of neuroimmunology*, 77, 195-203.

Vogt, A.B., Kropshofer, H., Kalbacher, H., Kalbus, M., Rammensee, H.G., Coligan, J.E., and Martin, R. (1994). Ligand motifs of HLA-DRB5*0101 and DRB1*1501 molecules delineated from self-peptides. *J Immunol*, 153, 1665-1673.

Wraith, D.C., Smilek, D.E., Mitchell, D.J., Steinman, L., and mcdevitt, H.O. (1989). Antigen recognition in autoimmune encephalomyelitis and the potential for peptide-mediated immunotherapy. *Cell*, 59, 247-255.

Wucherpfennig, K.W., Sette, A., Southwood, S., Oseroff, C., Matsui, M., Strominger, J.L., and Hafler, D.A. (1994a). Structural requirements for binding of an immunodominant myelin basic protein peptide to DR2 isotypes and for its recognition by human T cell clones. *The Journal of experimental medicine*, 179, 279-290.

Wucherpfennig, K.W., Zhang, J., Witek, C., Matsui, M., Modabber, Y., Ota, K., and Hafler, D.A. (1994b). Clonal expansion and persistence of human T cells specific for an immunodominant myelin basic protein peptide. *J Immunol*, 152, 5581-5592.

In: Myelin Basic Protein
Editor: Joan M. Boggs

ISBN: 978-1-60456-699-4

Chapter VI

INTERACTIONS OF THE 18.5 KDA MYELIN BASIC PROTEIN WITH LIPID BILAYERS: STUDIES BY ELECTRON PARAMAGNETIC RESONANCE SPECTROSCOPY AND IMPLICATIONS FOR GENERATION OF AUTOIMMUNITY IN MULTIPLE SCLEROSIS

Joan M. Boggs[1*], Ian R. Bates[2,3], Abdiwahab A. Musse[2] and George Harauz[2*]

[1] Department of Molecular Structure and Function, Research Institute, Hospital for Sick Children, Toronto, Ontario, Canada, and Department of Laboratory Medicine and Pathobiology, University of Toronto, Toronto, Ontario, Canada.

[2] Department of Molecular and Cellular Biology, and Biophysics Interdepartmental Group, University of Guelph, Guelph, Ontario, N1G 2W1, Canada.

[3] Currently at McGill University, Montréal, Canada.

[*] Joan Boggs: Department of Molecular Structure and Function, Hospital for Sick Children, 555 University Avenue, Toronto, Ontario, M5G 1X8, Canada, Telephone: 1-416-813-5919; FAX: 1-416-813-5022; E-mail: jmboggs@sickkids.ca. George Harauz: Department of Molecular and Cellular Biology, University of Guelph, 50 Stone Road East, Guelph, Ontario, Canada. N1G 2W1. Telephone: 1-519-824-4120, Ext. 52535; FAX: 519-837-1802; E-mail: gharauz@uoguelph.ca.

ABSTRACT

The central role of the 18.5 kDa splice isoform of classic myelin basic protein (MBP) is to maintain the structural integrity of the myelin sheath of the adult central nervous system, by holding together the apposing cytoplasmic leaflets of the oligodendrocyte membrane in a tight, multilamellar arrangement. Electron paramagnetic resonance (EPR) spectroscopy is a powerful tool for probing the association of proteins such as MBP with lipid membranes. We review here the basic principles of the methodology, including the reconstitution of the MBP isoform into a physiologically relevant environment, and describe our work using EPR spectroscopy to probe 18.5 kDa MBP's depth of penetration into myelin-like lipid bilayers, as well as MBP-protein interactions. We have used two recombinant variants, rmC1 and rmC8, of murine 18.5 kDa MBP corresponding to the most cationic, C1, and least cationic, C8, variants of the protein, respectively. When interacting with large unilamellar vesicles, the N-terminal and C-terminal sites in rmC1 were located below the plane of the phospholipid headgroups. One of MBP's dominant antigenic epitopes is segment Pro85 to Pro96 (human sequence numbering, corresponding to Pro82 to Pro93 in the mouse). When bound to a myelin-like membrane surface, this segment exhibited a sinusoidal depth profile that was characteristic of an amphipathic α-helix, penetrating up to 12 Å into the bilayer. The α-helix was tilted slightly, and the central lysyl residue was in an ideal snorkelling position for side chain interaction with the negatively-charged phospholipid head groups. A proline-rich segment adjacent to this helix was exposed to the external milieu, accessible to modifying proteins such as kinases, or signalling proteins containing SH3-domains. Upon reduction in net charge of the whole protein by citrulline-mimicking substitutions at various sites throughout it, this α-helix was disrupted, had a greater surface exposure, and was more susceptible to proteolysis. The C-terminal domain (a secondary epitope) dissociated from the membrane entirely. Thus, in addition to decreased myelin compaction, deimination of MBP may also contribute to its increased proteolytic degradation and autoantigenic presentation, *in vivo*, during the early stages of multiple sclerosis.

ABBREVIATIONS

C1-C8 – MBP charge isomers, or components, 1 to 8; CaM – calmodulin; DPPG – dipalmitoylphosphatidylglycerol; EPR – electron paramagnetic resonance; Golli – genes of oligodendrocyte lineage; IDP – intrinsically disordered protein; MAP kinase – mitogen-activated protein kinase; MARCKS – myristoylated alanine-rich C kinase substrate; MS – multiple sclerosis; MTS – methanethiosulfonate; NMR – nuclear magnetic resonance; R1 – spin label; rmMBP – recombinant murine 18.5 kDa classic MBP isoform; rmC1, rmC8 – recombinant murine MBP charge isomers, or components, 1 and 8; SDSL – site-directed spin-labelling; SH3 – Src homology 3; SL – spin label; TID – (3-(trifluoromethyl)-3-(*m*-[^{125}I]iodophenyl)diazirine).

INTRODUCTION

Myelin basic protein – a structural, peripheral membrane protein of the central nervous system

In mammals, the myelin basic protein (MBP) gene encodes numerous splice variants, including the "classic" ones ranging in molecular mass from 14 kDa to 21.5 kDa (Campagnoni and Skoff, 2001) (see Chapter I). The major classic MBP isoform is 18.5 kDa in adult central nervous system myelin in humans and cattle, and we henceforth refer to this member of the family simply as MBP (Boggs, 2006; Harauz et al., 2004). In electron micrographs of sectioned myelin, MBP is localised in the major dense lines of the myelin sheath (Arroyo and Scherer, 2000). These electron dense lamellae are formed by the tight apposition of the cytoplasmic leaflets of the oligodendrocyte membrane, wherein MBP serves as an adhesive molecule. In addition, the *in vitro* reconstitution of MBP with unilamellar vesicles containing anionic lipids results in the formation of multilamellar structures similar to those seen in myelin sheaths (Sedzik et al., 1984).

The protein has an extreme net positive charge of +19 at neutral pH and interacts strongly with the cytoplasmic membrane. As shown using force measurements and atomic force microscopy on model myelin membranes, the synergistic MBP-lipid interactions rely on a balance of interactions between the basic residues of MBP and the acidic headgroups of the lipid bilayer, in order to assemble the proper multilamellar structure seen in myelin sheaths (Hu et al., 2004). Small changes to this balance could result in significant changes in myelin adhesion or stability (Jo and Boggs, 1995). Consequently, alteration of the cationicity of MBP may represent a regulatory mechanism for normal myelin assembly, defining microdomains of altered microstructure or composition (DeBruin et al., 2005; DeBruin and Harauz, 2007; Maggio et al., 2005; Rosetti and Maggio, 2007), or a degradative mechanism in neurodegenerative diseases such as multiple sclerosis (MS), as we shall discuss below.

In this last regard, MBP that is isolated from brain shows extensive post-translational modifications with varying degrees of deimination, phosphorylation, deamidation, methylation, and N-terminal acylation (reviewed in (Harauz et al., 2004; Kim et al., 2003); also see Chapters II and III). These modifications give rise to multiple charge variants denoted as C1 to C8, when separated on carboxymethyl (CM52) cellulose resin. The C1 component represents the least modified and the most cationic component with net charge +19 at neutral pH; it is the most abundant form in healthy adult humans, and the most effective lipid-aggregating charge variant of MBP (Moscarello et al., 1994). The remaining components, known as C2-C8, differ by the successive additional loss of one unit of positive charge as a result of combinations of the aforementioned post-translational modifications. The effectiveness of these charge components in inducing multilamellar membrane structures is reduced as a function of their overall net positive charge, as shown by various studies, *e.g.*, (Boggs et al., 1997; Cheifetz and Moscarello, 1985). Similar qualitative results with respect to the membrane competency of these charge components, as a function of their net positive charge, have also been obtained from a molecular-level membrane adhesion study using optical waveguide spectrometry (Shanshiashvili et al., 2003).

Deimination of MBP is correlated with early development, and with demyelination and disease severity in multiple sclerosis

The most severe post-translational modification of MBP, in terms of its reduction in overall cationicity, is the extensive arginine deimination of the protein, which gives rise to the C8 charge component (Moscarello et al., 2007) (see Chapter III). There is a reduction in net positive charge of the protein by one unit for each arginyl residue that is converted to citrulline. In general, the C8 charge component of MBP contains citrulline predominantly at 6 sites (residues Arg25, Arg31, Arg122, Arg130, Arg159, and Arg170, using human 18.5 kDa MBP sequence numbering) as assessed by amino acid sequencing (Wood and Moscarello, 1989), resulting in a net positive charge of only +13 compared to the +19 of the C1 charge component. Other sites are also deiminated, as well as phosphorylated or otherwise modified, and strictly speaking, the C8 component really comprises a heterogeneous mixture of modified proteins with net charge $\leq +13$.

The C8 component is increased in normal children relative to adult levels, but is also increased in adults with MS (Moscarello et al., 1994), indicating it may have both a physiological and a pathological role (Harauz and Musse, 2007). Because of this overall charge reduction, deiminated MBP is less effective, compared to the most highly cationic C1 charge component, in organising lipids into a multilamellar structure (Boggs et al., 1997; Ishiyama et al., 2001; Wood and Moscarello, 1989). The most highly deiminated component (18 of 19 arginyl residues, on average), isolated from a patient with a severe form of MS (Wood et al., 1996), caused fragmentation of lipid membranes, similar to the effects of apolipoproteins (Boggs et al., 1999b). Immunoelectron microscopical studies have shown epitopes of C8 to be present in the (extracellular) intraperiod line of myelin, whereas C1 is found only in the (cytoplasmic) major dense line (McLaurin et al., 1993). These and other observations suggest that the extent of MBP deimination may have a profound impact on the stability of myelin sheaths in the central nervous system, both in normal development and in MS, as reviewed in (Harauz and Musse, 2007; Moscarello et al., 2007).

Electron paramagnetic resonance (EPR) spectroscopy to study lipid-reconstituted MBP

As reviewed in other chapters in this book, MBP is an intrinsically disordered protein, whose conformational variability is best studied by spectroscopic means, *cf.*, (Harauz et al., 2004). Most forms of spectroscopy (fluorescence, circular dichroism, solution NMR) require the sample to be in soluble form, without aggregates or large complexes that would cause excessive scattering or otherwise diminish the usable signal. When reconstituted with lipids to mimic its natural environment in myelin, MBP forms large assemblies that are difficult or impossible to study by these approaches. Solid-state NMR spectroscopy is a developing technique that will eventually yield atomic level information on MBP in a native milieu (see Chapter X; also Zhong et al., 2007), but in the meantime, other approaches are required to obtain details of MBP's interactions with lipids. One of these is EPR spectroscopy, which can be used to study semi-solid material, and which was initially applied to human MBP-lipid complexes by spin-labelling its two methionines, Met20 and Met167 (corresponding to murine Met19 and Met165) (Boggs et al., 1980; Stollery et al., 1980). This procedure has the disadvantages that the sulfur becomes doubly alkylated and positively-charged, and two residues are labelled. Nevertheless, the results revealed features of MBP-lipid interactions which agreed with observations using other techniques, and increased our understanding of how MBP interacts with the lipid bilayer.

Although MBP is a water-soluble protein without long stretches of hydrophobic amino acids, it had been found to have a perturbing effect on lipid bilayers. In particular, it has been shown to affect the motion of the fatty acid chains even at locations deep within the bilayer (Boggs et al., 1981; Sankaram et al., 1989), and to decrease the lipid phase transition temperature (Papahadjopoulos et al., 1975; reviewed in Boggs et al., 1982). The spin-labelled methionines of MBP sensed the same change in phase transition temperature (Figure 1) of a lipid with saturated fatty acid chains such as dipalmitoylphosphatidylglycerol (DPPG), as was detected by differential scanning calorimetry of samples with a large MBP/lipid ratio, including similar hysteresis effects between heating and cooling scans (Boggs et al., 1980).

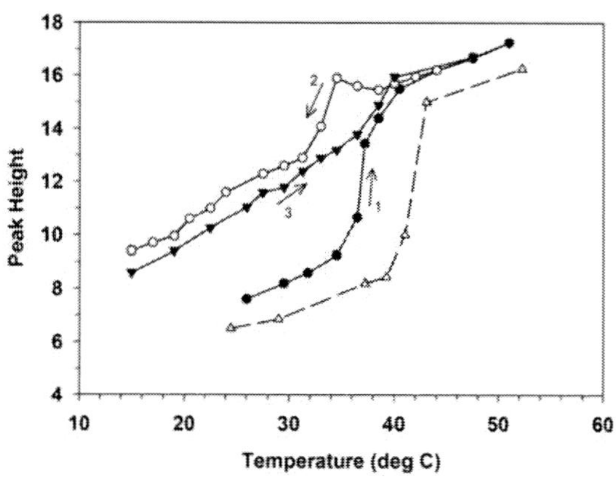

Figure 1. Effect of lipid environment on the motion of a spin-label covalently bound to MBP (Boggs et al., 1980). Temperature dependence of the height of the center line of the EPR spectrum of iodo-acetamide spin-label, covalently bound to Met20 and Met167 of human MBP, reconstituted with DPPG (solid lines). Curve 1, 1st heating scan (closed circles); curve 2, cooling scan (open circles); curve 3, 2nd heating scan (closed inverted triangles). The height is proportional to mobility. The height of the center line of the EPR spectrum of a fatty acid spin label in pure DPPG is also plotted against temperature to show the phase transition temperature of the pure lipid (dashed line, open triangles). The MBP-lipid sample shows hysteresis between heating and cooling, similar to that seen by differential scanning calorimetry. There is no hysteresis for the pure lipid sample. On the 1st heating scan of the MBP-DPPG sample, the spin label bound to MBP senses a sharp lipid phase transition but at a lower temperature than that of pure DPPG, indicating some perturbation of the lipid by the protein. The motion is greater in the liquid crystalline phase than in the gel phase indicating the spin labelled side chain can detect the degree of order and packing of the lipid. On the cooling scan, the phase transition sensed by the protein spin label is broader and at a lower temperature than on heating, indicating a larger perturbing effect of the protein on the lipid in the vicinity of the bound spin label after the protein has interacted with the fluid liquid crystalline phase lipid. Protein side chains are able to penetrate into the bilayer more in the fluid phase. On reheating, the phase transition sensed by the protein spin label is very broad but at a higher temperature than on cooling. This result indicates that after the lipid froze back into the gel phase, the protein side chains were extruded from the bilayer to some extent, but less than on the 1st heating scan. Figure adapted from reference (Boggs et al., 1980) and used with permission from the American Chemical Society.

This hysteresis indicated greater lipid perturbation after the protein interacted with the fluid phase of the lipid than when it interacted with the ordered gel phase of the lipid, suggesting greater penetration of the side chains into the fluid lipid and extrusion from the ordered lipid. Furthermore, the mobility of the side chains increased when the lipid was in the disordered liquid crystalline phase, and decreased when the lipid was in the gel phase. These experiments showed that the spin-labelled side chains could detect the degree of lipid order and also the lipid-perturbing effect of the protein molecule to which they were attached. Subsequent NMR studies indicated that, of the two methionines, it was primarily Met20 (murine Met19) that penetrated into the bilayer (Hughes et al., 1982). However, spin-labelled Met20 sensed effects due to penetration of other side chains of MBP into the lipid, which then caused greater disordering of the lipid and decreased its phase transition temperature.

Site-directed spin-labelling (SDSL) and EPR spectroscopy of soluble and membrane-associated proteins

We have subsequently used site-directed spin-labelling (SDSL) in combination with EPR spectroscopy (Biswas et al., 2001; Cornish et al., 1994; Feix and Klug, 1998; Hubbell and Altenbach, 1994; Oh et al., 2000) to map out which residues of MBP penetrate into the bilayer, and to determine the topology and structure of MBP on the bilayer surface. The SDSL procedure involves first introducing cysteinyl residues at positions of interest in the protein (Cornish et al., 1994; Hubbell and Altenbach, 1994). The second step is to mark the site with a sulfhydryl-specific spin label; the most common is a methanethiosulfonate spin label (MTS-SL), which is preferred because of its relatively small size (Figure 2). The last step is studying the spin-labelled sample using EPR spectroscopy. A simple spectrum can reveal details on probe mobility from the height and width of the center-line of the recorded spectrum (panels A-C in Figure 2); as the mobility decreases, the spectrum broadens and the height decreases.

The most powerful EPR measurements involve determining the solvent accessibility of the spin-labelled site, through the collisional interactions of the nitroxide radical with quickly relaxing paramagnetic reagents with differential solubilities. In the most common solvent accessibility experiments, the environment of each spin label is studied in the presence of NiEDDA or O_2 (Figure 2C). Equilibrating a membrane system with air will result in a gradient of O_2 that will be high in the middle of the bilayer, and low in the outer solution due to the hydrophobicity of oxygen. Conversely, NiEDDA in solution will result in a gradient of NiEDDA at a higher concentration in the aqueous milieu, going to a lower concentration in the center of the bilayer. Separate continuous wave power saturation experiments with NiEDDA and O_2 can be used to determine Ångstrom level penetration of the spin label into the bilayer, after distance calibration with spin-labelled lipids of known depth penetration. Secondary structure is revealed by the periodicity of the label's accessibility to these reagents (Nelson et al., 2005; Qin et al., 1996).

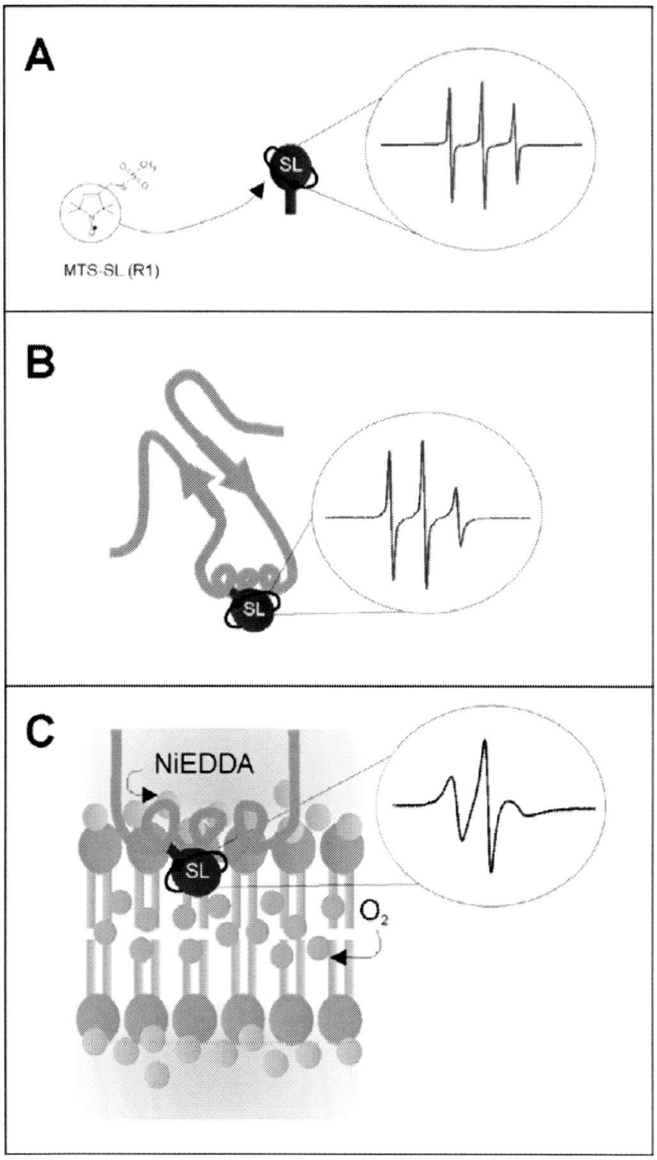

Figure 2. The effect of different environments on the methanethiosulfonate spin label (MTS-SL). The motion of the spin label translates into different EPR (first-order derivative) spectra, going from very sharp when the spin label is free in solution (A), to broader, due to the slight immobilisation when bound to protein (B), to much broader when embedded in the bilayer (C). The spectra are shown normalized to the same center peak height. However, the height also decreases when the spectrum becomes broader and the mobility decreases. In addition to the mobility information that can be garnered from analysing the spectrum (peak heights, widths, *etc.*), continuous wave power saturation is used to determine Ångstrom level depth of penetration of the residue into the bilayer. This technique involves two paramagnetic species: O_2 (hydrophobic) is at a higher concentration within the bilayer, whereas NiEDDA (hydrophilic) is at a higher concentration outside the bilayer (C). An examination of the broadening effect of these two reagents with the spin label, and calibration with spin-labelled lipids, can provide an accurate depth measurement. Figure adapted from reference (Harauz et al., 2004) and used with permission from Elsevier Ltd.

The methodology is well established. In an SDSL/EPR study of lysozyme, the motions of the spin labels were found to correlate very well with their positions (interior or exterior) on the crystallographic structure (Langen et al., 2000; Mchaourab et al., 1996; Mchaourab et al., 1997). Similarly, an SDSL/EPR study of rhodopsin confirmed that spin labels were accurate probes of protein-membrane interactions and did not disrupt the structure significantly (Farahbakhsh et al., 1992; Langen et al., 1999). Studies of influenza hemagglutinin fusion peptides by both SDSL/EPR and NMR confirmed the conservation of global structural features (Tamm et al., 2003), and studies on intrinsically disordered proteins such as amyloid peptides and α–synuclein have revealed details of their self-assembly (Der-Sarkissian et al., 2003; Török et al., 2002). Determination of the exposure of different residues to paramagnetic reagents has been used to provide high resolution structural maps for a number of other proteins (Altenbach et al., 1996; Kersten et al., 2000; Macosko et al., 1997; Perozo et al., 1998; Perozo et al., 2001; Rabenstein and Shin, 1995a; Rabenstein and Shin, 1995b; Voss et al., 1996). This technique has also been used to determine the structure of spin-labelled peptides bound to another protein (Qin et al., 1996; Qin and Cafiso, 1996).

SDSL/EPR spectroscopy of MBP in a myelin-like membrane environment

The SDSL/EPR approach is ideally suited to MBP for several reasons. First of all, the 18.5 kDa isoform of MBP is natively Cys-less (Figure 3A). Therefore, cysteinyl residues could be introduced for the purposes of spin-labelling (Figure 3B) without having first to undergo the laborious process of removing the native cysteines. Secondly, MBP causes the aggregation of vesicles, to which EPR is insensitive. In fact, sedimenting the aggregate by centrifugation is an excellent way to increase the sample concentration for EPR. Thirdly, SDSL/EPR can give accurate, Ångstrom level depths of spin label penetration, which is ideal for comparing the variations in membrane-association of different MBP isoforms. Finally, SDSL/EPR can give secondary structure information about MBP in a native environment, by Cys-scanning through each residue in a region with periodic secondary structure.

In order to understand further the molecular basis of myelin breakdown due to MBP deimination, we have used *in vitro* systems and two recombinant forms of the 18.5 kDa protein which we denote as rmC1 and rmC8, corresponding to the respective natural isoforms (Bates et al., 2000; Bates et al., 2002) (Figure 3A). The results of the SDSL in conjunction with EPR spectroscopy, and the accessibility to paramagnetic reagents, showed that there was very little immobilisation in solution as indicated by the sharp hyperfine lines in the spectrum, as in Figure 2A, consistent with the absence of a defined tertiary structure for MBP (Biswas et al., 2001). When the spin-labelled proteins were incubated with cytoplasmic large unilamellar vesicles (Cyt-LUVs, of composition mimicking the cytoplasmic leaflet of the myelin sheath: 44 mol% cholesterol, the rest being zwitterionic or negatively-charged phospholipids, with some sphingomyelin), there was significant broadening of the spectra of all the spin-labelled mutants, as in Figure 2C, indicative of the sites binding to the membrane bilayer. Power saturation experiments were used to determine the depth of each spin-labelled residue and revealed some notable divergence between rmC1 and rmC8 (Figure 4A).

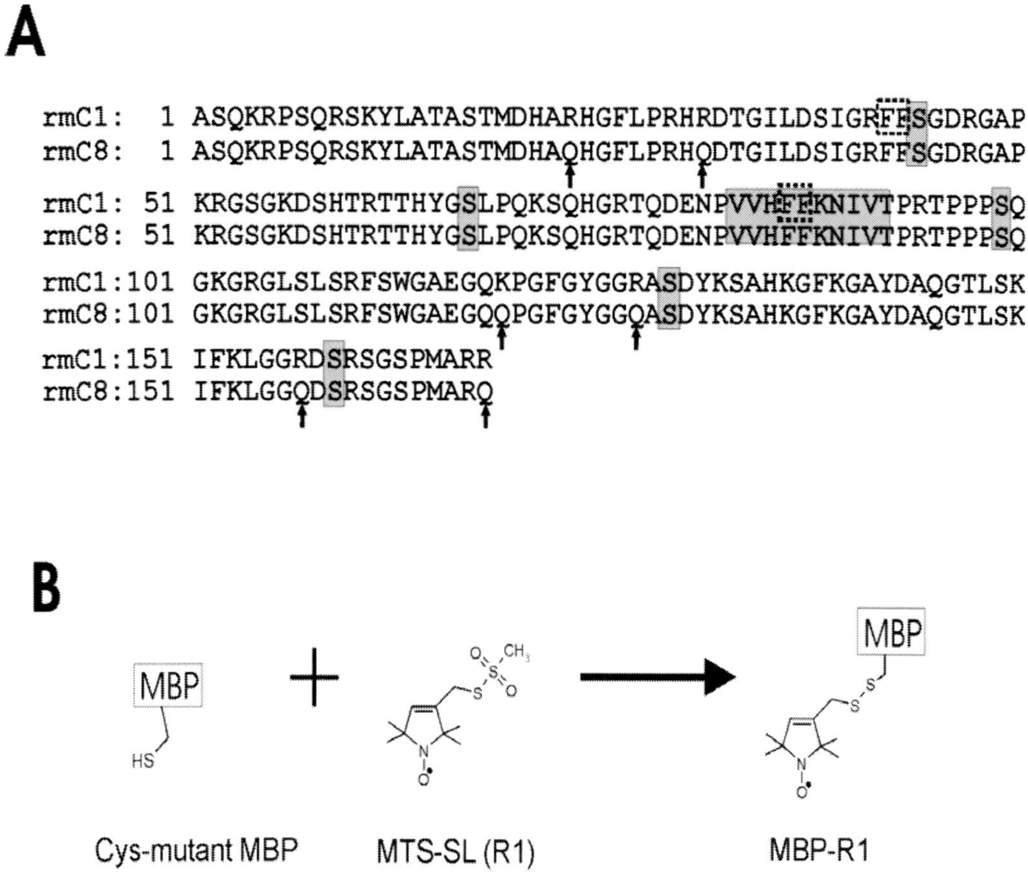

Figure 3. (A) Amino acid sequences of rmC1 and rmC8. Shaded squares indicate residues that were mutated to Cys and spin-labelled. The basic residues in rmC1 that were converted to Gln to yield rmC8 are indicated with an arrow, and the two Phe-Phe sites that were mutated to Ala-Ala are indicated by dotted squares. The hexahistidine tag used for purification has been omitted for clarity. (B) Scheme for the spin-labelling reaction of cysteinyl residues. The Cys-containing proteins were reacted with the MTS-SL while bound to the Ni^{2+}-nitrilotriacetic acid resin during the purification procedure. Figure adapted from reference (Bates et al., 2003) and used with permission from the American Society for Biochemistry and Molecular Biology.

In particular, rmC8 was significantly more exposed in the C-terminal half than rmC1, consistent with the fact that four of the six Arg/Lys → Gln substitutions were in this region. This greater accessibility of the C-terminal half of rmC8 in MS would potentially facilitate its interactions with other proteins, such as proteases or calmodulin (CaM, see below), which may be involved in myelinogenesis during development, but may result in deleterious interactions in adults with MS.

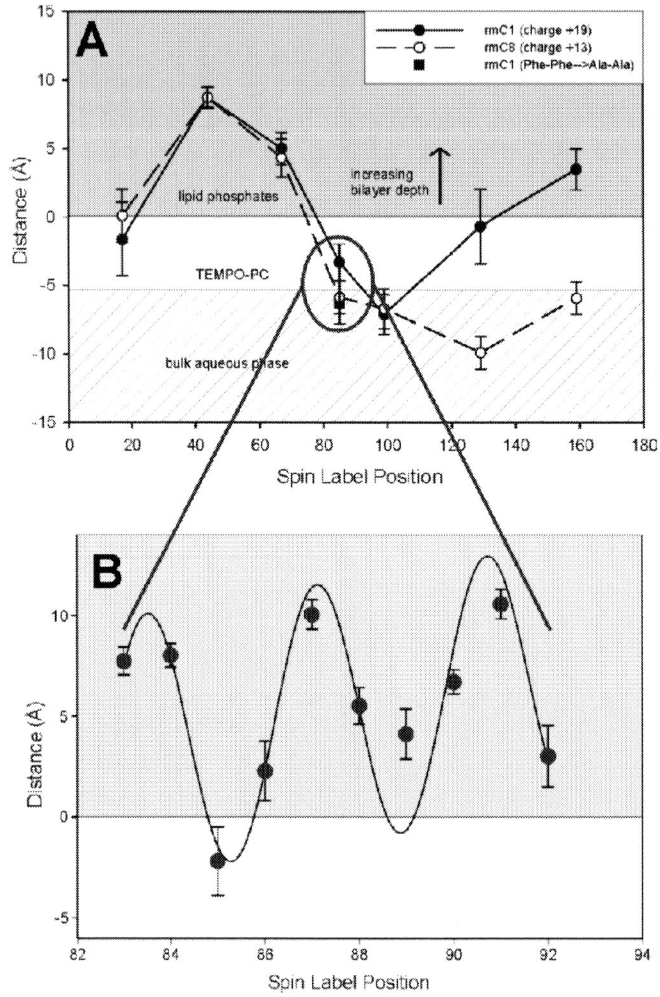

Figure 4. (A) Depths of penetration of all rmC1 and rmC8 spin labels into the lipid bilayer of the Cyt-LUVs. The gray shading indicates values below the lipid phosphates (in the bilayer), whereas the cross-hatching specifies the tentative region of distance determination. The horizontal line at 5 Å indicates the location of the nitrogen atom of the TEMPO-phosphatidylcholine nitroxide (Farahbakhsh et al., 1992). (B) The region of rmC1 from Val83 to Thr92 was subsequently studied by Cys-substitution of each residue, followed by SDSL/EPR of each protein species in turn. The periodicity of the penetration depth was fitted to a sine function. The resulting fit revealed a periodicity of 3.6 residues, and amplitude of 10 Å, indicative of an amphipathic α-helix. Moreover, the helix was tilted by ~9° with respect to the plane of the bilayer, similar to fusogenic peptides. Figure adapted from references (Bates et al., 2003; Bates et al., 2004; Harauz, 2004) and used with permission from the American Society for Biochemistry and Molecular Biology and Elsevier Ltd.

Conformation of an immunodominant epitope of MBP *in situ*

The segment Pro85-Pro96 of human MBP, corresponding to the murine residues Pro82-Pro93 (Figure 3A), is the minimal B cell epitope of MBP, and is also an important T cell recognition site. The amino acid sequence of this region of the protein is highly conserved amongst mammals (Farès et al., 2006; Harauz et al., 2004), and is post-translationally *un*modified save for the threonyl 92 residue, which is phosphorylated (as discussed below). We have probed the structure of this region by Cys-scanning through rmMBP(Val83-Thr92), and examining the depth of each residue in the lipid bilayer (Bates et al., 2004).

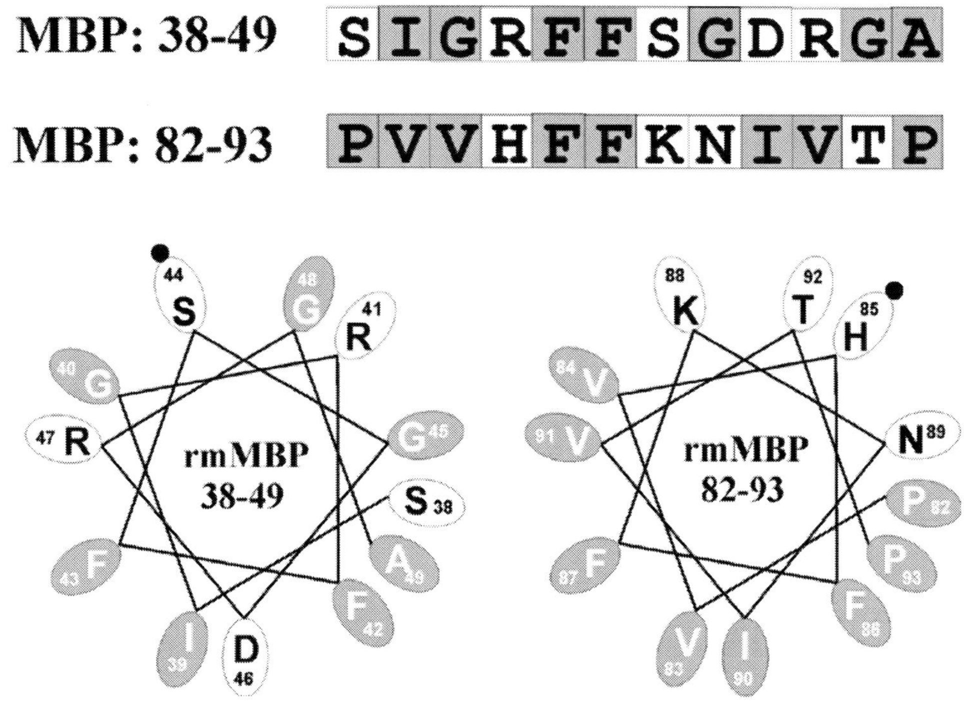

Figure 5. Helical wheel representations of the Phe42-Phe43 and Phe86-Phe87 regions of rmMBP, including 5-6 residues on either side of each Phe-Phe pair. The gray shading represents apolar and hydrophobic residues, and the white represents polar residues. The sites of the cysteinyl substitutions are indicated by closed circles. Figure adapted from reference (Bates et al., 2003) and used with permission from the American Society for Biochemistry and Molecular Biology.

The results demonstrated conclusively that this portion of the protein formed an amphipathic α-helix *in situ* (Figure 5), since it had a depth profile that could be modelled by a sine wave function, with a periodicity of approximately 3.6 residues (Figure 4B). The α-helix was embedded into the lipid bilayer, and tilted at an angle of 9° with respect to the membrane surface. This tilt was a result of the larger hydrophobic potential profile around the C-terminal end of the helix. Amphipathic peptides with fusogenic properties are also tilted with respect to the membrane plane, but at greater angles (Brasseur, 1991; Macosko et al., 1997). The smaller tilt angle for this MBP domain may explain why MBP is able to cause hemifusion of lipid bilayers without causing mixing of the vesicle compartments (Cajal et al., 1997). Thus, the degree of tilt for this segment of MBP is sufficient for deep penetration and firm membrane adhesion, but is not large enough to destabilise the bilayer.

The constituent hydrophobic side chains of rmMBP(Val83-Thr92) penetrated at a depth of up to 12 Å into the bilayer, thus explaining the ability of MBP to perturb the lipid packing. Substitution of AlaAla for the Phe86-Phe87 pair caused this segment to be less deeply embedded in the bilayer (Bates et al., 2003) (Figure 4A), similar to the results of (Victor et al., 1999) for the basic effector region of MARCKS (myristoylated alanine-rich protein kinase C substrate). Thus, both hydrophobic and electrostatic interactions anchor MBP to the membrane. This study provided the first experimental evidence of specific, local secondary structure in MBP when bound to a lipid bilayer, which was later confirmed by solution NMR spectroscopy (Farès et al., 2006) (see Chapter X).

Another significant consequence of this work concerns the role of Lys88Cys-R1 (murine numbering, Figure 3A), which was the deepest penetrating residue on the polar face of the helix in this study (5.5 Å into the bilayer). Thus, this residue was in an ideal position for snorkelling, i.e., positioning the positively-charged group of the amino acid in the polar region, while the aliphatic part was in the hydrophobic portion of the bilayer (Figure 6).

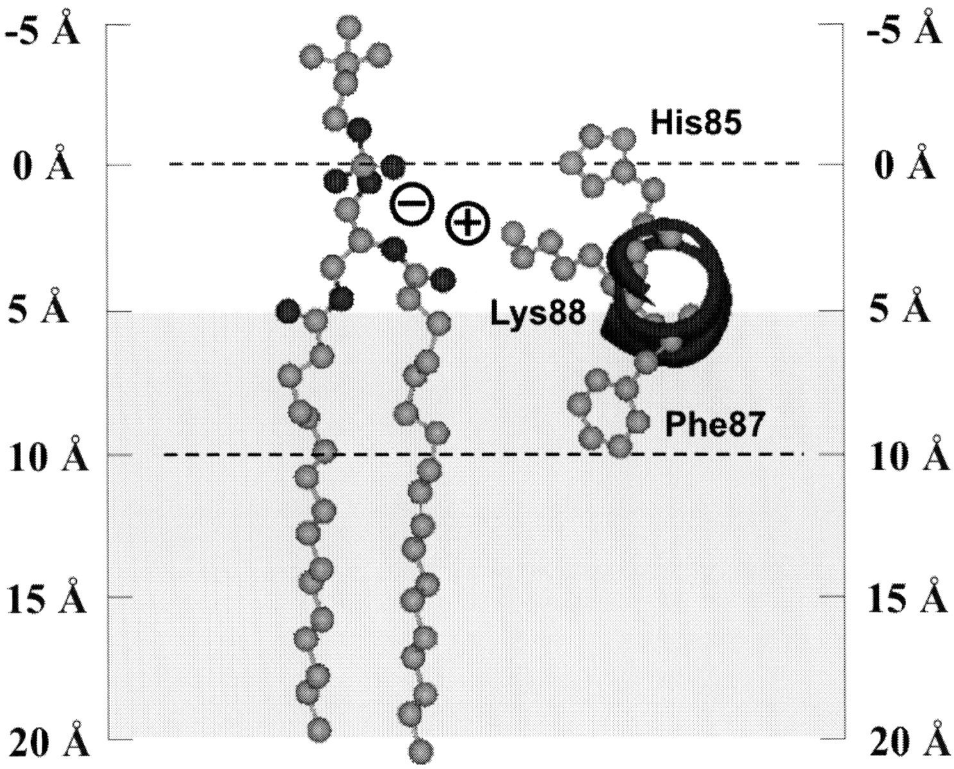

Figure 6. A schematic representation of the depth of penetration of the α-helical immunodominant epitope with the most exposed (His85) and deepest penetrating residues (Phe87) indicated. The hydrocarbon region starts ~7 Å from the lipid-water interface. Here, Lys88 is shown in a snorkelling orientation interacting with the negatively-charged phosphate group of a phospholipid. Figure adapted from reference (Bates et al., 2004) and used with permission from the American Society for Biochemistry and Molecular Biology.

In peripheral membrane proteins, snorkelling is thought to allow the long and bendable side chain of lysine to place the charged amino group in the more polar interface region, while keeping the hydrocarbon part of the side chain inside the hydrophobic part of the membrane, thus resulting in stronger binding (Mishra et al., 1994; Strandberg and Killian, 2003).

Our SDSL/EPR studies also showed that residue Ser44Cys-R1, next to the other Phe-Phe pair in MBP, was as deeply embedded in the bilayer as the side chains on the hydrophobic side of helix Val83-Thr92 (Figure 4A). However, substitution of AlaAla for this Phe42-Phe43 pair did not alter the accessibility of Ser44Cys-R1 in contrast to the Phe86-Phe87 pair, indicating that the hydrophobic location of Ser44Cys-R1 may be due to the overall structure of this region. If this region were to form a helix, it would not be amphipathic (Figure 5). Therefore, we posit that it must acquire some other type of conformation when inserted into the lipid bilayer.

Deimination of MBP exposes the central immunodominant epitope

In contrast to rmC1, the primary immunodominant epitope of rmC8, the less cationic recombinant murine MBP variant, forms a more highly surface-exposed and shorter amphipathic α-helix (Musse et al., 2006) (Figure 7). Whereas the entire Val83-Thr92 sequence of the membrane-bound rmC1 variant formed an α-helix, the observed EPR parameters for the rmC8 mutants showed only a short α-helix consisting of residues His85-Thr92. This phenomenon occurs despite the fact that none of the Arg/Lys → Gln substitutions are in this region, indicating a global effect of deimination on the MBP structure. We confirmed the topological difference of this epitope in these two variants of MBP by performing proteolytic digestion of the membrane-bound proteins. Cathepsin D digested membrane-bound rmC8 3-fold faster than rmC1, and cleavage at Phe86-Phe87 occurred much more readily in rmC8 than rmC1 (Musse et al., 2006). How the topologies with respect to the membrane of other known epitopes of MBP, primarily the C-terminal region (Figure 4A), are affected by protein deimination, is presently under investigation.

The SH3-ligand domain of MBP

Downstream from the amphiphathic helix is a TPRTPPPS (Thr92-Ser99) motif (Figure 3A) that has been demonstrated to bind to SH3-domain (Src homology 3) containing proteins, such as non-receptor tyrosine kinases (Polverini et al., 2008). In addition, the threonyl residues within this motif are MAP (mitogen-activated protein) kinase targets. Phosphorylation at Thr92, close to the C-terminus of the α-helix, may serve to destabilise this structure (Andrew et al., 2002), stabilise the adjacent polyproline type II helix (Polverini et al., 2008; Rath et al., 2005), modulate this region's interactions with the membrane and/or other proteins (Boggs et al., 2006; Cheifetz et al., 1985; Cheifetz and Moscarello, 1985; Hill and Harauz, 2005; Polverini et al., 2008), or target it to specific membrane microdomains (DeBruin et al., 2005; DeBruin et al., 2006). Spin-labelled Ser99Cys-R1 was found by SDSL/EPR to be exposed to the aqueous phase, in membrane-associated rmC1 and rmC8 (Bates et al., 2003) (Figure 4A). The accessibility of this residue suggests that the entire Pro-rich region C-terminal to the amphipathic helix of both variants should be accessible to enzymes and other proteins, including those with SH3-domains, even when MBP is bound to the membrane. Indeed, lipid-associated rmC1 and rmC8 both bound the SH3-domain of Fyn to the lipid bilayer (Polverini et al., 2008) (see Chapter IX).

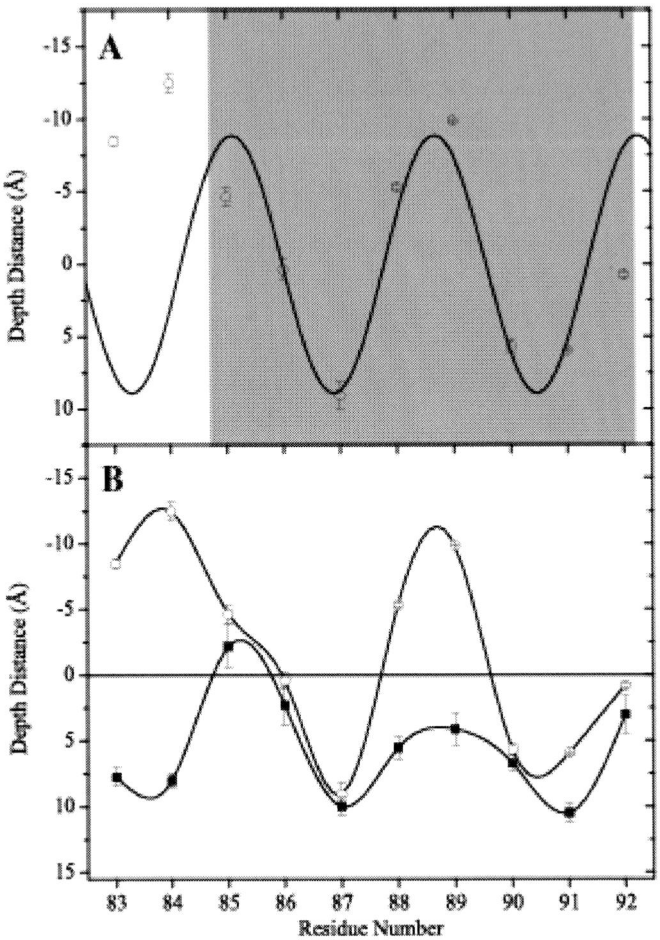

Figure 7. (A) The depth immersion of *rmC8* mutants into the lipid bilayer of the Cyt-LUVs (○) fitted to a harmonic wave function. Positive and negative values are in reference to distances below and above the lipid phosphate headgroups of the membrane, respectively. The shaded area highlights the region predicted to form an amphipathic α-helix. (B) Membrane depth measurements of the primary immunodominant epitope of MBP in normal (rmC1) (open circles) and disease-associated (rmC8) (closed circles) charge variants. In the normal myelin-associated charge variant of MBP (rmC1), this entire epitope (Val83-Thr92, murine 18.5 kDa numbering) forms a highly membrane-sequestered amphipathic α-helix, with a 9° tilt with respect to the plane of the bilayer (Figure 4B). In contrast, in the disease-associated and less cationic charge variant (rmC8), this epitope is found to be more highly surface-exposed, and to form a shorter α-helix than in rmC1. Figure adapted from reference (Musse et al., 2006) and used with permission from the National Academy of Sciences USA.

Other studies have shown that tryptic sites near this Ser99Cys-R1 position, *viz.*, Arg94-Thr95 and Arg104-Gly105, were cleaved when MBP was bound to a lipid membrane, although two other tryptic sites closer to the N-terminus, Arg22-His23 and Arg62-Thr63, were protected (Medveczky et al., 2006). The latter observation was consistent with studies using the hydrophobic photolabel TID (3-(trifluoromethyl)-3-(*m*-[^{125}I]iodophenyl)diazirine), showing that the N-terminal half of MBP is more deeply embedded into the bilayer than the

C-terminal half (Boggs et al., 1999a), as well as with our SDSL/EPR mapping of the penetration of MBP into the membrane (Bates et al., 2003).

Protein-protein interactions

In addition to SH3-domains, actin, and tubulin (see Chapters VIII and IX), MBP binds to calmodulin (CaM) in a specific, Ca^{2+}-dependent manner, with a 1:1 stoichiometry and sub-micromolar affinities. It also has a significantly different interaction when deiminated, possibly indicating two binding sites (Libich et al., 2003a; Libich et al., 2003b). Regions of MBP with hydrophobic residues in the center, basic residues at one or both ends, and thought to be capable of forming α-helices, are Arg41-Arg52, Pro82-Arg94 (the region studied above), and Ser149-Arg160. These regions were modelled as α-helices, and docking simulations were performed to investigate their interactions with the CaM peptide-binding tunnel (Polverini et al., 2004) (see Chapter IX). Almost equally favourable CaM-binding modes were found for each of them, indicating that there are several plausible CaM-binding sites in MBP. These predictions were supported by our EPR results for all three regions, using rmC1 variants with one spin-labelled residue in the region under study. The EPR results further suggested that in rmC8, the region Ser149-Arg160 (containing an Arg157Gln substitution) inserted into the CaM binding tunnel in the opposite direction from rmC1, indicating a major effect of deimination on the interaction of MBP with CaM. This possibility may partly explain our finding that CaM dissociated C8 from actin more readily than C1 (Boggs et al., 2005), and the apparently biphasic CaM titration curve (Libich et al., 2003a).

Conclusions and Future Directions

Knowledge of the conformational repertoire of the various isoforms and modified variants of MBP is essential to understanding the mechanism by which these proteins stabilise compacted myelin multilayers and interact with other proteins, potentially in events coupling cellular signals to cytoskeletal organisation during myelination or remyelination. The structural analyses described here – mapping the depth of penetration of MBP along its length, and the identification of a central amphipathic α-helix – indicate that deimination renders an otherwise partially-sequestered immunodominant epitope (Val86-Thr95, human 18.5 kDa MBP numbering) of this protein highly surfaced-exposed (Figure 8).

This important observation is consistent with earlier findings which showed the unusual accessibility of this epitope, *in vivo*, in degenerating myelin of MS lesions in close proximity to activated microglia and myelin-laden macrophages, as reviewed in (Musse and Harauz, 2007). Proteolytic processing of such a surface-exposed epitope by myelin-associated proteases, and/or phagocytosis of the damaged myelin by immune-derived cells, could result in the release of this encephalitogenic epitope from the central nervous system, and its processing for antigen presentation to the peripheral immune system. Deimination of MBP may thus, in addition to reducing the degree of myelin compaction, also participate in the autoimmune pathogenesis of the disease by revealing otherwise inaccessible antigens of myelin to the surveillance of immune cells.

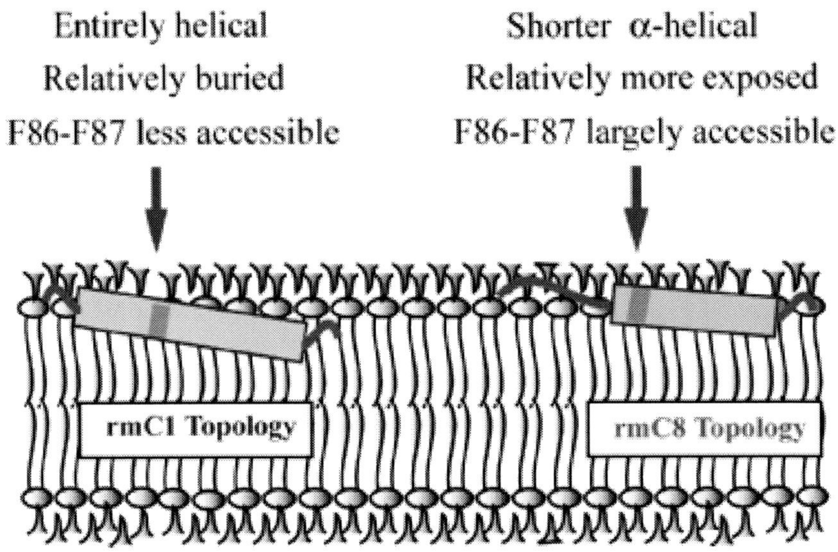

Figure 8. Schematic topology of the primary immunodominant epitope (Val83–Thr92, using murine 18.5 kDa numbering) of rmMBP in normal (rmC1) and disease-associated (rmC8) charge components. Reduction in global net positive charge of MBP by deimination reduces the overall helicity of this epitope, and renders it highly surface-exposed and susceptible to proteolysis. Figure adapted from reference (Musse et al., 2006) and used with permission from the National Academy of Sciences USA.

ACKNOWLEDGEMENTS

Our EPR spectroscopic work on myelin basic protein has been supported primarily by the Canadian Institutes of Health Research (CIHR, Operating Grants MOP #6506 to JMB, and MOP #43982 to GH), and the Natural Sciences and Engineering Research Council of Canada (NSERC, Discovery Grant RG121541 to GH). Drs. Ian Bates and Abdiwahab Musse were both recipients of Doctoral Studentships from the Multiple Sclerosis Society of Canada. We are grateful to Dr. Jimmy Feix, Medical College of Wisconsin, for assistance with some of the EPR measurements described in this chapter.

REFERENCES

Altenbach, C., Yang, K., Farrens, D.L., Farahbakhsh, Z.T., Khorana, H.G., and Hubbell, W.L. (1996). Structural features and light-dependent changes in the cytoplasmic interhelical E-F loop region of rhodopsin: a site-directed spin-labeling study. *Biochemistry*, 35, 12470-12478.

Andrew, C.D., Warwicker, J., Jones, G.R., and Doig, A.J. (2002). Effect of phosphorylation on alpha-helix stability as a function of position. *Biochemistry*, 41, 1897-1905.

Arroyo, E.J., and Scherer, S.S. (2000). On the molecular architecture of myelinated fibers. *Histochem. Cell Biol*, 113, 1-18.

Bates, I.R., Boggs, J.M., Feix, J.B., and Harauz, G. (2003). Membrane-anchoring and charge effects in the interaction of myelin basic protein (MBP) with lipid bilayers studied by spin-labelling. *J. Biol. Chem*, 278, 29041-29047.

Bates, I.R., Feix, J.B., Boggs, J.M., and Harauz, G. (2004). An immunodominant epitope of myelin basic protein is an amphipathic α-helix. *J. Biol. Chem*, 279, 5757-5764.

Bates, I.R., Libich, D.S., Wood, D.D., Moscarello, M.A., and Harauz, G. (2002). An Arg/Lys→Gln mutant of recombinant murine myelin basic protein as a mimic of the deiminated form implicated in multiple sclerosis. *Protein Expr. Purif*, 25, 330-341.

Bates, I.R., Matharu, P., Ishiyama, N., Rochon, D., Wood, D.D., Polverini, E., Moscarello, M.A., Viner, N.J., and Harauz, G. (2000). Characterization of a recombinant murine 18.5-kDa myelin basic protein. *Protein Expr. Purif*, 20, 285-299.

Biswas, R., Kuhne, H., Brudvig, G.W., and Gopalan, V. (2001). Use of EPR spectroscopy to study macromolecular structure and function. *Sci. Prog*, 84, 45-67.

Boggs, J.M. (2006). Myelin basic protein: a multifunctional protein. *Cell. Mol. Life Sci*, 63, 1945-1961.

Boggs, J.M., Moscarello, M.A., and Papahadjopoulos, D. (1982). *Structural organization of myelin: Role of lipid-protein interactions determined in model systems*. Pages 1-51, in Jost, P.C., and Griffith, O.H. (eds.), Lipid-Protein Interactions, Wiley-Interscience, New York, USA.

Boggs, J.M., Rangaraj, G., Gao, W., and Heng, Y.M. (2006). Effect of phosphorylation of myelin basic protein by MAPK on its interactions with actin and actin binding to a lipid membrane *in vitro*. *Biochemistry*, 45, 391-401.

Boggs, J.M., Rangaraj, G., Hill, C.M.D., Bates, I.R., Heng, Y.M., and Harauz, G. (2005). Effect of arginine loss in myelin basic protein, as occurs in its deiminated charge isoform, on mediation of actin polymerization and actin binding to a lipid membrane *in vitro*. *Biochemistry*, 44, 3524-3534.

Boggs, J.M., Rangaraj, G., and Koshy, K.M. (1999). Analysis of the membrane-interacting domains of myelin basic protein by hydrophobic photolabeling. *Biochim. Biophys. Acta*, 1417, 254-266.

Boggs, J.M., Rangaraj, G., Koshy, K.M., Ackerley, C., Wood, D.D., and Moscarello, M.A. (1999). Highly deiminated isoform of myelin basic protein from multiple sclerosis brain causes fragmentation of lipid vesicles. *J. Neurosci. Res*, 57, 529-535.

Boggs, J.M., Stollery, J.G., and Moscarello, M.A. (1980). Effect of lipid environment on the motion of a spin-label covalently bound to myelin basic protein. *Biochemistry*, 19, 1226-1234.

Boggs, J.M., Wood, D.D., and Moscarello, M.A. (1981). Hydrophobic and electrostatic interactions of myelin basic proteins with lipid. Participation of N-terminal and C-terminal portions. *Biochemistry*, 20, 1065-1073.

Boggs, J.M., Yip, P.M., Rangaraj, G., and Jo, E. (1997). Effect of posttranslational modifications to myelin basic protein on its ability to aggregate acidic lipid vesicles. *Biochemistry*, 36, 5065-5071.

Brasseur, R. (1991). Differentiation of lipid-associating helices by use of three-dimensional molecular hydrophobicity potential calculations. *J. Biol. Chem*, 266, 16120-16127.

Cajal, Y., Boggs, J.M., and Jain, M.K. (1997). Salt-triggered intermembrane exchange of phospholipids and hemifusion by myelin basic protein. *Biochemistry*, 36, 2566-2576.

Campagnoni, A.T., and Skoff, R.P. (2001). The pathobiology of myelin mutants reveal novel biological functions of the MBP and PLP genes. *Brain Pathol*, 11, 74-91.

Cheifetz, S., Boggs, J.M., and Moscarello, M.A. (1985). Increase in vesicle permeability mediated by myelin basic protein: effect of phosphorylation of basic protein. *Biochemistry*, 24, 5170-5175.

Cheifetz, S., and Moscarello, M.A. (1985). Effect of bovine basic protein charge microheterogeneity on protein-induced aggregation of unilamellar vesicles containing a mixture of acidic and neutral phospholipids. *Biochemistry*, 24, 1909-1914.

Cornish, V.W., Benson, D.R., Altenbach, C.A., Hideg, K., Hubbell, W.L., and Schultz, P.G. (1994). Site-specific incorporation of biophysical probes into proteins. *Proc. Natl. Acad. Sci*, USA, 91, 2910-2914.

DeBruin, L.S., Haines, J.D., Bienzle, D., and Harauz, G. (2006). Partitioning of myelin basic protein into membrane microdomains in a spontaneously demyelinating mouse model for multiple sclerosis. *Biochem. Cell Biol*, 84, 993-1005.

DeBruin, L.S., Haines, J.D., Wellhauser, L.A., Radeva, G., Schonmann, V., Bienzle, D., and Harauz, G. (2005). Developmental partitioning of myelin basic protein into membrane microdomains. *J. Neurosci. Res*, 80, 211-225.

DeBruin, L.S., and Harauz, G. (2007). White matter rafting - Membrane microdomains in myelin. *Neurochem. Res*, 32, 213-228.

Der-Sarkissian, A., Jao, C.C., Chen, J., and Langen, R. (2003). Structural organization of alpha-synuclein fibrils studied by site-directed spin labeling. *J. Biol. Chem*, 278, 37530-37535.

Farahbakhsh, Z.T., Altenbach, C., and Hubbell, W.L. (1992). Spin labeled cysteines as sensors for protein-lipid interaction and conformation in rhodopsin. *Photochem. Photobiol*, 56, 1019-1033.

Farès, C., Libich, D.S., and Harauz, G. (2006). Solution NMR structure of an immunodominant epitope of myelin basic protein. Conformational dependence on environment of an intrinsically unstructured protein. *FEBS J*, 273, 601-614.

Feix, J.B., and Klug, C.S. (1998). *Site-directed spin labeling of membrane proteins and peptide-membrane interactions*. Pages 251-281, in Berliner, X. (ed.), Spin Labeling: The Next Millenium, Plenum, New York.

Harauz, G., Ishiyama, N., Hill, C.M.D., Bates, I.R., Libich, D.S., and Farès, C. (2004). Myelin basic protein-diverse conformational states of an intrinsically unstructured protein and its roles in myelin assembly and multiple sclerosis. *Micron*, 35, 503-542.

Harauz, G., and Musse, A.A. (2007). A tale of two citrullines - Structural and functional aspects of myelin basic protein deimination in health and disease. *Neurochem. Res*, 32, 137-158.

Hill, C.M.D., and Harauz, G. (2005). Charge effects modulate actin assembly by classic myelin basic protein isoforms. *Biochem. Biophys. Res. Commun*, 329, 362-369.

Hu, Y., Doudevski, I., Wood, D., Moscarello, M., Husted, C., Genain, C., Zasadzinski, J.A., and Israelachvili, J. (2004). Synergistic interactions of lipids and myelin basic protein. *Proc. Natl. Acad. Sci*, USA, 101, 13466-13471.

Hubbell, W.L., and Altenbach, C. (1994). Investigation of structure and dynamics in membrane proteins using site-directed spin labeling. *Curr. Opin. Struct. Biol*, 4, 566-573.

Hughes, D.W., Stollery, J.G., Moscarello, M.A., and Deber, C.M. (1982). Binding of myelin basic protein to phospholipid micelles. *J. Biol. Chem*, 257, 4698-4700.

Ishiyama, N., Bates, I.R., Hill, C.M.D., Wood, D.D., Matharu, P., Viner, N.J., Moscarello, M.A., and Harauz, G. (2001). The effects of deimination of myelin basic protein on structures formed by its interaction with phosphoinositide-containing lipid monolayers. *J. Struct. Biol*, 136, 30-45.

Jo, E., and Boggs, J.M. (1995). Aggregation of acidic lipid vesicles by myelin basic protein: dependence on potassium concentration. *Biochemistry*, 34, 13705-13716.

Kersten, M.V., Dunn, S.D., Wise, J.G., and Vogel, P.D. (2000). Site-directed spin-labeling of the catalytic sites yields insight into structural changes within the F_0F_1-ATP synthase of *Escherichia coli. Biochemistry*, 39, 3856-3860.

Kim, J.K., Mastronardi, F.G., Wood, D.D., Lubman, D.M., Zand, R., and Moscarello, M.A. (2003). Multiple sclerosis: an important role for post-translational modifications of myelin basic protein in pathogenesis. *Mol. Cell. Proteomics*, 2, 453-462.

Langen, R., Cai, K., Altenbach, C., Khorana, H.G., and Hubbell, W.L. (1999). Structural features of the C-terminal domain of bovine rhodopsin: a site-directed spin-labeling study. *Biochemistry*, 38, 7918-7924.

Langen, R., Oh, K.J., Cascio, D., and Hubbell, W.L. (2000). Crystal structures of spin labeled T4 lysozyme mutants: implications for the interpretation of EPR spectra in terms of structure. *Biochemistry*, 39, 8396-8405.

Libich, D.S., Hill, C.M.D., Bates, I.R., Hallett, F.R., Armstrong, S., Siemiarczuk, A., and Harauz, G. (2003). Interaction of the 18.5-kDa isoform of myelin basic protein with Ca^{2+}-calmodulin: effects of deimination assessed by Trp fluorescence spectroscopy, dynamic light scattering, and circular dichroism. *Protein Sci*, 12, 1507-1521.

Libich, D.S., Hill, C.M.D., Haines, J.D., and Harauz, G. (2003). Myelin basic protein has multiple calmodulin-binding sites. *Biochem. Biophys. Res. Commun*, 308, 313-319.

Macosko, J.C., Kim, C.H., and Shin, Y.K. (1997) The membrane topology of the fusion peptide region of influenza hemagglutinin determined by spin-labeling EPR. *J. Mol. Biol*, 267, 1139-1148.

Maggio, B., Rosetti, C.M., Borioli, G.A., Fanani, M.L., and Del Boca, M. (2005). Protein-mediated surface structuring in biomembranes. *Braz. J. Med. Biol. Res*, 38, 1735-1748.

Mchaourab, H.S., Lietzow, M.A., Hideg, K., and Hubbell, W.L. (1996). Motion of spin-labeled side chains in T4 lysozyme. Correlation with protein structure and dynamics. *Biochemistry*, 35, 7692-7704.

Mchaourab, H.S., Oh, K.J., Fang, C.J., and Hubbell, W.L. (1997). Conformation of T4 lysozyme in solution. Hinge-bending motion and the substrate-induced conformational transition studied by site-directed spin labeling. *Biochemistry*, 36, 307-316.

McLaurin, J., Ackerley, C.A., and Moscarello, M.A. (1993). Localization of basic proteins in human myelin. *J. Neurosci. Res*, 35, 618-628.

Medveczky, P., Antal, J., Patthy, A., Kekesi, K., Juhasz, G., Szilagyi, L., and Graf, L. (2006). Myelin basic protein, an autoantigen in multiple sclerosis, is selectively processed by human trypsin 4. *FEBS Lett*, 580, 545-552.

Mishra, V.K., Palgunachari, M.N., Segrest, J.P., and Anantharamaiah, G.M. (1994). Interactions of synthetic peptide analogs of the class A amphipathic helix with lipids. Evidence for the snorkel hypothesis. *J. Biol. Chem*, 269, 7185-7191.

Moscarello, M.A., Mastronardi, F.G., and Wood, D.D. (2007). The role of citrullinated proteins suggests a novel mechanism in the pathogenesis of multiple sclerosis. *Neurochem. Res*, 32, 251-256.

Moscarello, M.A., Wood, D.D., Ackerley, C., and Boulias, C. (1994). Myelin in multiple sclerosis is developmentally immature. *J. Clin. Invest*, 94, 146-154.

Musse, A.A., Boggs, J.M., and Harauz, G. (2006). Deimination of membrane-bound myelin basic protein in multiple sclerosis exposes an immunodominant epitope. *Proc. Natl. Acad. Sci*, USA, 103, 4422-4427.

Musse, A.A., and Harauz, G. (2007). Molecular "negativity" may underlie multiple sclerosis: role of the myelin basic protein family in the pathogenesis of MS. *Int. Rev. Neurobiol*, 79, 149-172.

Nelson, W.D., Blakely, S.E., Nesmelov, Y.E., and Thomas, D.D. (2005). Site-directed spin labeling reveals a conformational switch in the phosphorylation domain of smooth muscle myosin. *Proc. Natl. Acad. Sci*, USA, 102, 4000-4005.

Oh, K.J., Altenbach, C., Collier, R.J., and Hubbell, W.L. (2000). *Site-directed spin labeling of proteins: Applications to diphtheria toxin.* Pages 147-169, in Holst, O. (ed.), Bacterial Toxins: Methods and Protocols, Humana Press Inc., Totowa, NJ.

Papahadjopoulos, D., Moscarello, M., Eylar, E.H., and Isac, T. (1975). Effects of proteins on thermotropic phase transitions of phospholipid membranes. *Biochim. Biophys. Acta*, 401, 317-335.

Perozo, E., Cortes, D.M., and Cuello, L.G. (1998). Three-dimensional architecture and gating mechanism of a K+ channel studied by EPR spectroscopy. *Nat. Struct. Biol*, 5, 459-469.

Perozo, E., Kloda, A., Cortes, D.M., and Martinac, B. (2001). Site-directed spin-labeling analysis of reconstituted Mscl in the closed state. *J. Gen. Physiol*, 118, 193-206.

Polverini, E., Boggs, J.M., Bates, I.R., Harauz, G., and Cavatorta, P. (2004). Electron paramagnetic resonance spectroscopy and molecular modelling of the interaction of myelin basic protein (MBP) with calmodulin (CaM)-diversity and conformational adaptability of MBP CaM-targets. *J. Struct. Biol*, 148, 353-369.

Polverini, E., Rangaraj, G., Libich, D.S., Boggs, J.M., and Harauz, G. (2008). Binding of the proline-rich segment of myelin basic protein to SH3-domains - Spectroscopic, microarray, and modelling studies of ligand conformation and effects of post-translational modifications. *Biochemistry*, 47, 267-282.

Qin, Z., and Cafiso, D.S. (1996). Membrane structure of protein kinase C and calmodulin binding domain of myristoylated alanine rich C kinase substrate determined by site-directed spin labeling. *Biochemistry,* 35, 2917-2925.

Qin, Z., Wertz, S.L., Jacob, J., Savino, Y., and Cafiso, D.S. (1996). Defining protein-protein interactions using site-directed spin- labeling: the binding of protein kinase C substrates to calmodulin. *Biochemistry*, 35, 13272-13276.

Rabenstein, M., and Shin, Y.K. (1995). A peptide from the heptad repeat of human immunodeficiency virus gp41 shows both membrane binding and coiled-coil formation. *Biochemistry*, 34, 13390-13397.

Rabenstein, M.D., and Shin, Y.K. (1995). Determination of the distance between two spin labels attached to a macromolecule. *Proc. Natl. Acad. Sci*, USA, 92, 8239-8243.

Rath, A., Davidson, A.R., and Deber, C.M. (2005). The structure of "unstructured" regions in peptides and proteins: role of the polyproline II helix in protein folding and recognition. *Biopolymers*, 80, 179-185.

Rosetti, C.M., and Maggio, B. (2007). Protein-induced surface structuring in myelin membrane monolayers. *Biophys. J*, 93, 4254-4267.

Sankaram, M.B., Brophy, P.J., and Marsh, D. (1989). Spin-label ESR studies on the interaction of bovine spinal cord myelin basic protein with dimyristoylphosphatidylglycerol dispersions. *Biochemistry*, 28, 9685-9691.

Sedzik, J., Blaurock, A.E., and Höchli, M. (1984). Lipid/myelin basic protein multilayers. A model for the cytoplasmic space in central nervous system myelin. *J. Mol. Biol*, 174, 385-409.

Shanshiashvili, L.V., Suknidze, N.Ch., Machaidze, G.G., Mikeladze, D.G., and Ramsden, J.J. (2003). Adhesion and clustering of charge isomers of myelin basic protein at model myelin membranes. *Arch. Biochem. Biophys*, 419, 170-177.

Stollery, J.G., Boggs, J.M., and Moscarello, M.A. (1980). Variable interaction of spin-labeled human myelin basic protein with different acidic lipids. *Biochemistry* 19, 1219-1226.

Strandberg, E., and Killian, J.A. (2003). Snorkeling of lysine side chains in transmembrane helices: how easy can it get? *FEBS Lett*, 544, 69-73.

Tamm, L.K., Abildgaard, F., Arora, A., Blad, H., and Bushweller, J.H. (2003). Structure, dynamics and function of the outer membrane protein A (OmpA) and influenza hemagglutinin fusion domain in detergent micelles by solution NMR. *FEBS Lett*, 555, 139-143.

Török, M., Milton, S., Kayed, R., Wu, P., McIntire, T., Glabe, C.G., and Langen, R. (2002). Structural and dynamic features of Alzheimer's Abeta peptide in amyloid fibrils studied by site-directed spin labeling. *J. Biol. Chem*, 277, 40810-40815.

Victor, K., Jacob, J., and Cafiso, D.S. (1999). Interactions controlling the membrane binding of basic protein domains: phenylalanine and the attachment of the myristoylated alanine-rich C- kinase substrate protein to interfaces. *Biochemistry*, 38, 12527-12536.

Voss, J., He, M.M., Hubbell, W.L., and Kaback, H.R. (1996). Site-directed spin labeling demonstrates that transmembrane domain XII in the lactose permease of *Escherichia coli* is an alpha-helix. *Biochemistry*, 35, 12915-12918.

Wood, D.D., Bilbao, J.M., O'Connors, P., and Moscarello, M.A. (1996). Acute multiple sclerosis (Marburg type) is associated with developmentally immature myelin basic protein. *Ann. Neurol*, 40, 18-24.

Wood, D.D., and Moscarello, M.A. (1989). The isolation, characterization, and lipid-aggregating properties of a citrulline containing myelin basic protein. *J. Biol. Chem*, 264, 5121-5127.

Zhong, L., Bamm, V.V., Ahmed, M.A., Harauz, G., and Ladizhansky, V. (2007). Solid-state NMR spectroscopy of 18.5 kDa myelin basic protein reconstituted with lipid vesicles: spectroscopic characterisation and spectral assignments of solvent-exposed protein fragments. *Biochim. Biophys. Acta (Biomembranes),* 1768, 3193-3205.

In: Myelin Basic Protein
Editor: Joan M. Boggs

ISBN: 978-1-60456-699-4

Chapter VII

INSIGHTS INTO THE INTERACTION OF MYELIN BASIC PROTEIN WITH MICROTUBULES

Mauricio R. Galiano, Cecilia Lopez Sambrooks and Marta E. Hallak

ABSTRACT

One of the cellular elements of the glia in the central nervous system (CNS) is the oligodendrocyte. Lamella from these specialized cells, which wraps the neuronal axons, provides and maintains the myelin sheath that serves to speed up the action potentials by saltatory conduction. In oligodendrocytes, a large class of proteins, predominantly cytoplasmic, called myelin basic proteins (MBP) have a central role in myelin formation. During myelinogenesis, these cells extend branched processes toward axons; the extension of these processes as well as the formation of the myelin sheaths are supported by the oligodendrocyte cytoskeleton assembled by microtubules and microfilaments (Wilson and Brophy, 1989). Several studies have recognized the relevance of cytoskeletal arrangement during the morphological differentiation of oligodendrocytes (Lunn et al., 1997a; Lunn et al., 1997b; Song et al., 1999; Richter-Landsberg, 2000) as well as the reorganization of microtubules and microfilaments during the formation of processes and branches (Song et al., 2001). The present review will emphasize the interaction between MBP and the microtubules. Moreover among the putative functions derived from this interaction, and also between MBP and microfilaments, an active participation of MBP in oligodendroglial differentiation has been suggested. Thus, besides the known role of MBP in myelin compaction, another function of MBP is in the ordering of the oligodendrocyte's cytoskeletal elements during CNS axonal myelination.

* Centro de Investigaciones en Química Biológica de Córdoba, CIQUIBIC, (UNC-CONICET), Departamento de Química Biológica, Facultad de Ciencias Químicas, Universidad Nacional de Córdoba, 5000-Córdoba, Argentina. Telephone: 54-351-4334171; Fax: 54-351-4334074; E-mail: mhallak@mail.fcq.unc.edu.ar

INTRODUCTION

In general the cytoskeleton participates in several functions of eukaryotic cells. It is responsible for maintenance of cell morphology, intracellular vesicular transport and cell surface receptor modulation. It is also vital for the processes of cell motility, differentiation and mitosis (Rodriguez et al., 2003). The microtubules are actively implicated in these functions. The dynamic characteristics of the microtubules are dependent not only on their intrinsic dynamic properties but also on the interactions of microtubules with several cytoplasmic proteins. The putative role of MBP as a microtubule-interacting protein involved in the coordination of cytoskeleton changes during oligodendrocyte differentiation, and how its post-translational modifications are involved in these interactions, will be the focus of discussion in this chapter.

Implication of MBP in the signal transduction derived from membrane alterations

One important aspect of the MBP-microtubules relationship is the characterization of the role of MBP in the transduction of signals derived from lipid alterations at the cell surface promoting cytoskeleton changes during myelin compaction (Dyer and Benjamins, 1988; Dyer et al., 1994; Bansal et al., 1999; Boggs and Wang, 2001; Boggs and Wang, 2004). Several studies have analyzed the role of the interactions between myelin glycosphingolipids at oligodendrocyte membranes, which could favor the compaction and formation of mature myelin. Through these interactions many changes are triggered at the membrane concomitantly with a reorganization of the underlying cytoskeleton. MBP should be mediating these cytoskeletal changes as a transmitter of environmental information through the cellular membrane, participating therefore in the regulation of oligodendrocyte differentiation and myelination. This has been suggested from studies where oligodendrocyte cultures treated with anti-glycosphingolipid antibodies (Dyer and Benjamins, 1988; Dyer and Benjamins 1989a; Dyer et al., 1994) or with glycosphingolipid-containing liposomes (Boggs and Wang, 2001; Boggs and Wang, 2004) showed an altered distribution of many glycosphingolipids and, in the case of antibodies, extensive contraction of membrane sheets. Both treatments mimic the lipid arrangement during normal myelination. In addition, both effectors induced the microtubule depolymerization of the lacy veins; a more extensive effect on microfilaments has also been reported for the cytoskeleton of oligodendrocytes exposed to liposomes (Boggs and Wang, 2001).

Recently, Boggs and Wang (2004) reported that cytoskeleton depolymerization was necessary to induce the redistribution of glycosphingolipids and myelin-specific proteins, including MBP. These authors found that the stabilization of actin microfilaments by jasplakinolide (a cyclic peptide isolated from the marine sponge *Jaspis johnstoni*) inhibited the clustering of membrane constituents induced by liposomes. From these studies it was concluded that the depolymerization of the cytoskeleton is an early event necessary for the glycosphingolipid redistribution induced by liposomes (Boggs and Wang, 2004). Moreover, it was established that the microtubule depolymerization induced by glycosphingolipid-containing liposomes takes place once actin microfilaments are depolymerized. Even though the mechanism of these changes is still not fully understood, it has been determined that the signals conferred either by anti-glycosphingolipid antibodies or by glycosphingolipid-containing liposomes are transmitted across the membrane to the cytoskeleton. Among the different effects triggered that could induce cytoskeleton changes, it was reported that the anti-galactocerebroside antibody leads to an influx of Ca^{2+} (Dyer and Benjamins, 1990) and phospholipid turnover (Dyer, 1993), and a decrease in the levels of MBP phosphorylation (Dyer et al., 1994). Myelin oligodendrocyte glycoprotein (MOG) and proteolipid protein (PLP) may be the membrane receptors that transmit the out-coming signal that may coordinate the oligodendrocyte differentiation (Boggs and Wang, 2004). MBP seems to be one of the endogenous targets of the signals elicited by these receptors since no effects were observed in oligodendrocytes derived from MBP-deficient Shiverer mice (Dyer et al., 1994). These studies indicate that MBP influences the cytoskeletal reorganization induced by anti-galactocerebroside antibody in oligodendrocytes. Nevertheless, it still remains to be determined whether MBP mediates these changes by itself or through its interaction with one or more proteins, whose association with MBP indirectly promotes the cytoskeletal changes.

In the course of myelinogenesis the presence of an intact cytoskeleton is necessary for extension of oligodendrocyte processes at initial stages. In turn, oligodendrocytes concomitantly produce and expand the membrane sheets wrapping the axons. The signals derived from the interactions between glycosphingolipids expressed in apposed surfaces may elicit the cascade of events that probably coordinate the cytoskeletal reorganization within the oligodendroglial membrane sheets during myelin formation and compaction. In view of the different findings described for myelin production in MBP-deficient Shiverer mice (Privat et al., 1979; Barbarese et al., 1983; Inoue et al., 1983; Kimura et al., 1998), it was suggested that the absence of MBP may influence the abnormal myelin development found in these mutants. Despite the fact that the contact between glycosphingolipids takes place, transiently or in localized domains, the lack of MBP probably does not allow the synchronized display of cytoskeleton that mediates the optimal myelin development. It is expected that future studies will elucidate if MBP is the main protein that modulates the signal transmission across the membrane that induces depolymerization of the cytoskeleton followed by clustering of membrane constituents.

The interaction between myelin basic protein and the cytoskeleton

As we described above, in a number of studies it was shown that MBP participates in the transduction of extracellular signals that modify the cytoskeletal structures (Dyer and Benjamins, 1988; Dyer et al., 1994; Dyer and Benjamins, 1989a; Dyer et al., 1990; Staugaitis et al., 1996; Lintner and Dyer, 2000; Arvanitis et al., 2002). Concurrently, other studies showed that MBP is predominantly associated with microtubules (Wilson and Brophy, 1989; Dyer and Benjamins, 1989b; Richter-Landsberg, 2000). Using primary glial cultures, a large amount of information was obtained about the morphological changes that the oligodendroglial cells undergo during differentiation. As is known, in the CNS this cell type acquires a complex star-like morphology. Throughout maturation, oligodendrocytes extend several processes that contact and spiral around axons, forming later the compact myelin sheath responsible for the saltatory conduction of nerve impulses. These changes require a well coordinated reorganization of the cytoskeleton. The differentiation of bipolar precursor progenitor cells to immature oligodendrocytes begins with the extension of multiple processes. After further differentiation, mature oligodendrocytes develop large arborized processes which are supported by the coordinated action of microtubules and microfilaments (Wilson and Brophy, 1989; Lunn et al., 1997a; Song et al., 2001). Diverse cytoskeletal-associated proteins influence this cytoskeletal organization (Richter-Landsberg, 2000; Jiang et al., 2005; Lee et al., 2005; Brockschnieder et al., 2006; Bacon et al., 2007).

The understanding of which elements coordinate these two cytoskeletal arrays and how both structures interact with each other during oligodendrocyte development, is limited. Possible candidates for mediating the linkage of these two polymer systems are the structural microtubule-associated proteins (MAPs), which were suggested to play this role in neuronal cells (Rodriguez et al., 2003; Tögel et al., 1998; Noiges et al., 2002; Dehmelt et al., 2003; Dehmelt and Halpain, 2004; Cueille et al., 2007). The expression of microtubule-associated proteins (MAPs) has been described in oligodendrocytes (Vouyiouklis and Brophy, 1995; LoPresti, et al, 1995; Müller et al., 1997; Richter-Landsberg and Gorath 1999; Lopresti, 2002; Galiano et al., 2004), but their function as structural linker protein remains unclear. Taking into account the specialized functions of oligodendrocytes, perhaps directly related to its complex morphology (Murtie et al., 2007), several studies have suggested a close association between myelin constituents and cytoskeletal elements (Wilson and Brophy, 1989; Dyer and Benjamins, 1989b; Gillespie et al., 1989; Lee et al., 2005; Galiano et al., 2006). As we already mentioned, MBP and also CNP (2',3'-cyclic nucleotide 3'-phosphodiesterase), two specific oligodendroglial proteins, may be implicated in the coordinated reorganization of microtubules and actin microfilaments, also mediating the anchorage of cytoskeletal elements to the oligodendrocyte plasma membrane (Boggs and Rangaraj, 2000; Lee et al., 2005; Hill et al., 2005). Both proteins are known to colocalize with microtubules in cultured oligodendrocytes (Dyer and Benjamins, 1989a; Dyer and Benjamins, 1989b). Wilson and Brophy (1989) partially copurified MBP and CNP with tubulin and actin after detergent extraction of myelin membranes from cultured oligodendrocytes. This is a strong indication that these proteins may interact with the cytoskeleton.

Indeed several biochemical investigations with isolated proteins have described the putative MBP interaction with the cytoskeleton (Dobrowolski et al., 1986; Modesti and Barra, 1986; Pirollet et al., 1992; Saoudi et al., 1995; Boggs and Rangaraj, 2000; Hill et al., 2005; Hill and Harauz, 2005). An actin polymerizing property of MBP was shown by Dobrowolski et al. (1986). These authors, using electron microscopy, observed that MBP binds to G-actin inducing the formation of actin filaments. In addition, these actin-binding and polymerizing activities described for MBP were associated with its lipidic interaction (Boggs and Rangaraj, 2000). MBP was found to be able to interact with actin and to simultaneously bind acidic lipids. The association of MBP with actin or lipids was regulated by calmodulin in the presence of Ca^{2+} (Dobrowolski et al., 1986; Boggs and Rangaraj, 2000). These ultrastructural findings established that MBP could mediate the binding of actin filaments to the membrane, probably regulating the interactions between the cytoskeleton and the oligodendrocyte membrane during differentiation and myelin formation (Boggs and Rangaraj, 2000).

There is also evidence for the interaction of MBP with tubulin (Modesti and Barra, 1986; Pirollet et al., 1992; Saoudi et al., 1995; Hill et al., 2005). From these studies, many functions have been attributed to MBP as a microtubule-associated protein. Modesti and Barra (1986) first reported that the *in vitro* interaction of MBP with tubulin causes the inhibition of tubulin carboxypeptidase activity. Pirollet et al. (1992) identified two proteins, MBP and STOP (stable tubule only polypeptide), as two soluble brain proteins capable of inducing microtubule cold-stability. This microtubule-stabilizing activity of both proteins is also regulated by Ca^{2+}-calmodulin. A putative microtubule bundling activity of MBP was described using immunofluorescence analysis.

Through a comparative study to define the structural elements implicated in MAP-tubulin polymers interaction, Saoudi et al. (1995) found that, after treatment of microtubules with subtilisin (to produce microtubules composed of tubulin subunits lacking carboxy termini), Histone H1 and MBP, two highly basic proteins, shared with MAP2 the capability to induce similar polymer stabilization of control and subtilisin-treated microtubules, in contrast to tau protein, which enhanced the stabilization of subtilisin-treated microtubules relative to control microtubules. The interactions between subtilisin-treated microtubules and MAPs were observed by immunofluorescence, and MBP showed an *in vitro* microtubule-stabilizing activity similar to that described for structural MAPs (Saoudi et al., 1995), suggesting that the binding of MBP onto microtubules involves diverse tubulin-interacting domains.

Understanding of the molecular events that regulate the MBP-microtubule interaction was aided by an important contribution made by studies to clarify the interaction of MBP with calmodulin (Chan et al., 1990; Libich et al., 2003; Polverini et al., 2004). The main calmodulin-binding site of the 18.5 kDa isoform of MBP was localized at its C-terminal sequence. Analyzing MBP mutants with deleted terminal sequences, the presence of a second calmodulin-binding site in a central segment of the MBP sequence was described (Libich et al., 2003). Further studies characterized the presence of a third calmodulin-binding site localized at the N-terminal sequence (Polverini et al., 2004). These multiple calmodulin-binding sites described for MBP could be alternatively implicated in the multiple interactions established by this protein with different cellular elements. An example corresponds to the sequence of the second calmodulin-binding site, which could be involved in the association of MBP with membranes (Bates et al., 2003). We could speculate that one of the remaining calmodulin-binding sites of MBP could overlap with amino acid sequences also implicated in MBP-microtubule interactions similar to the bifunctional calmodulin-binding and microtubule-stabilizing domains that have been described for STOP proteins (Bosc et al., 2001; Bosc et al., 2003). Even though MBP does not have sequence homology with STOP proteins, a functional analysis to characterize the MBP capability to interact with microtubules, using MBP with mutations in its calmodulin-binding sites, may contribute to a better understanding of the Ca^{2+}-calmodulin regulation of MBP–microtubule interactions. We must take into consideration that the interactions of MBP proteins with cytoskeletal elements and other biological macromolecules are also modulated by post-translational modifications of MBP (see review, Boggs, 2006 and Chapter VIII). This issue will be evaluated below.

Recently, Hill et al. (2005) have provided new findings consistent with a role for MBP in microtubule regulation. These authors characterized by light scattering and transmission electron microscopy the biochemical interaction of highly purified MBP isoforms with tubulin. They found that MBP is able to induce tubulin polymerization and also to bundle the generated microtubules. From these results it was suggested that MBP may present different tubulin-binding motifs which could be comprised of the amino acid sequences derived from exons III and IV of the classic MBP gene (now called exons 7 and 8 of the total MBP gene, which expresses both golli and classic MBP isoforms – see Chapters I and X).

As we mentioned above, the association of MBP with microtubules was established in oligodendroglial cell cultures (Wilson and Brophy, 1989; Dyer and Benjamins, 1989b) and it was shown that MBP colocalizes with microtubules in the cellular sheet extensions (Richter-Landsberg, 2000). Similarly, as was established for MBP, CNP also showed a cytoskeleton-associated distribution, presenting a close association with actin microfilaments of immature cells (Wilson and Brophy, 1989). It is known that CNP expression precedes MBP expression (Pfeiffer et al., 1981; Dubois-Dalcq et al., 1986; Reynolds and Wilkin, 1988). It was shown that immature oligodendrocytes express high levels of CNP during differentiation. When these cells differentiate into mature oligodendrocytes, the distribution of CNP was observed through large arborized processes in numerous subsets of branches and at the cortical edge of the cell body, colocalizing with tubulin (Lee et al., 2005). Furthermore, it was described that CNP preferentially interacts with tubulin and promotes microtubule assembly, causing the reorganization of microtubules and actin filaments. From these studies, it was suggested that CNP plays also a vital role during process outgrowth and branching in myelinating oligodendrocytes. In accordance with the CNP functions proposed by Lee et al. (2005), we may conjecture that both proteins (CNP and MBP) could perform a coordinated activity on the microtubule cytoskeleton. In regard to the sequential expression of CNP and MBP (Pfeiffer et al., 1981; Dubois-Dalcq et al., 1986; Reynolds and Wilkin, 1988), it is possible that at early stages of oligodendrocyte differentiation, CNP induces microtubule disruption by depleting tubulin subunits through its interaction with tubulin heterodimers (Lee et al., 2005). In turn, with further differentiation, cells must reorganize a stable microtubule network. This is a scenario wherein MBP, in addition to structural MAPs, may play an important role through interaction with tubulin. Future studies directed to elucidate the participation of both proteins together on cytoskeletal dynamics may contribute to a better understanding of how these proteins influence oligodendroglial morphology and functions.

Oligodendroglial distribution of microtubules and microfilaments

Initially it was described that microtubules were distributed throughout all regions of the cell and that actin microfilaments were enriched in the cellular processes (Wood and Bunge, 1984; Wilson and Brophy, 1989). Later Simpson and Armstrong (1999) showed that microfilaments are also present in the oligodendroglial cell body (see Figure 1 and 2 for examples of the microfilament and microtubule distribution in oligodendrocytes). However a more complete description of the distribution of microtubules and microfilaments was made by Song et al. (2001) that arrived at a better understanding of the coordinated reorganization of both structures during oligodendroglial differentiation and also established a vital role for actin microfilaments during development. Microfilaments lead the rearrangement of microtubules at branching points and regions of process outgrowth. It was described that the enrichment of microfilaments in the leading edges and branching sites was distinct at different levels of differentiation. Such an enrichment was predominantly marked in immature oligodendrocytes compared to mature ones (Figure 2) (Song et al., 2001).

Furthermore there is a different distribution of labile and stable microtubules that may also reflect the functions of the different cellular regions of the oligodendrocytes (Lunn et al., 1997a; Song et al., 2001). The C-terminal tyrosine of α-tubulin is removed by a tubulin-carboxypeptidase (TCP), leaving Glu as the C-terminal residue, and re-added by a tubulin-tyrosine-ligase, as a cyclic modification called the cycle of detyrosination/tyrosination

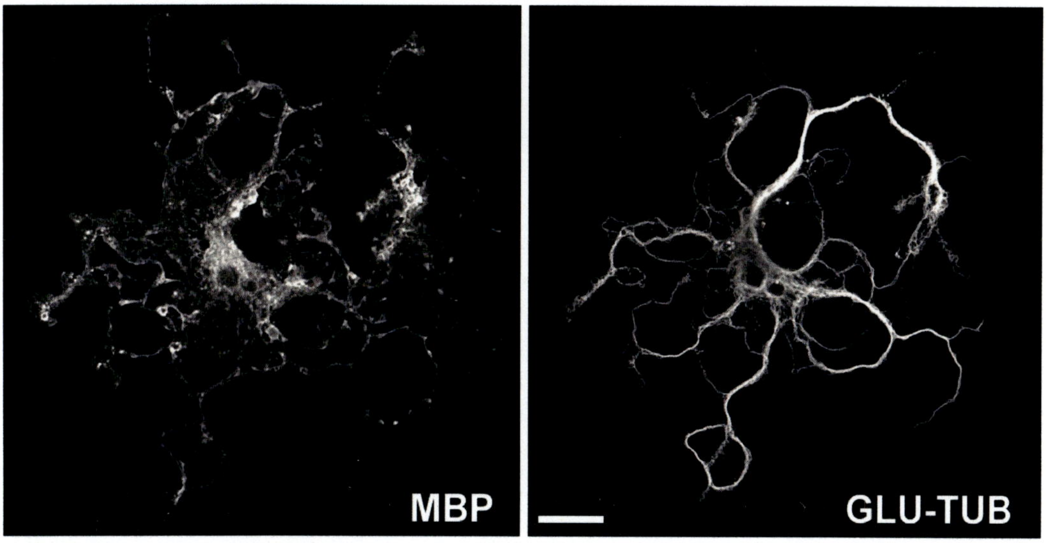

Figure 1. Microtubule cytoskeleton of MBP+ oligodendrocyte.
Cells were fixed in 4% paraformaldehyde in PBS for 30 min, washed with PBS and then permeabilized with 0.01% Triton X-100 in PBS for 10 min. Then, cells were doubly stained with anti-MBP antibody and anti-Glu-tubulin antibody. (Bar= 20 μm).

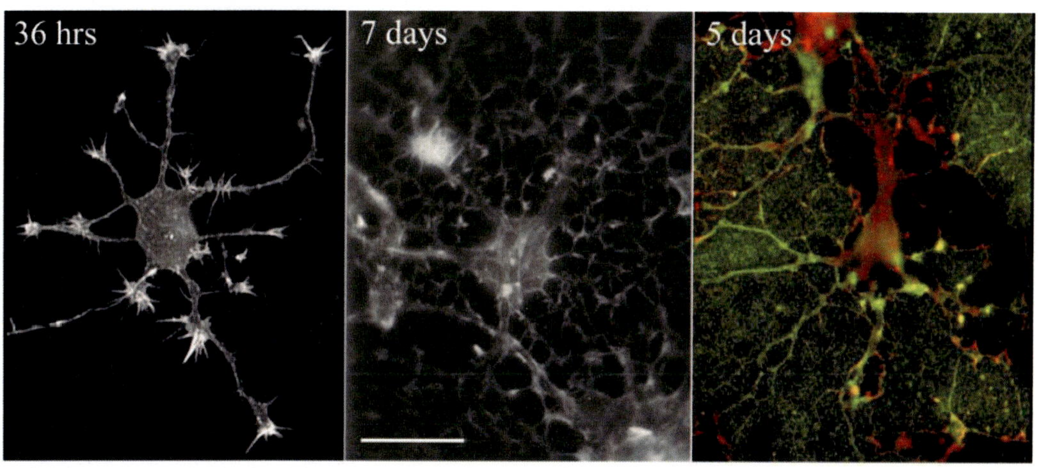

Figure 2. Actin cytoskeleton of MBP+ oligodendrocyte.
Microfilaments in the developing and mature oligodendrocytes. Oligodendrocytes were either stained with Texas red phalloidin for microfilaments (left and middle panels) or double-labeled for microfilaments and MBP (right panel). Left panel shows a young oligodendrocyte (36 h in culture). The middle panel shows a mature oligodendrocyte (7 days in vitro). In the right panel (5 days in vitro), microfilaments are in red and MBP is in green. Microfilaments were enriched in the peripheral cortical region just beneath the plasma membrane, both in cell bodies and processes. (Bar= 50 μm). [From Song et al., (2001) by copyright permission of Elsevier Limited (license number: 1833220954734)].

(Hallak et al., 1977; Barra et al., 1988; Erck et al., 2005). This cycle generates two tubulin pools: the tyrosinated α-tubulin (Tyr-tubulin) displayed by dynamic microtubules, and detyrosinated α-tubulin (Glu-tubulin) which lacks a C-terminal tyrosine derived from the preferential activity of TCP on stable microtubules. Thus, Glu-tubulin is enriched in stable microtubules exhibiting little dynamic behavior (Gundersen et al., 1984; Wehland and Weber 1987). In oligodendrocytes, for example the cell body and main processes contain more stable microtubules, being enriched in Glu-tubulin (also known as microtubules enriched with acetylated tubulin) (Figure 1). In contrast, the newly formed processes and branches are enriched in Tyr-tubulin (which implicate more dynamic microtubules). This differential distribution was appreciated as a vital property of the oligodendroglial cytoskeleton involved in its myelinating function. Thus, the detyrosinated microtubule subpopulation mainly observed in primary processes, may be essential to support the highly branched morphology of oligodendrocyte, also mediating the transport of organelles and macromolecules. In contrast, the enrichment of tyrosinated microtubules observed at the distal tips of small processes, which are rapidly depolymerized by nocodazole, may reflect the dynamic microtubule arrays of cellular extensions that *in vivo* contact the axons for myelination. Besides we must recall that the microfilaments coordinate the rearrangement of microtubules during myelin formation, similar to that described for neurite initiation (Dent and Kalil, 2001; Rodriguez et al., 2003; Dehmelt and Halpain, 2004).

Although the mechanisms that regulate the distribution of these microtubule subpopulations are as yet unclear, it is assumed that the microtubule stabilization is representative of the physiological effect of MAPs on microtubules (Dustin, 1984). The putative effects resulting from the interaction of MBP or CNP with tubulin may account for the microtubule distribution observed in different regions of the oligodendrocyte during differentiation (Wilson and Brophy, 1989; Dyer and Benjamins, 1989b; Staugaitis et al., 1990; Allinquant et al., 1991; Richter-Landsberg, 2000; Lee et al., 2005). The strong interaction between MBP and microtubules (Wilson and Brophy, 1989) was suggested to be responsible for the enhanced microtubule stability found in more mature oligodendrocytes, probably derived from the temporally regulated expression of MBP. However, it has been recently shown that the expression of several STOP isoforms in oligodendroglial cultures, aside from MAPs, might also contribute to this microtubule stability observed in oligodendrocytes (Galiano et al., 2004). As was mentioned before, STOP proteins are able to induce cold-microtubule stability in different cell types (Bosc et al., 2003). It is interesting that the structural MAP proteins (such as MAP2 or tau) showed a differentially regulated expression throughout oligodendrocyte maturation (LoPresti et al., 1995; Müller et al., 1997; Richter-Landsberg and Gorath, 1999; Song et al., 1999; Gorath et al., 2001; Lopresti, 2002; Galiano et al., 2004). A similar expression pattern was described for MBP proteins (Jordan et al., 1989). By immunofluorescence localization and confocal microscopy it was characterized that the 14 and 18.5 kDa MBP isoforms had a plasma membrane distribution, while the 17 and 21.5 kDa MBP isoforms (which share the sequence codified by exon II of MBP (exon 6 of the total MBP gene)) presented a diffuse distribution throughout the cytoplasm and also showed nuclear accumulation (Staugaitis et al., 1990; Allinquant et al., 1991). Pedraza et al. (1997) determined that these MBP isoforms are transported from cytosol to nucleus by an active process apparently regulated by the phosphorylation state of MBP. These MBP isoforms (17 and 21.5 kDa) showed high levels of expression at early stages of oligodendrocyte development (Barbarese et al., 1978; Jordan et al., 1989) and thus both may play regulatory roles in the myelination program within the nucleus (Pedraza et al., 1997). The association of these MBP isoforms with microtubules may be involved in these regulatory roles in oligodendrocyte differentiation.

It is known that the events required by the oligodendroglial cells for their maturation and subsequent specialization into myelin-producing cells are carefully coordinated. The MBP expression by oligodendrocytes is a clear example of how these cells achieve their vital function (Jordan et al., 1989; Allinquant et al., 1991; Staugaitis et al., 1996; Dyer et al., 1997; Baumann and Pham-Dinh, 2001). In addition to the putative nuclear regulating activities of the exon II (exon 6)-containing MBP isoforms (Pedraza et al., 1997), a cytoskeletal-associated mechanism that mediates the transport of organelles and macromolecules (Bizzozero et al., 1982; Sato et al., 1986; Lunn et al., 1997a; Song et al., 1999), including transport of MBP mRNAs, is activated before development of the oligodendroglial arborized morphology (Trapp et al., 1987; Ainger et al., 1993; Carson et al., 1998).

Also noteworthy is the fact that the exon II-containing MBP isoforms (the 17 and 21.5 kDa isoforms) are known to be closely associated with tubulin *in vivo* (Gillespie et al., 1989; Karthigasan et al., 1996). Although the previously mentioned microtubule-stabilizing activity could not be completely assigned to these MBP isoforms, MBP, in addition to STOP proteins, was recently shown to be capable of inducing microtubule-cold stability in differentiated oligodendrocytes (Galiano et al., 2006). The participation of both proteins as microtubule-stabilizing effectors was independently shown in oligodendrocyte cultures derived from MBP-deficient Shiverer mice, or from STOP knock out mice using specific siRNAs directed to knock down the expression of STOP and MBP proteins, respectively. After the induction of microtubule depolymerization by cold exposure, it was observed for each condition that the absence of MBP and STOP expression correlated with the depletion of cold-stable microtubules in transfected oligodendrocytes (Galiano et al., 2006). From these assays, it was concluded that the *in vitro* microtubule-stabilizing activity of MBP previously reported by Pirollet et al. (1992) also occurred in cultured oligodendrocytes (see Figure 3).

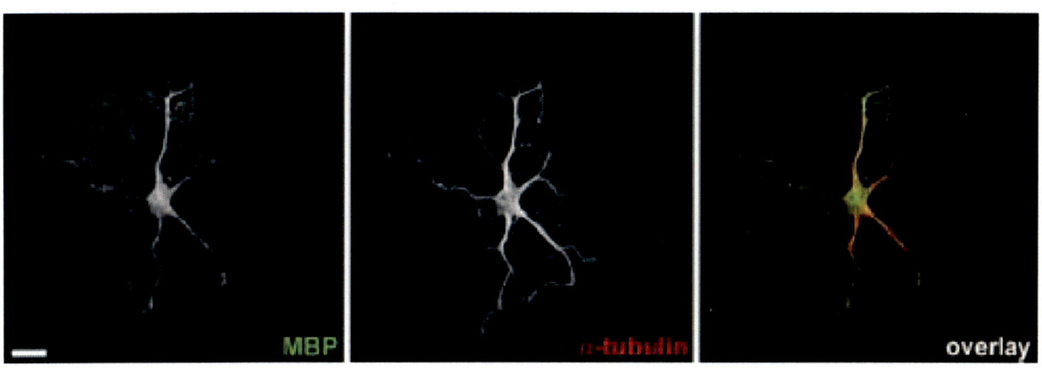

Figure 3. Colocalization of MBP and tubulin in cold-stable microtubules.
Single confocal planes were collected to analyze the localization of MBP and tubulin in oligodendrocyte cultures derived from mice, as carried out in Galiano et al. (2006). Confocal microscopy analysis was done in mature oligodendrocytes that were exposed to 4°C for 45 min and then permeabilized, fixed and double labeled with anti-MBP and anti-α-tubulin antibodies followed by Alexa488-conjugated anti-rat and Cy3-conjugated anti-mouse secondary antibodies. Confocal images of double-labeled cells were taken as z-series. Confocal images of MBP (left), α-tubulin (center) and merge (right) in a single focal plane. In the right panel, colocalized pixels are pseudocolored yellow. (Bar = 20 μm).

This function assigned to MBP was observed in mature oligodendrocytes where MBP is expressed and localized in cell extensions (Galiano et al., 2006), suggesting a physiological activity resulting from MBP-microtubule interactions throughout oligodendrocyte differentiation.

Effects of Post-Translational Modifications on Tubulin-MBP interactions

The oligodendroglial cytoskeleton requires continuous microtubule turnover and protein synthesis in order to maintain the specialized architecture of the processes and myelin sheath (Benjamins and Nedelkoska, 1994). However, the dynamics of these polymer systems, and the functions that depend on them, are also highly regulated by post-translational modifications of their constituents and of several cytoskeletal interacting proteins (Valetti et al., 1999; Nakamura et al., 2001; Cassimeris, 2002; Coquelle et al., 2002; Klein et al., 2002; Liu et al., 2003; Jiang et al., 2005; Kim et al., 2006; Bacon et al., 2007; Li et al., 2007). MBP is susceptible to many post-translational modifications, such as methylation, deamidation, N-terminal acylation, ADP ribosylation, deimination and phosphorylation (Harauz et al., 2004; Hill et al., 2005; Boggs, 2006) (see Chapters II, III, and VI). Some of these modifications change the positive net charge of the MBP molecule and can alter its association with lipids (Wood and Moscarello, 1989; Boggs et al., 1997; Bates et al., 2003) or with proteins such as calmodulin, actin or tubulin (Libich et al., 2003; Polverini et al., 2004; Hill et al., 2005; Hill and Harauz, 2005; Boggs et al., 2005) (see Chapters VIII and IX). With regard to the changes observed in MBP-tubulin interactions after MBP modification the exhaustive study performed by Hill et al. (2005), using diverse purified bovine MBP charge variants and recombinant murine MBP variants, has allowed to arrival at a better understanding of how these modifications regulate the interaction of MBP with tubulin. They characterized the capability of MBP to *in vitro* polymerize tubulin and bundle microtubules in a dose-dependent manner. By comparison with the classic 18.5 kDa MBP, the exon II (exon 6)-containing MBP isoforms do not show a significant difference in tubulin-polymerizing capacity or microtubule bundling. With respect to the MBP post-translational modifications, they found that modifications that reduce the net positive charge of MBP, such as deamidation or phosphorylation, enhanced the ability of MBP to assemble microtubules.

The reduced ability of modified MBP to assemble microtubules differs from that observed for MBP-actin interactions. In this case, the modifications that reduce the net positive charge of MBP reduce its ability to assemble actin (Boggs et al., 2005; Hill and Harauz, 2005; Boggs et al., 2006). These observations suggest that modifications such as phosphorylation can mediate the dissociation of MBP from actin filaments leading to its association with tubulin polymers. A similar phosphorylation-dependent localization has been described in neuronal cells for other MAPs, such as MAP2c, tau proteins or STOP (Ozer and Halpain, 2000; Biernat et al., 2002; Baratier et al., 2006). However, the phosphorylation of these MAPs promotes their dissociation from microtubules and localizes them in actin structures. In addition to the opposite effects on microtubule interaction after phosphorylation of MBP and structural MAPs, these studies raised the role of MAP phosphorylation as a significant protein modification that modulates the dynamics of MAP interaction with microtubules, altering the distribution of MAPs among microtubules, affecting microtubule stability and other functions, such as microtubule-associated transport (Olmsted et al., 1989; Li and Black, 1996; Mandell and Banker, 1996; Ebneth et al., 1998; Bulinski et al., 2001; Mandelkow et al., 2004). Additionally, the regulation of MAPs' association to microtubules by phosphorylation could be involved in the local dissociation of MAPs which may allow access of severing proteins to microtubules. These microtubule depolymerizing proteins could be relevant at sites where microtubules require enhanced dynamics, which favours process elongation or branching formation (Baas and Qiang, 2005).

From studies with isolated oligodendrocytes in culture, Vartanian et al. (1986) established that oligodendrocyte-substratum interaction activates protein kinase C (PKC)-mediated phosphorylation of MBP, as well as promoting the synthesis of MBP, suggesting that this modification of MBP may represent an important event in subsequent steps involved in myelin formation (Vartanian et al., 1986). Potentially, this phosphorylated state of MBP may also be required for its participation as a microtubule-stabilizing protein during early stages of oligodendrocyte differentiation. In contrast, as was mentioned before, a decrease in MBP phosphorylation could be part of the different events that mediate microtubule depolymerization upon association of anti-glycosphingolipid antibodies with the oligodendrocyte membrane (Dyer et al., 1994; Dyer et al., 1997) although we still do not fully understand how the oligodendroglial cytoskeleton could be affected once MBP is phosphorylated. The phosphorylation of MBP at different sites may result in local secondary structural transitions that modify its interaction with tubulin (Hill et al., 2005).

We have evaluated the effects on MBP distribution in primary oligodendroglial cultures after its phosphorylation (MG and MH unpublished observations) with agents that activate protein kinase C (tumor-promoting phorbol ester) 12-O-tetradecanoyl phorbol-13-acetate (TPA) (Vartanian et al., 1986; Pedraza et al., 1997). A few such observations are detailed below. Oligodendroglial cells incubated with TPA presented a higher level of colocalization of MBP with tubulin than control cells. Although we still cannot assign this difference to a direct effect on the MBP phosphorylation state, this enhanced MBP association with tubulin is in agreement with the *in vitro* analysis of the different MBP isoforms (Hill et al., 2005).

The degree of phosphorylation of the different structural MAPs might govern their redistribution among microtubules in a highly dynamic way even when the microtubule polymerization does not show a massive alteration of its dynamics (Olmsted et al., 1993; Bulinski et al., 2001). In the case of MBP, it has been demonstrated that its interactions with the cytoskeleton might be regulated, in addition to the Ca^{2+}-calmodulin binding, by post-translational modifications (Boggs, 2006). Thus, these modifications seem to have a central role in site-specific modulation of the MBP-tubulin interactions probably derived from concomitant changes of MBP secondary structure (Hill et al., 2005). A question that remains open is the likelihood that MBP functions as a factor which controls cytoskeletal architecture and stability in oligodendrocytes.

Because of the diverse circumstances related to the MBP-tubulin interaction, MBP seems to play more than one role as a consequence of these associations. MBP is capable of inducing microtubule cold stability (Pirollet et al., 1992; Galiano et al., 2006) and also of promoting *in vitro* tubulin polymerization (Hill et al., 2005), both activities being related to the progression of oligodendrocyte development. Besides these activities, it has also been mentioned that MBP mediates the reorganization of the microtubule cytoskeleton after alterations of the oligodendrocyte membrane (Dyer and Benjamins, 1988; Dyer and Benjamins, 1989a; Dyer et al., 1994). In addition, it has also been postulated from *in vitro* assays that MBP inhibits TCP activity (Modesti and Barra, 1986). These two reputed activities are related to an effect of MBP on microtubule depolymerization, a phenomenon directly associated with regulation of oligodendrocyte differentiation and myelin formation.

With regard to the putative inhibitory effect of MBP on TCP activity, we have evaluated if the association of MBP with microtubules, favored by its phosphorylation, regulates TCP activity. We found that TPA-treated cells showed a higher ratio of Tyr-/Glu-tubulin than control cells (determined from quantitative analysis of Tyr-/Glu-tubulin fluorescence intensity in oligodendrocytes) (MG and MH unpublished results). It should be mentioned that the tubulin distribution of untreated oligodendrocytes determined from the staining of tyrosinated microtubules (Tyr-tubulin) and detyrosinated microtubules (Glu-tubulin) presented similar patterns (see Figure 4), both showing a notable distribution of tubulin at the distal tips of processes (Lunn et al., 1997a; Song et al., 2001).

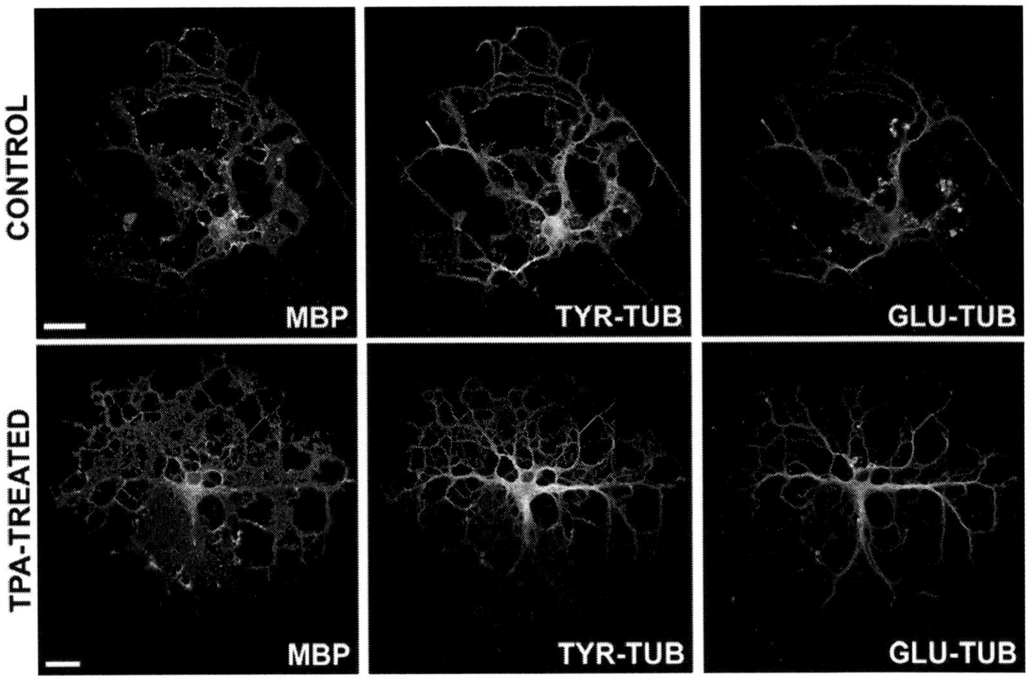

Figure 4. Distribution of tyrosinated and detyrosinated microtubules of MBP^+ oligodendrocytes. Oligodendrocyte cultures were untreated (CONTROL) or treated with TPA (300 ng/ml) for 2 h (TPA-TREATED) before fixation in 4% paraformaldehyde in PBS for 30 min, washed with PBS and then permeabilized with 0.01% Triton X-100 in PBS for 10 min. Then, cells were stained with anti-MBP antibody plus anti-Tyr- and anti-Glu-tubulin antibodies to analyze the distribution of both microtubule subpopulations for each condition. (Bar= 20 μm).

Even though the difference of staining intensity found in treated cells does not seem easily appreciable (see Figure 4), the enhancement of Tyr-tubulin after TPA treatment is in agreement with the enhanced tubulin association found for MBP (MG and MH unpublished results). Thus, despite the fact that we must confirm that the enhanced Tyr/Glu-tubulin ratio is derived from the inhibition of TCP activity by phosphorylated MBP, these results suggest that the association of phosphorylated MBP with microtubules could be mediating the inhibition of TCP activity, which has been described to be mainly active on microtubules and associated with microtubules (Arce et al., 1978; Contin et al., 1999).

The TCP activity has been characterized in different cell types (Contin et al., 1999). However, it has not been studied in oligodendrocytes, where the MBPs are naturally expressed. This enzyme is still not fully characterized, but it is known that the association of TCP with microtubules seems to be modulated by phosphorylation (Sironi et al., 1997). Moreover, the microtubule-associated activity of TCP has been detected in the distal regions of neural processes after culture for a few days (Contin and Arce, 2000), suggesting that the enzyme is associated with microtubules in limited regions of the cells, or is selectively associated with a subpopulation of microtubules which are susceptible to detyrosination. In view of these facts, the observations mentioned above may account for a localized effect of phosphorylated MBP on TCP activity, probably inhibiting TCP upon association of phosphorylated MBP with microtubules. Similar to the pattern reported in oligodendroglial cells (Lunn et al., 1997a; Song et al., 2001), myoblasts undergoing morphogenesis (Gundersen et al., 1989), or already differentiated neuroblastoma cells (Wehland and Weber, 1987), present elevated levels of Glu-tubulin. Thus, the putative regulation of TCP activity by MBP in oligodendrocyte processes may correspond to another event affecting cytoskeletal elements during differentiation. We must also consider a putative pro-differentiating effect assigned to TPA (Yong et al., 1988; Tint et al., 1992; Stariha et al., 1997; Kabir et al., 2001). Enhanced extension of processes in cells after TPA treatment has been described. This TPA-induced morphological differentiation was achieved through diverse proteins susceptible to PKC phosphorylation. Thus, the relative enhancement of tyrosinated tubulin in TPA-treated oligodendrocytes could derive from many activated factors. Further studies will elucidate the relevance of this apparent inhibitory effect of MBP on TCP activity. Probably, the characterization of cytosolic carboxypeptidases in oligodendroglial cells, similar to that recently described by Kalinina et al. (2007) for various tissues, will be valuable for testing this hypothesis.

CONCLUSION

The current knowledge of oligodendroglial differentiation certainly suggests that the development of this specialized process requires a multi-step sequence of coordinated events. The inter-relationship of cytoskeletal elements leading to the extension of cellular processes may be essential for the myelination of axons by oligodendrocytes (Richter-Landsberg, 2000, Song et al., 2001). Among a broad spectrum of cytoskeletal-interacting proteins, we emphasize the putative function of MBP as a microtubule-associated protein whose temporal interaction with tubulin throughout oligodendrocyte differentiation may be necessary for optimal myelin production and maintenance (Dyer et al., 1994; Dyer et al., 1997; Hill et al., 2005; Galiano et al., 2006). Probably, during early stages of differentiation, MBP contributes to the maintenance of stable microtubules required by the oligodendrocyte to achieve their arborized morphology. Through the progression of this cellular specialization, the different outside-in signals received by the cell could in turn alter the dynamics of microtubules, mainly at the edge of extended processes. These alterations could be partially mediated by MBP, whose association with tubulin is regulated by post-translational modifications. Hence, variations of MBP-tubulin association may account for the cytoskeletal reorganization required for the morphological changes observed during oligodendrocyte maturation. Further studies will reveal the physiological functions of MBP and its post-translational modifications related to the oligodendroglial cytoskeleton and cell morphology.

ACKNOWLEDGEMENTS

The authors are grateful to Drs. H. Barra, G. Pilar, C. Bosc and D. Job for reading the manuscript and providing helpful comments. M.E.H. is a Career Investigator and M.R.G. a CONICET fellow.

REFERENCES

Allinquant, B., Staugaitis, S.M., D'Urso, D., and Colman, D.R. (1991). The ectopic expression of myelin basic protein isoforms in Shiverer oligodendrocytes: Implications for myelinogenesis. *J. Cell Biol*, 113, 393-403.

Ainger, K., Avossa, D., Morgan, F., Hill, S.J., Barry, C., Barbarese, E., and Carson, J.H. (1993). Transport and localization of exogenous myelin basic protein mRNA microinjected into oligodendrocytes. *J. Cell Biol*, 123, 431– 441.

Arce, C.A., Hallak, M.H., Rodriguez, J.A., Barra, H.S., and Caputto, R. (1978). Capability of tubulin and microtubules to incorporate and to release tyrosine and phenylalanine and the effect of the incorporation of these amino acids on tubulin assembly. *J. Neurochem*, 31, 205-210.

Arvanitis, D.N., Yang, W., and Boggs, J.M. (2002). Myelin proteolipid protein, basic protein, the small isoform of myelin associated glyprotein, and p42MAPK are associated in the Triton X-100 extract of central nervous system myelin. *J. Neurosci. Res.*, 70, 8-23.

Baas, P.W., and Qiang, L. (2005). Neuronal microtubules: when the MAP is the roadblock. *Trends Cell Biol*, 15, 183-187.

Bacon, C., Lakics, V., Machesky, L., and Rumsby, M. (2007). N-WASP regulates extension of filopodia and processes by oligodendrocyte progenitors, oligodendrocytes, and Schwann cells-implications for axon ensheathment at myelination. *Glia*, 55, 844-858.

Bansal, R., Winkler, S., and Bheddah, S. (1999). Negative regulation of oligodendrocyte differentiation by galactosphingolipids. *J. Neurosci*, 19, 7913-7924.

Baratier, J., Peris, L., Brocard, J., Gory-Fauré, S., Dufour, F., Bosc, C., Fourest-Lieuvin, A., Blanchoin, L., Salin, P., Job, D., and Andrieux, A. (2006). Phosphorylation of Microtubule-associated Protein STOP by Calmodulin Kinase II. *J. Biol. Chem*, 281, 19561–19569.

Barbarese, E., Carson, J.H., and Braun, P.E. (1978). Accumulation of the four myelin basic proteins in mouse brain during development. *J. Neurochem.*, 31, 779-782.

Barbarese, E., Nielson, M.L., and Carson, J.H. (1983). The effect of the shiverer mutation on myelin basic protein expression in homozygous and heterozygous mouse brain. *J. Neurochem*, 40, 1680-1686.

Barra, H.S., Arce, C.A., and Argaraña, C.E. (1988). Posttranslational tyrosination/ detyrosination of tubulin. *Mol. Neurobiol*, 2, 133-153.

Bates, I.R., Boggs, J.M., Feix, J.B., and Harauz, G. (2003). Membrane-anchoring and charge effects in the interaction of myelin basic protein with lipid bilayers studied by site-directed spin labeling. *J. Biol. Chem.*, 278, 29041-29047.

Baumann, N., and Pham-Dinh, D. (2001). Biology of oligodendrocyte and myelin in the mammalian central nervous system. *Physiol. Rev*, 81, 871-927.

Benjamins, J.A., and Nedelkoska, L. (1994). Maintenance of membrane sheets by cultured oligodendrocytes requires continuous microtubule turnover and Golgi transport. *Neurochem. Res*, 19, 631-639.

Biernat, J., Wu, Y.Z., Timm, T., Zheng-Fischhofer, Q., Mandelkow, E., Meijer, L., and Mandelkow, E-M. (2002). Protein kinase MARK/PAR-1 is required for neurite outgrowth and establishment of neuronal polarity. *Mol. Biol. Cell*, 13, 4013–4028.

Bizzozero, O.A., Pasquini, J.M., and Soto, E.F. (1982). Differential effect of colchicine upon the entry of proteins into myelin and myelin related membranes. *Neurochem. Res*, 7, 1415–1425.

Boggs, J.M., Yip, P.M., Rangaraj, G., and Jo, E. (1997). Effect of posttranslational modifications to myelin basic protein on its ability to aggregate acidic lipid vesicles. *Biochemistry*, 36, 5065-5071.

Boggs, J.M., and Rangaraj, G. (2000). Interaction of lipid-bound myelin basic protein with actin filaments and calmodulin. *Biochemistry*, 39, 7799-7806.

Boggs, J.M., and Wang, H. (2001). Effect of liposomes containing cerebroside and cerebroside sulfate on cytoskeleton of cultured oligodendrocytes. *J. Neurosci. Res.*, 66, 242-253.

Boggs, J.M., and Wang, H. (2004). Co-clustering of galactosylceramide and membrane proteins in oligodendrocyte membranes on interaction with polyvalent carbohydrate and prevention by an intact cytoskeleton. *J. Neurosci. Res*, 76, 342-355.

Boggs, J.M., Rangaraj, G., Hill, C.M., Bates, I.R., Heng, Y.M., and Harauz, G. (2005). Effect of arginine loss in myelin basic protein, as occurs in its deiminated charge isoform, on mediation of actin polymerization and actin binding to a lipid membrane in vitro. *Biochemistry,* 44, 3524-3534.

Boggs, J.M., Rangaraj, G., Gao, W., and Heng, Y.M. (2006). Effect of phosphorylation of myelin basic protein by MAPK on its interactions with actin and actin binding to a lipid membrane in vitro. *Biochemistry*, 45, 391-401.

Boggs, J.M. (2006). Myelin basic protein: a multifunctional protein. *Cell Mol. Life Sci*, 63, 1945-1961.

Bosc, C., Frank, R., Denarier, E., Ronjat, M., Schweitzer, A., Wehland, J., and Job, D. (2001). Identification of novel bifunctional calmodulin-binding and microtubule-stabilizing motifs in STOP proteins. *J. Biol. Chem*, 276, 30904–30913.

Bosc, C., Andrieux, A., and Job, D. (2003). STOP proteins. *Biochemistry*, 42, 12125-12132.

Brockschnieder, D., Sabanay, H., Riethmacher, D., and Peles, E. (2006). Ermin, a myelinating oligodendrocyte-specific protein that regulates cell morphology. *J. Neurosci.*, 26, 757–762.

Bulinski, J.C., Odde, D.J., Howell, B.J., Salmon, T.D., and Waterman-Storer, C.M. (2001). Rapid dynamics of the microtubule binding of ensconsin in vivo. *J. Cell Sci.*, 114, 3885-3897.

Carson, J.H., Kwon, S., and Barbarese, E. (1998). RNA trafficking in myelinating cells. *Curr. Opin. Neurobiol*, 8, 607–612.

Cassimeris, L. (2002). The oncoprotein 18/stathmin family of microtubule destabilizers. *Curr. Opin. Cell Biol*, 14, 18–24.

Chan, K.F., Robb, N.D., and Chen, W.H. (1990). Myelin basic protein: Interaction with calmodulin and gangliosides. *J. Neurosci. Res*, 25, 535-544.

Contin, M.A., Sironi, J.J., Barra, H.S., and Arce, C.A. (1999). Association of tubulin carboxypeptidase with microtubules in living cells. *Biochem. J*, 339, 463-471.

Contin, M.A., and Arce, C.A. (2000). Tubulin carboxypeptidase/microtubules association can be detected in the distal region of neural processes. *Neurochem. Res.*, 25, 27-36.

Coquelle, F.M., Caspi, M., Cordelieres, F.P., Dompierre, J.P., Dujardin, D.L., Koifman, C., Martin, P., Hoogenraad, C.C., Akhmanova, A., Galjart, N., De Mey, J.R., and Reiner, O. (2002). LIS1, CLIP-170's key to the dynein/dynactin pathway. *Mol. Cell Biol*, 22, 3089–3102.

Cueille, N., Blanc, C.T., Popa-Nita, S., Kasas, S., Catsicas, S., Dietler, G., and Riederer, B.M. (2007). Characterization of MAP1B heavy chain interaction with actin. *Brain Res. Bull.*, 71, 610-618.

Dehmelt, L., Smart, F.M., Ozer, R.S., and Halpain, S. (2003). The role of microtubule-associated protein 2c in the reorganization of microtubules and lamellipodia during neurite initiation. *J. Neurosci.,* 23, 9479-9490.

Dehmelt, L., and Halpain, S. (2004). Actin and microtubules in neurite initiation: are MAPs the missing link? *J. Neurobiol*, 58, 18-33.

Dent, E.W., and Kalil, K. (2001). Axon branching requires interactions between dynamic microtubules and actin filaments. *J. Neurosci.*, 21, 9757-9769.

Dobrowolski, Z., Osinska, H., Mossakowska, M., and Barylko, B. (1986). Ca^{2+}-calmodulin-dependent polymerization of actin by myelin basic protein. *Eur. J. Cell Biol*, 42, 17-26.

Dubois-Dalcq, M., Behar, T., Hudson, L. and Lazzarini, R.A. (1986). Emergence of three myelin proteins in oligodendrocytes cultured without neurons. *J. Cell Biol.*, 102, 384–392.

Dustin, P. (1984). *Microtubules* (2nd edition). New York, NY, USA: Springer-Verlag.

Dyer, C.A., and Benjamins, J.A. (1988). Antibody to galactocerebroside alters organization of oligodendroglial membrane sheets in culture. *J. Neurosci*, 8, 4307-4318.

Dyer, C.A., and Benjamins, J.A. (1989a). Organization of oligodendroglial membrane sheets: II. Galactocerebroside:antibody interactions signal changes in cytoskeleton and myelin basic protein. *J. Neurosci. Res*, 24, 212-221.

Dyer, C.A., and Benjamins, J.A. (1989b). Organization of oligodendroglial membrane sheets. I: Association of myelin basic protein and 2',3'-cyclic nucleotide 3'-phosphohydrolase with cytoskeleton. *J. Neurosci. Res.*, 24, 201-211.

Dyer, C.A., and Benjamins, J.A. (1990). Glycolipids and transmembrane signaling: Antibodies to galactocerebroside cause an influx of calcium in oligodendrocytes. *J. Cell Biol.*, 111, 626-633.

Dyer, C.A. (1993). Novel oligodendrocyte transmembrane signaling systems. *Mol. Neurobiol*, 7, 1–22.

Dyer, C.A., Philibotte, T.M., Wolf, M.K., and Billings-Gagliardi, S. (1994). Myelin Basic Protein mediates extracellular signals that regulate microtubule stability in oligodendrocyte membrane sheets. *J. Neurosci. Res*, 39, 97-107.

Dyer, C.A., Philibotte, T.M., Wolf, M.K., and Billings-Gagliardi, S. (1997). Regulation of cytoskeleton by myelin components: studies on shiverer oligodendrocytes carrying an Mbp transgene. *Dev. Neurosci*, 19, 395-409.

Ebneth, A., Godemann, R., Stamer, K., Illenberger, S., Trinczek, B., and Mandelkow, E. (1998). Overexpression of tau protein inhibits kinesin-dependent trafficking of vesicles, mitochondria, and endoplasmic reticulum: implications for Alzheimer's disease. *J. Cell Biol.*, 143, 777-794.

Erck, C., Peris, L., Andrieux, A., Meissirel, C., Gruber, A.D., Vernet, M., Schweitzer, A., Saoudi, Y., Pointu, H., Bosc, C., Salin, P.A., Job, D., and Wehland, J. (2005). A vital role of tubulin-tyrosine-ligase for neuronal organization. *Proc. Natl. Acad. Sci*, USA, 102, 7853-7858.

Galiano, M.R., Bosc, C., Schweitzer, A., Andrieux, A., Job, D., and Hallak, M.E. (2004). Astrocytes and oligodendrocytes express different STOP protein isoforms. *J. Neurosci. Res*, 78, 329-337.

Galiano, M.R., Andrieux, A., Deloulme, J.C., Bosc, C., Schweitzer, A., Job, D., and Hallak, M.E. (2006). Myelin Basic Protein functions as a microtubule stabilizing protein in differentiated oligodendrocytes. *J. Neurosci. Res*, 84, 534–541.

Gillespie, C.S., Wilson, R., Davidson, A., and Brophy, P.J. (1989). Characterization of a cytoskeletal matrix associated with myelin from rat brain. *Biochem. J*, 260, 689-696.

Gorath, M., Stahnke, T., Mronga, T., Goldbaum, O., and Richter-Landsberg, C. (2001). Developmental changes of tau protein and mRNA in cultured rat brain oligodendrocytes. *Glia*, 36, 89-101.

Gundersen G.G., Kalnoski M.H., and Bulinski J.C. (1984). Distinct populations of microtubules: tyrosinated and nontyrosinated alpha tubulin are distributed differently in vivo. *Cell,* 38, 779-89.

Gundersen, G.G., Khawaja, S., and Bulinski, J.C. (1989). Generation of a stable, posttranslationally modified microtubule array is an early event in myogenic differentiation. *J. Cell Biol.*, 109, 2275-2288.

Hallak M.E., Rodriguez J.A., Barra H.S., and Caputto R. (1977). Release of tyrosine from tyrosinated tubulin. Some common factors that affect this process and the assembly of tubulin. *FEBS Lett*, 73, 147-150.

Harauz, G., Ishiyama, N., Hill, C.M., Bates, I.R., Libich, D.S., and Farès, C. (2004). Myelin basic protein: Diverse conformational states of an intrinsically unstructured protein and its roles in myelin assembly and multiple sclerosis. *Micron*, 35, 503-542.

Hill, C.M., and Harauz, G. (2005). Charge effects modulate actin assembly by classic myelin basic protein isoforms. *Biochem. Biophys. Res. Commun*, 329, 362-369.

Hill, C.M., Libich, D.S., and Harauz, G. (2005). Assembly of tubulin by classic Myelin Basic Protein isoforms and regulation by post-translational modification. *Biochemistry*, 44, 16672-16683.

Inoue, Y., Inoue, K., Terashima, T., Mikoshiba, K., and Tsukada, Y. (1983). Developmental changes of oligodendroglia in the posterior funiculus of "Shiverer" mutant mouse spinal cord, with special reference to myelin formation. *Anat. Embryol. (Berl),* 168, 159-171.

Jiang, S., Avraham, H.K., Park, S.Y., Kim, T.A., Bu, X., Seng, S., and Avraham, S. (2005). Process elongation of oligodendrocytes is promoted by the Kelch-related actin-binding protein Mayven. *J. Neurochem*, 92, 1191–1203.

Jordan, C., Friedrich, V.Jr., and Dubois-Dalcq, M. (1989). In situ hybridization analysis of myelin gene transcripts in developing mouse spinal cord. *J. Neurosci.*, 9, 248-257.

Kabir, N., Schaefer, A.W., Nakhost, A., Sossin, W.S., and Forscher, P. (2001). Protein kinase C activation promotes microtubule advance in neuronal growth cones by increasing average microtubule growth lifetimes. *J. Cell Biol.*, 152, 1033-1044.

Kalinina, E., Biswas, R., Berezniuk, I., Hermoso, A., Aviles, F.X., and Fricker, L.D. (2007). A novel subfamily of mouse cytosolic carboxypeptidases. *FASEB J.*, 21, 836-850.

Karthigasan, J., Garvey, J.S., Ramamurthy, G.V. and Kirschner, D.A. (1996). Immunolocalization of 17 and 21.5 kDa MBP isoforms in compact myelin and radial component. *J. Neurocytol*, 25, 1-7.

Kim, H.J., DiBernardo, A.B., Sloane, J.A., Rasband, M.N., Solomon, D., Kosaras, B., Kwak, S.P., and Vartanian, T.K. (2006). WAVE1 is required for oligodendrocyte morphogenesis and normal CNS myelination. *J. Neurosci.*, 26, 5849-5859.

Kimura, M., Sato, M., Akatsuka, A., Saito, S., Ando, K., Yokoyama, M., and Katsuki, M. (1998). Overexpression of a minor component of myelin basic protein isoform (17.2 kDa) can restore myelinogenesis in transgenic shiverer mice. *Brain Res.*, 785, 245-252.

Klein, C., Krämer, E-M., Cardine, A.M., Schraven, B., Brandt, R., and Trotter, J. (2002). Process outgrowth of oligodendrocytes is promoted by interaction of Fyn kinase with the cytoskeletal protein Tau. *J. Neurosci*, 22, 698–707.

Lee, J., Gravel, M., Zhang, R., Thibault, P., and Braun, P.E. (2005). Process outgrowth in oligodendrocytes is mediated by CNP, a novel microtubule assembly myelin protein. *J. Cell Biol*, 170, 661–673.

Li, Y., and Black, M.M. (1996). Microtubule assembly and turnover in growing axons. *J. Neurosci*, 16, 531–544.

Li, W., Zhang, B., Tang, J., Cao, Q., Wu, Y., Wu, C., Guo, J., Ling, E.A., and Liang, F. (2007). Sirtuin 2, a mammalian homolog of yeast silent information regulator-2 longevity regulator, is an oligodendroglial protein that decelerates cell differentiation through deacetylating α-tubulin. *J. Neurosci*, 27, 2606 –2616.

Libich, D.S., Hill, C.M., Haines, J.D., and Harauz, G. (2003). Myelin basic protein has multiple calmodulin-binding sites. *Biochem. Biophys. Res. Commun.*, 308, 313-319.

Lintner, R.N., and Dyer, C.A. (2000). Redistribution of cholesterol in oligodendrocyte membrane sheets after activation of distinct signal transduction pathways. *J. Neurosci. Res*, 60, 437-449.

Liu, A., Muggironi, M., Marin-Husstege, M., and Casaccia-Bonnefil, P. (2003). Oligodendrocyte process outgrowth in vitro is modulated by epigenetic regulation of cytoskeletal severing proteins. *Glia*, 44, 264-274.

LoPresti, P., Szuchet, S., Papasozomenos, S.C., Zinkowski, R.P., and Binder, L.I. (1995). Functional implications for the microtubule associated protein tau: Localization in oligodendrocytes. *Proc. Natl. Acad. Sci*, USA, 92, 10369–10373.

LoPresti, P. (2002). Regulation and differential expression of Tau mRNA isoforms as oligodendrocytes mature in vivo: Implications for Myelination. *Glia*, 37, 250-257.

Lunn, K.F., Baas, P.W., and Duncan, I.D. (1997a). Microtubule organization and stability in the oligodendrocyte. *J. Neurosci*, 17, 4921-4932.

Lunn, K.F., Clayton, M.K., and Duncan, I.D. (1997b). The temporal progression of the myelination defect in the taiep rat. *J. Neurocytol*, 26, 267-281.

Mandelkow, E-M., Thies, E., Trinczek, B., Biernat, J., and Mandelkow, E. (2004). MARK/PAR1 kinase is a regulator of microtubule-dependent transport in axons. *J. Cell Biol*, 167, 99–110.

Mandell, J.W., and Banker, G. (1996). A spatial gradient of tau protein phosphorylation in nascent axons. *J. Neurosci*, 16, 5727–5740.

Modesti, N.M., and Barra, H.S. (1986). The interaction of myelin basic protein with tubulin and the inhibition of tubulin carboxypeptidase activity. *Biochem. Biophys. Res. Commun*, 136, 482-489.

Müller, R., Heinrich, M., Heck, S., Blohm, D., and Richter-Landsberg, C. (1997). Expression of microtubule-associated proteins MAP2 and tau in cultured rat brain oligodendrocytes. *Cell Tissue Res*, 288, 239-249.

Murtie, J.C., Macklin, W.B., and Corfas, G. (2007). Morphometric analysis of oligodendrocytes in the adult mouse frontal cortex. *J. Neurosci. Res*. 85, 2080-2086.

Nakamura, M., Zhou, X.Z., and Lu, K.P. (2001). Critical role for the EB1 and APC interaction in the regulation of microtubule polymerization. *Curr. Biol*, 11, 1062-1067.

Noiges, R., Eichinger, R., Kutschera, W., Fischer, I., Nemeth, Z., Wiche, G., and Propst, F. (2002). Microtubule-associated protein 1A (MAP1A) and MAP1B: light chains determine distinct functional properties. *J. Neurosci.*, 22, 2106-2114.

Olmsted, J.B., Stemple, D.L., Saxton, W.M., Neighbors, B.W., and McIntosh, J.R. (1989). Cell cycle-dependent changes in the dynamics of MAP 2 and MAP 4 in cultured cells. *J. Cell Biol.*, 109, 211-223.

Ozer, R.S., and Halpain, S. (2000). Phosphorylation-dependent localization of microtubule-associated protein MAP2c to the actin cytoskeleton. *Mol. Biol. Cell*, 11, 3573-3587.

Pedraza, L., Fidler, L., Staugaitis, S., and Colman, D. (1997). The active transport of Myelin Basic Protein into the nucleus suggests a regulatory role in myelination. *Neuron*, 18, 579-589.

Pfeiffer, S.E., Barbarese, E., and Bhat, S. (1981). Noncoordinate regulation of myelinogenic parameters in primary cultures of dissociated fetal rat brain. *J. Neurosci. Res.*, 6, 369-380.

Pirollet, F., Derancourt, J., Haiech, J., Job, D., and Margolis, R.L. (1992). Ca^{2+}-calmodulin regulated effectors of microtubule stability in bovine brain. *Biochemistry*, 31, 8849–8855.

Polverini, E., Boggs, J.M., Bates, I.R., Harauz, G., Cavatorta, P. (2004). Electron paramagnetic resonance spectroscopy and molecular modelling of the interaction of myelin basic protein (MBP) with calmodulin (CaM)-diversity and conformational adaptability of MBP CaM-targets. *J. Struct. Biol.*, 148, 353-369.

Privat, A., Jacque, C., Bourre, J.M., Dupouey, P. and Baumann, N. (1979). Absence of the major dense line in myelin of the mutant mouse shiverer. *Neurosci. Lett.*, 12, 107-112.

Reynolds, R. and Wilkin, G.P. (1988). Development of macroglial cells in rat cerebellum. II. An in situ immunohistochemical study of oligodendroglial lineage from precursor to mature myelinating cell. *Development*, 102, 409-425.

Richter-Landsberg, C., and Gorath, M. (1999). Developmental regulation of alternative spliced isoforms of mRNA encoding MAP2 and Tau in rat brain oligodendrocytes during culture maturation. *J. Neurosci. Res*, 56, 259-270.

Richter-Landsberg, C. (2000). The oligodendroglia cytoskeleton in health and disease. *J. Neurosci. Res*, 59, 11-18.

Rodriguez, O.C., Schaefer, A.W., Mandato, C.A., Forscher, P., Bement, W.M., and Waterman-Storer, C.M. (2003). Conserved microtubule-actin interactions in cell movement and morphogenesis. *Nat. Cell Biol.*, 5, 599-609.

Saoudi, Y., Paintrand, I., Multigner, L., and Job, D. (1995). Stabilization and bundling of subtilisin-treated microtubules induced by microtubule associated proteins. *J. Cell Sci.*, 108, 357-367.

Sato, C., Schriftman, M., and Larocca, J.N. (1986). Transport of sulfatides towards myelin. Effect of colchicine, monensin and calcium on their intracellular traslocation. *Neurochem. Int*, 9, 265–271.

Simpson, P.B., and Armstrong, R.C. (1999). Intracellular signals and cytoskeletal elements involved in oligodendrocyte progenitor migration. *Glia*, 26, 22–35.

Sironi, J.J., Barra, H.S., and Arce, C.A. (1997). The association of tubulin carboxypeptidase activity with microtubules in brain extracts is modulated by phosphorylation/dephosphorylation processes. *Mol. Cell Biochem.*, 170, 9-16.

Song, J., O'Connor, L.T., Yu, W., Baas, P.W., and Duncan, I.D. (1999). Microtubule alterations in cultured taiep rat oligodendrocytes lead to deficits in myelin membrane formation. *J. Neurocytol*, 28, 671-683.

Song, J., Goetz, B.D., Baas, P.W., and Duncan, I.D. (2001). Cytoskeletal reorganization during the formation of oligodendrocyte processes and branches. *Mol. Cell Neurosci*, 17, 624-636.

Stariha, R.L., Kikuchi, S., Siow, Y.L., Pelech, S.L., Kim, M., and Kim, S.U. (1997). Role of extracellular signal-regulated protein kinases 1 and 2 in oligodendroglial process extension. *J. Neurochem*, 68, 945-953.

Staugaitis, S.M., Smith, P.R., and Colman, D.R. (1990). Expression of myelin basic proteins isoforms in nonglial cell lines. *J. Cell Biol*, 110, 1719–1727.

Staugaitis, S.M., Colman, D.R., and Pedraza, L. (1996). Membrane adhesion and other functions for the myelin basic proteins. *Bioessays,* 18, 13-18.

Tint, I., Bondertu, E.M., Federt, H.H., Reboulleaut, C.P., Vasiliev, J.M., and Gelfand, I.M. (1992). Reversible structural alterations of undifferentiated and differentiated human neuroblastoma cells induced by phorbol ester. *Proc. Natl. Acad. Sci*, USA, 89, 8160-8164.

Tögel, M., Wiche, G., and Propst, F. (1998). Novel Features of the Light Chain of Microtubule-associated Protein MAP1B: Microtubule Stabilization, Self Interaction, Actin Filament Binding, and Regulation by the Heavy Chain. *J. Cell Biol.*, 143, 695-707.

Trapp, B.D., Moench, T., Pulley, M., Barbosa, E., Tennekoon, G., and Griffin, J. (1987). Spatial segregation of mRNA encoding myelin-specific proteins. *Proc. Natl. Acad. Sci,* USA, 84, 7773-7777.

Valetti, C., Wetzel, D.M., Schrader, M., Hasbani, M.J., Gill, S.R., Kreis, T.E., and Schroer, T.A. (1999). Role of dynactin in endocytic traffic: effects of dynamitin overexpression and colocalization with CLIP-170. *Mol. Biol. Cell*, 10, 4107-4120.

Vartanian, T., Szuchet, S., Dawson, G., and Campagnoni, A.T. (1986). Oligodendrocyte adhesion activates protein kinase C mediated phosphorylation of myelin basic protein. *Science*, 234, 1395–1398.

Vouyiouklis, D.A., and Brophy, P.J. (1995). Microtubule-associated proteins in developing oligodendrocytes: transient expression of a MAP2c isoform in oligodendrocyte precursors. *J. Neurosci. Res.*, 42, 803-817.

Wehland, J., and Weber, K. (1987). Turnover of the carboxy-terminal tyrosine of alpha-tubulin and means of reaching elevated levels of detyrosination in living cells. *J. Cell Sci.*, 88, 185-203.

Wilson, R., and Brophy, P.J. (1989). Role for the oligodendrocyte cytoskeleton in myelination. *J. Neurosci. Res*, 22, 439-448.

Wood, P.M., and Bunge, R.P. (1984). *The biology of the oligodendrocyte.* In Norton, W.T. (Ed.), Oligodendroglia (pp. 1–46). New York, NY,USA: Plenum Press.

Wood, D.D., and Moscarello, M.A. (1989). The isolation, characterization, and lipid-aggregating properties of a citrulline containing myelin basic protein. *J. Biol. Chem*, 264, 5121-5127.

Yong, V.W., Sekiguchi, S., Kim, M.W., and Kim, S.U. (1988). Phorbol ester enhances morphological differentiation of oligodendrocytes in culture. *J. Neurosci. Res*, 19, 187-194.

In: Myelin Basic Protein
Editor: Joan M. Boggs

ISBN: 978-1-60456-699-4

Chapter VIII

MYELIN BASIC PROTEIN INTERACTIONS WITH ACTIN AND TUBULIN *IN VITRO*: BINDING, ASSEMBLY, AND REGULATION

Joan M. Boggs[*]

ABSTRACT

The sole function of MBP has long been thought to be adhesion of the cytosolic surfaces of the oligodendrocyte membranes of central nervous system (CNS) myelin. However, MBP is known to bind to a number of proteins, including actin and tubulin, and can cause their polymerization and bundling *in vitro*. It can also tether actin filaments and bundles to a membrane surface. These interactions can be regulated by physiological post-translational modifications of MBP (primarily Ser/Thr phosphorylation and Arg deimination to citrulline), by Ca^{2+}-calmodulin binding to MBP, and by physiological changes in membrane surface potential. *In vivo*, both actin and tubulin are isolated together with MBP in a low density, detergent-insoluble, glycosphingolipid-enriched fraction of myelin, which also contains caveolin and kinases, and thus may be a membrane signaling domain. Some MBP is colocalized with actin *in vivo* at the membrane edges of cultured oligodendrocytes and with cytoskeletal veins of microfilaments and microtubules. In OLs from the shiverer mutant mouse, which lacks MBP, actin microfilaments do not form bundles and are not colocalized with microtubules. Moreover, microtubular structures are abnormal in size, and cell processes are usually smaller than normal with a larger cell body. In addition, extracellular signaling via glycosphingolipids to the cytoskeleton does not occur. These observations and changes in shiverer OLs suggest that interactions of MBP with actin microfilaments

[*] Department of Molecular Structure and Function, Research Institute, Hospital for Sick Children, 555 University Ave., Toronto, Ontario, Canada M5G 1X8 and Department of Laboratory Medicine and Pathobiology, University of Toronto, Toronto, Ontario, Canada M5G 1L5. Telephone 1-416-813-5919; Fax 1-416-813-5022; E-mail: jmboggs@sickkids.ca

and microtubules occur also *in vivo* in OLs, and that these interactions are important for cell function, membrane process formation, and myelination.

INTRODUCTION

Myelin basic protein (MBP) was first shown to bind to other proteins such as actin, tubulin, calmodulin, and tropomyosin in the 1980s (Barylko and Dobrowolski, 1984; Dobrowolski et al., 1986a; Dobrowolski et al., 1986b; Modesti and Barra, 1986; Grand and Perry, 1980; Chan et al., 1990). MBP not only bound to actin and tubulin, but polymerized them into actin filaments and microtubules, and bundled the filaments and microtubules together into larger assemblies. The binding of Ca^{2+}-calmodulin to MBP dissociated it from actin and tubulin, and resulted in depolymerization of the filaments and microtubules. MBP was found to be one of two proteins in rat brain which could stabilize microtubules from depolymerizing in the cold, called STOP (Stable Tubule Only Polypeptide) activity (Pirollet et al., 1992) (see Chapter VII). These early intriguing observations were largely ignored until recently, because these proteins are negatively-charged, whereas MBP is highly positively-charged. It thus appeared at first glance that the interaction was merely electrostatic, and therefore non-specific and not of physiological consequence.

It is now known that polycationic proteins can interact with a number of polyanionic ligands, proteins and surfaces in functional ways (Jones et al., 2004). Some proteins, such as MARCKS (myristoylated alanine-rich C kinase substrate), gravin, K-Ras4B, and Src have an unstructured basic cluster which binds polyanionic ligands (Murray et al., 1997; McLaughlin et al., 2005), but also have other domains which are acidic. This clustered arrangement of amino acid types is different from MBP, whose basic and acidic residues are distributed throughout its sequence. The high net charge and low hydrophobicity of basic proteins such as MBP, microtubule-associated proteins (MAP) such as tau and MAP2, α-synuclein (Tompa, 2005), or proteins with a basic effector domain such as MARCKS (Harauz et al., 2000), maximize intramolecular electrostatic repulsion leading to an extended disordered structure (Hill et al., 2002; Harauz et al., 2004). Such proteins have sufficient flexibility to bind to various charged surfaces and ligands, and to acquire whatever local conformation is necessary to optimize binding to several different targets (Dyson and Wright, 2002; Fuxreiter et al., 2004). Moreover, they are often multifunctional regulatory proteins associated with signal transduction, adhesion, cell-cycle regulation, gene expression, alternative splicing of pre-mRNAs, or chaperone action (Benmerah et al., 2003; Tompa et al., 2005; Tompa, 2005; Pandey et al., 2004; Haynes and Iakoucheva, 2006). Disordered proteins with high surface charge which are involved in multiple interactions may integrate hubs of various biological activities (Uversky et al., 2005; Patil and Nakamura, 2006).

In this regard, MBP also interacts with several other proteins in addition to binding to negatively-charged proteins such as actin and tubulin. Its binding to calmodulin has been studied extensively, and three different domains of MBP may be involved (Libich et al., 2003a; Libich et al., 2003b; Polverini et al., 2004) (see Chapter IX). In addition, MBP has a domain that has been predicted to be a PXXP SH3-target consensus sequence (Moscarello, 1997), *viz.*, TPRTPPP (residues 92-98) (murine sequence). This domain has recently been shown to form a poly-proline helix and to bind SH3 domains of several proteins (Polverini et al., 2008; see Chapter IX). MBP has also been shown to bind to clathrin and catalyze its polymerization into artificial baskets that appear structurally similar to the baskets assembled by other assembly proteins (Prasad et al., 1995). Thus, MBP has also been suggested to be a multifunctional protein (Boggs, 2006; Dyer, 1997; Martenson, 1980; Pedraza et al., 1997). In this Chapter, we will focus on its interactions with actin and tubulin.

MBP-Mediated Interactions with Actin and Tubulin, and Effect of Post-Translational Modifications

MBP binds to G-actin in solution at an MBP/actin mole ratio of 1:2 and causes its polymerization into filaments under otherwise non-polymerizing, low ionic strength conditions (Dobrowolski et al., 1986a; Roth et al., 1993) (Figure 1A).

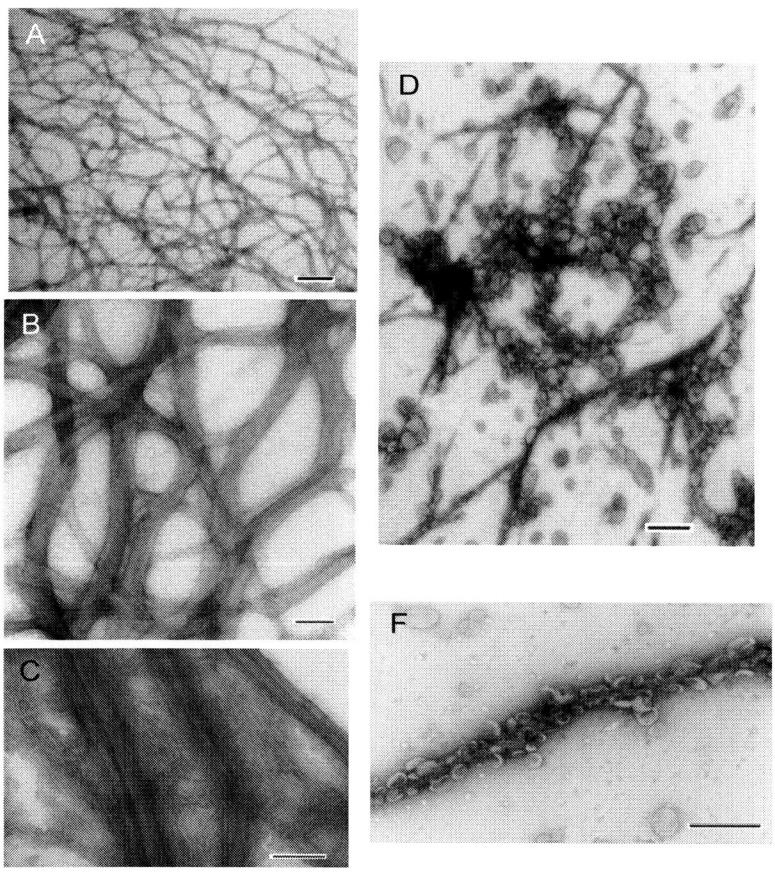

Figure 1. Electron micrographs of (A) F-actin, bar = 100 nm; (B) F-actin in the presence of equimolar MBP/C1 in F-buffer, bar =100 nm; (C) F-actin in the presence of equimolar P-MBP/C1 in F-buffer, bar = 100 nm. Strands of unbundled filaments next to bundled filaments can be seen in panel C, in contrast to panel B where virtually all actin filaments are bundled; (D) F-actin in the presence of equimolar MBP/C1 and negatively-charged phosphatidylcholine/phosphatidylglycerol 4:1 (m/m) large unilamellar lipid vesicles, bar = 500 nm; (F) as in (D) but higher magnification, bar = 500 nm. The negatively-charged lipid vesicles are bound to F-actin bundles, via MBP.

It also binds to F-actin in a 1:1 mole ratio and induces the formation of ordered bundles of F-actin filaments (Barylko and Dobrowolski, 1984) (Figure 1B). Calcium-activated calmodulin (Ca^{2+}CaM), dissociates MBP from actin bundles (Barylko and Dobrowolski, 1984) and from actin filaments resulting in their depolymerization (Dobrowolski et al., 1986a). MBP may interact with the cytoskeleton in OLs, in cytosolic inclusions in myelin, and even in compact myelin. Actin and tubulin are present in compact myelin where they may be associated with MBP and the radial component, a series of tight junctions, containing tight junction proteins claudin-11 (Gow et al., 1999) and zonula occludens-1 ZO-1 (Li et al., 2004), that pass through many layers of myelin (Karthigasan et al., 1994; Yoshikawa, 2001). Colocalization studies by immunohistochemistry in immature cultured OLs indicated that MBP was closely associated with microtubules and actin microfilaments (Wilson and Brophy, 1989; LoPresti et al., 1995; Muller et al., 1997) (see Chapter VII). MBP has been co-immunoprecipitated with tubulin from brain tissue (Taketomi et al., 2002).

MBP interacts with G-actin and F-actin through electrostatic interactions and the interaction is inhibited by an increase in salt concentration (Boggs et al., 2005; Boggs et al., 2006; Hill and Harauz, 2005). Thus, the interaction of MBP with actin might be expected to be affected by post-translational modifications which decrease the net positive charge of MBP. It is modified by a variety of such post-translational modifications *in vivo*, including deamidation, Ser/Thr phosphorylation (Chou et al., 1976; Zand et al., 1998), and deimination of Arginine (Arg) to Citrulline (Cit) (Wood and Moscarello, 1989), also termed citrullination (see Chapters II and III). It can be isolated from myelin as a series of charge components (or isomers), termed C1, C2, C3, C4, C5, C6, and C8 due to various combinations of these modifications. Component C1 is the least modified, most positively-charged form of the protein, and C8 comprises the least positively-charged isomers due to citrullination. Recombinant murine forms of C1 (rmC1) and C8 (rmC8) have been prepared; the latter contains Gln substituted for Cit at 6 sites of deiminated Arg (Bates et al., 2002). The dissociation constant of the less charged isomer rmC8 for actin was a little greater than that of rmC1, and rmC8 had somewhat less ability to polymerize actin and bundle F-actin filaments than rmC1 (Boggs et al., 2005). Moreover, rmC8 was more readily dissociated from actin by Ca^{2+}-calmodulin than rmC1. In contrast to the effect of deimination, phosphorylation of C1 *in vitro* at Thr 94 and Thr 97 (bovine sequence numbering) using MAPK (Erickson et al., 1990; Hirschberg et al., 2003) significantly decreased the ability of MBP to polymerize actin and to bundle actin filaments (Figure 1C), despite having no effect on binding to F-actin or on the ability of Ca^{2+}-CaM to dissociate the complex (Boggs et al., 2006). The rate and extent of actin polymerization induced by naturally occurring MBP charge isomers C1-C6 decreased with reduction in positive charge (Hill and Harauz, 2005).

The greater effect of deimination of MBP on its binding to actin compared to that of phosphorylation may be due to the fact that the 6 deiminated Arg residues were distributed throughout the sequence, whereas the 2 phosphorylated Thr residues were in a localized region. Earlier studies indicated that MBP has at least two actin-binding sites (Roth et al., 1993). Phosphorylation at specific sites may interfere with only one of these regions, whereas deimination, which is spread over the entire sequence, may interfere with both but in a less severe way. The comparable rates of polymerization induced by rmC8 and rmC1, despite the reduced net charge and hence electrostatic interactions of rmC8, suggest that hydrophobic and/or hydrogen-bonding interactions may also contribute to actin binding by MBP. Indeed, the polymerizing activities of the different size 14-21.5 kDa recombinant isoforms also did not depend on their net charges (Hill and Harauz, 2005). Phenylalanyl residues in the MARCKS effector domain were found to contribute to its actin-binding affinity and rate of F-actin nucleation (Wohnsland et al., 2000). A common actin-binding motif is an amphipathic helix with hydrophobic side chains on one side, which binds to a hydrophobic pocket in actin between actin subdomains 1 and 3 (Dominguez, 2004). The two sets of Phe-Phe pairs at positions 42-43 and 86-87 in MBP (murine 18.5-kDa sequence numbering) may contribute to its actin binding.

MBP was found to be one of only two proteins in rat brains to be able to protect microtubules from cold-induced depolymerization (Pirollet et al., 1992; Galliano et al., 2006). It inhibits tubulin carboxypeptidase activity (Modesti and Barra, 1986) and it polymerizes tubulin and bundles microtubules in a calmodulin-dependent way (Hill et al., 2005) (Figure 2).

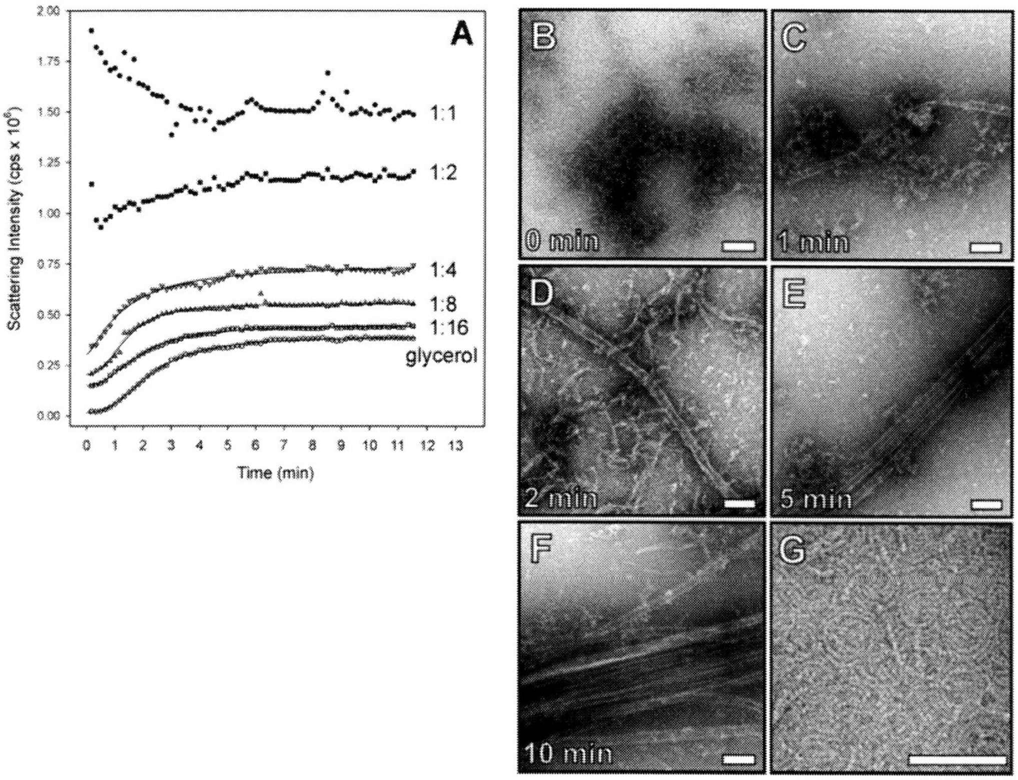

Figure 2. Assembly of tubulin by unmodified 18.5 kDa bovine MBP. (A) Time-dependence of increase in light scattering due to assembly of tubulin promoted by MBP at the molar ratios of MBP:tubulin indicated, or by 10% glycerol. (B-F) Electron micrographs of negatively-stained samples of tubulin after addition of MBP at a mole ratio to tubulin of 1:8, taken at the times indicated. Scale bars are 100 nm. A microtubule can be seen in panel D, and bundles of microtubules can be seen in panels E and F. (G) Tubulin ring structures observed 1 min after addition of MBP to tubulin at a 1:1 mole ratio. The scale bar is 50 nm. Reproduced from Hill et al. (2005) with permission from the American Chemical Society.

A reduction in positive charge due to deamidation in C2 decreased the ability of MBP to assemble tubulin, but further reductions in net positive charge in naturally occurring charge components C2-C6, enhanced the ability of MBP to assemble tubulin, in contrast to their effects on actin polymerization (Hill et al., 2005). The reduced-charge variant rmC8 had a similar ability to assemble tubulin as rmC1, also in contrast to its decreased interaction with actin. Thus, the interaction between MBP and tubulin seems less dependent on electrostatic effects than its interactions with actin. Site-specific phosphorylation of MBP may be the modification in these charge isomers which enhances the interaction with tubulin, as is the case with tau phosphorylated at certain sites (Feijoo et al., 2005). The rm21.5-, rm17.22- and rm18.5-kDa C1 had similar microtubule bundling and F-actin bundling activity, but that of rm14-kDa C1 was less (Hill and Harauz, 2005; Hill et al., 2005).

Cultured oligodendrocytes (OLs) produce large membrane sheets with large cytoskeletal veins and a lacy network of smaller veins, containing actin and tubulin (Dyer and Benjamins, 1989; Wilson and Brophy, 1989; Boggs and Wang, 2004). MBP is colocalized with these cytoskeletal veins. The cytoskeletal appearance in shiverer mouse OLs, which lack MBP, suggests that MBP interacts with actin and tubulin *in vivo*. In these mutant OLs, actin microfilaments did not form bundles (Dyer et al., 1995). The microtubular structures were also abnormal in size, and cell processes were usually smaller than normal with a larger cell body. In shiverer OLs transfected with the MBP gene, in which the average MBP expression was less than normal, a normal microtubule appearance and actin bundles were seen only in those membrane sheets where the MBP expression appeared normal. They were not seen in membrane sheets where the MBP expression was abnormally low (Dyer et al., 1997).

MBP-mediated binding of actin to membranes and effect of post-translational modifications

MBP can tether actin filaments to the surface of negatively-charged lipid vesicles (Figure 1 D,F), suggesting that it may be able to act as a membrane actin-binding protein (Boggs and Rangaraj, 2000). In the case of the MARCKS effector domain, theoretical calculations showed that the electrostatic potential was quite positive above the peptide, even when associated with acidic lipids (Murray et al., 1999). A similar effect for MBP would allow it to interact with actin while simultaneously binding to the lipid negative surface charge. MBP was observed to be partially colocalized with actin at the edge of membrane sheets of OLs where it might link actin to the membrane (Boggs et al., 2006) (Figure 3).

Many actin-binding proteins insert into the lipid bilayer, allowing them to anchor actin filaments to the membrane and possibly to also sense signals transmitted through the membrane. This insertion into the lipid bilayer has been detected by labeling with hydrophobic photolabels such as 3-(trifluoromethyl)-3-(*m*-[^{125}I]iodophenyl)diazirine (TID) (Isenberg and Goldmann, 1995), which react only with groups in the acyl chain region of the bilayer. The spin label studies, described in Chapter VI, indicate that hydrophobic side chains of MBP insert into the lipid bilayer; MBP can also be labeled by TID when bound to lipid (Jo and Boggs, 1995; Boggs et al., 1999). Like many actin-binding proteins which are myristoylated at the N-terminus (Behrisch et al., 1995; Isenberg and Goldmann, 1992), MBP is also acylated at the N-terminus, although in MBP the chain length is heterogeneous and no

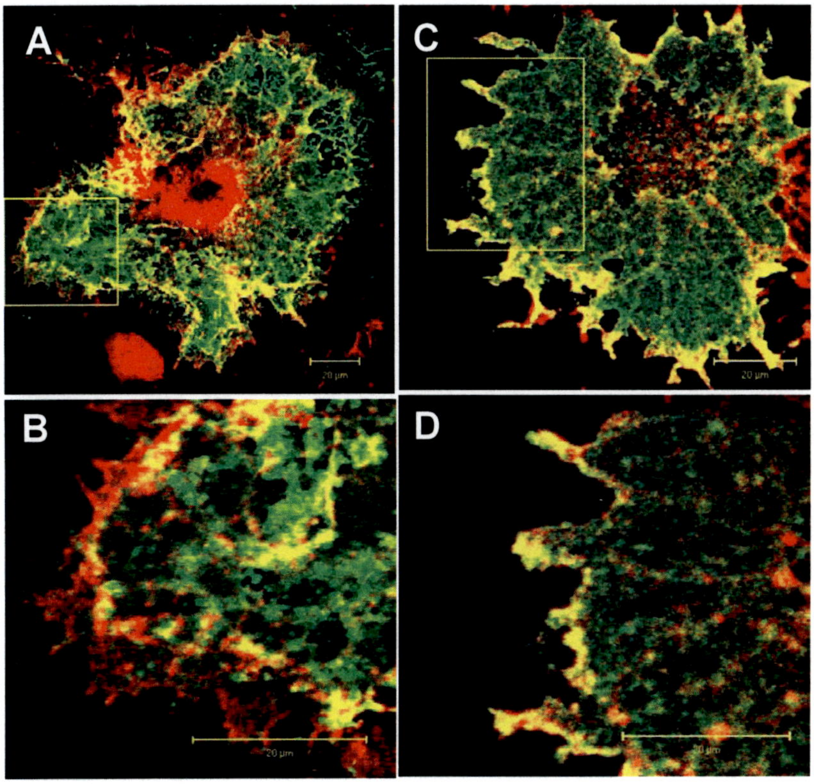

Figure 3. Confocal microscope images of cultured oligodendrocytes that have been fixed, permeabilized, and costained with rat anti-MBP Ab and Cy^{TM}2-conjugated-anti-rat IgG (green) and rhodamine-conjugated phalloidin (binds to F-actin; red). Merged images are shown. Panels B and D show enlarged images of the boxed sections marked in panels A and C, respectively. Colocalization of MBP and actin is seen at the edges of the membrane sheets as well as on cytoskeletal veins. In panel A, actin precedes MBP at the membrane edge, and in panel B, actin and MBP are both present together at the membrane edges and leading processes. The cell shown in panels C and D may be more mature than that in panels A and B. Scale bar = 20 μm in each case.

longer than 10 carbons (Moscarello et al., 1992). Other similarities, including analogous structural motifs, between MBP and the actin-binding MARCKS protein have been pointed out by Harauz et al. (2000).

The ability of MBP to bind actin to the myelin or oligodendrocyte membrane may allow it to participate in signaling (Dyer et al., 1994). Insertion of hydrophobic side chains or its N-terminal acyl group into the bilayer would allow it to sense mechanical signals transmitted through the membrane, *e.g.*, disordering effects on the lipid bilayer or bilayer curvature strain (McMahon and Gallop, 2005), and respond through a structural change, which could then impact on the cytoskeleton. Actin binding to MBP decreased the labeling of MBP by the hydrophobic photolabel TID, indicating that it decreased the hydrophobic interactions of MBP with the bilayer (Boggs and Rangaraj, 2000). This may be due to transfer of hydrophobic residues of MBP, such as the Phe-Phe pairs, from the lipid bilayer to the hydrophobic pocket in actin. This change in interaction of MBP with the bilayer could then create a cytosol-to-membrane signal caused by changes in interaction of the cytoskeleton with the membrane.

The amount of actin bound to the MBP-lipid vesicles decreased with increasing net negative surface charge of the lipid vesicles (Boggs and Rangaraj, 2000) (Figure 4).

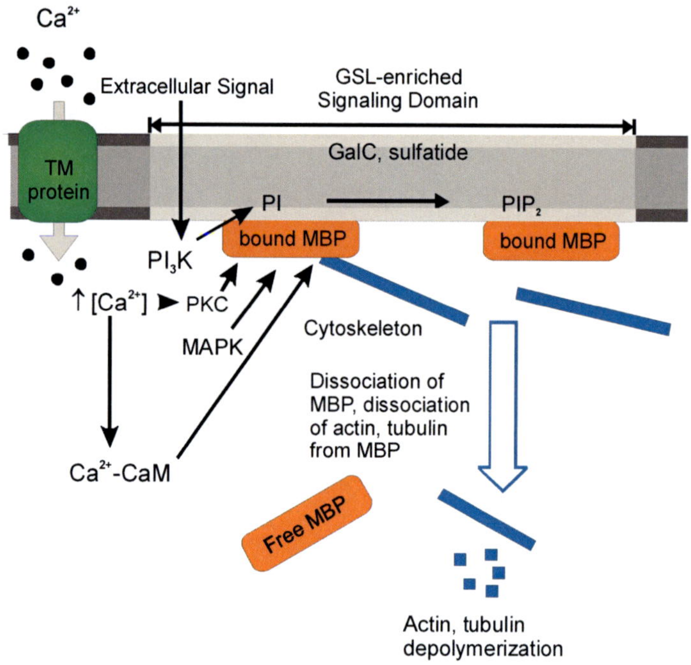

Figure 4. Schematic of mechanisms of regulation of binding of actin filaments and microtubules to MBP in oligodendrocytes and of transmission of extracellular signals to the cytoskeleton, via MBP. MBP bound to negatively charged lipids, phosphatidylserine and phosphatidylinositol (PI), on the cytosolic side of the membrane tethers actin filaments to the membrane. Extracellular signals impinging on GSL-enriched lipid rafts or membrane signaling domains in the OL membrane, such as those imparted by anti-GalC/sulfatide antibodies or by GalC/sulfatide-containing liposomes, as discussed in the text, cause Ca^{2+}-entry and/or activate kinases such as PI_3K and result in depolymerization of the cytoskeleton. The binding of Ca^{2+} to calmodulin causes dissociation of the actin from MBP, and of MBP from the membrane. It also causes dissociation of microtubules from MBP. Ca^{2+} activation of PKC causes phosphorylation of MBP which may decrease interactions of MBP with actin filaments and microtubules, as has been shown for phosphorylation of MBP with MAPK. PI_3K causes phosphorylation of PI and increased amounts of PIP_2, which are clustered by MBP, a "pipmodulin". The increased PIP_2 increases the membrane negative surface potential causing dissociation of negatively charged actin filaments from the membrane-bound MBP. The free actin filaments may then depolymerize, also causing depolymerization of microtubules, to which they may be bound, mediated by a number of cytoskeleton-binding proteins, and possibly also by MBP.

The binding of Ca^{2+}-CaM also caused dissociation of MBP and the MBP-actin complex from lipid vesicles (Boggs and Rangaraj, 2000; Boggs et al., 2005; Boggs et al., 2006) as it does for the basic domains of other proteins such as MARCKS (McLaughlin et al., 2005). Although basic residues of MBP are distributed throughout its sequence, in contrast to MARCKS where they are clustered, MBP also has several potential CaM binding sites (Polverini et al., 2004) (see Chapter IX), which may allow different regions of MBP to be dissociated from the bilayer by CaM. The deiminated protein component, C8, purified from human brain, and *in vitro*-phosphorylated component C1 (on Thr 94 and Thr 97 using MAPK) had significantly less ability to bind actin to lipid bilayers (Boggs et al., 2005; Boggs et al., 2006). The effect of phosphorylation was much greater than that of citrullination at six sites. Although average electrostatic forces are the primary determinant of the interaction of MBP with actin, phosphorylation may have an additional effect due to a site-specific electrostatic effect or to a conformational change. The 21.5 kDa isoform containing the exon II (now called exon 6, see Chapter I)-encoded sequence bound actin to the lipid bilayer less well than the 18.5 kDa form lacking this sequence (Boggs et al., 2005). The ability of other size isoforms to bind actin to the lipid bilayer has not been investigated.

Thus, deimination, a posttranslational modification of MBP, which normally occurs early in life but which is increased in adults with MS, and phosphorylation of MBP, which occurs in response to various extracellular signals in both myelin and OLs, both attenuate the ability of MBP to polymerize and bundle actin, and to bind actin to a membrane surface (Figure 4). Phosphorylation of a number of membrane actin-binding proteins such as MARCKS, regulates their interaction with the cytoskeleton and the membrane (Bubb et al., 1999).

An increase in negative surface potential of the lipid bilayer also decreased the binding of actin filaments to the lipid-bound MBP due to repulsion of the negatively-charged actin filaments. The increase in negative surface charge was produced by increasing the mole ratio of negatively-charged lipid to neutral lipid, phosphatidylcholine (PC) (Boggs and Rangaraj, 2000), or by incorporating PI, PIP, or PIP_2, at identical mole ratios of inositol lipid to PC (Boggs and Rangaraj, 2004) (Figure 4). We have shown that MBP can behave as a "PIP_2-modulin" similar to MARCKS, GAP43, and CAP23 (Laux *et al.* 2000) and cluster PIP_2 into localized domains (Harauz et al., 2008). Thus, changes in lipid composition resulting from signal transduction events could also regulate MBP-mediated binding of actin to the membrane.

MBP-cytoskeleton interactions *in vivo*

During the period of rapid myelination, in the first year of life in humans, the OL must produce large amounts of membrane (Simons and Trotter, 2007). Many events involved in myelination such as OL precursor migration in the CNS, differentiation, cell process extension, and membrane production may depend in part on the OL cytoskeleton (Kachar et al., 1986; Richter-Landsberg, 2007). Its interactions with the cytosolic side of the plasma membrane would allow it to participate in transmission of signals between the extracellular environment and the cytosol (Lunn et al., 1997). Such signals are received from extracellular matrix, growth hormones, and the axon. Continued interaction and communication between the axon and the mature compact myelin sheath occurs throughout life (Chakraborty et al., 1999; Witt and Brady, 2000; Simons and Trajkovic, 2006). Interactions between the adhered membranes of myelin in the multilayered myelin sheath may also provide signals which are transmitted across the membrane and across many layers of myelin.

Interactions of MBP with the cytoskeleton may allow it to play a role in signaling in OLs. Anti-galactosylceramide (GalC) antibody added to OLs mediates signals which cause Ca^{2+} entry, GalC and sulfatide redistribution in the membrane, phosphorylation changes of MBP, depolymerization of microtubules and other signal transduction events in cultured OLs (Dyer, 1993; Dyer and Benjamins, 1989) (Figure 4). GalC and sulfatide may be ligands for each other, which cause similar signaling effects (Boggs et al., 2004; Boggs et al., 2008a) as found for GSLs in other cells (Hakomori, 2002). The divalent cation Ca^{2+} can bridge sugars by binding sugar oxygens to its coordination sphere (Cook and Bugg, 1975), and it causes adhesion of liposomes containing GalC to liposomes containing sulfatide (Stewart and Boggs, 1993). Addition of GalC/sulfatide-containing liposomes to cultured OLs to interact with GalC and sulfatide in the OL membrane by these Ca^{2+}-mediated *trans* carbohydrate-carbohydrate interactions across apposed membranes had similar effects on lipid distribution and depolymerization of the cytoskeleton as anti-GalC antibody and also caused changes in protein distribution, including that of MBP (Boggs and Wang, 2001).

This extracellular signal impinging on GalC/sulfatide may be a signal for compaction of the MBP-containing domains (Dyer, 2002; Boggs et al., 2008a). MBP is required for these effects, since they do not occur in shiverer OLs, and are decreased in OLs in which MBP expression is inhibited with MBP siRNA, suggesting that MBP is involved in transduction of the signal (Dyer et al., 1994; Boggs et al., 2008b). Interactions of OLs with neurons also caused redistribution of OL lipids, in particular clustering of GalC (Fitzner et al., 2006). This neuron-induced GalC clustering also depended on the presence of MBP.

Stabilization of the actin cytoskeleton with jasplakinolide prevented all effects of the GalC/sulfatide-containing liposomes, including GalC and MBP redistribution and microtubule depolymerization, suggesting that the stability of microtubules in OLs depends on the integrity of the actin cytoskeleton (Boggs and Wang, 2004). The MBP redistribution caused by depolymerization of the actin cytoskeleton suggests that actin is linked directly or indirectly to MBP in OLs. In support of this idea, MBP was present in a Triton X-100 insoluble fraction from OLs, which resembled intact cytoskeleton and which contained actin and tubulin (Wilson and Brophy, 1989).

MBP might also bind actin to microtubules. In shiverer mouse OLs, actin filaments were not colocalized with microtubular structures as they were in wild type OLs (Dyer et al., 1995). MBP may also interact with actin and tubulin in the myelin sheath, in addition to OLs. Detergent extraction of myelin indicates that MBP is located in several different membrane domains or associated with different proteins. Some MBP distributes to two low density glycosphingolipid/cholesterol-enriched Triton-X-100 insoluble fractions, one of which is associated with actin, tubulin, caveolin, and kinases (Karthigasan et al., 1994; Arvanitis et al., 2005; Gillespie et al., 1989; Pereyra et al., 1988). Some MBP is also found in the Triton-X-100 supernatant associated with MAG, PLP, MAPK, and some phospholipid (Arvanitis et al., 2002). Although these isolated fractions do not necessarily come from different membrane domains in myelin *in situ* (Lichtenberg et al., 2005), the presence of MBP in several different fractions suggests that it can interact with different constituents or complexes in myelin, some of which interact with kinases. In this way, MBP may also participate in signaling in myelin. The p42/p44 MAPK in myelin is active and phosphorylates exogenous MBP (Arvanitis et al., 2005) and endogenous MBP, CNP, and tubulin (Y. Gong, W. Min and J.M. Boggs, unpublished). Different charge isomers and size isoforms of bovine MBP also are differentially associated with several detergent-insoluble fractions of myelin. A CHAPS-resistant low density fraction contained most of the phospho-Thr97-MBP whereas most of the citrullinated and methylated MBP were in a heavier CHAPS-insoluble fraction (DeBruin et al., 2005). The 18.5 kDa MBP was in the low density fraction and 21.5 kDa MBP was in the heavier CHAPS-insoluble fraction. This observation suggests that these different species of MBP associate preferentially with different myelin constituents and may perform different functions in these complexes.

In addition to kinases and phosphatases which can mediate rapid turnover of phosphate groups, myelin also contains mechanisms which control kinase activity, *e.g.*, phospholipase C to generate DAG, adenyl cyclase to generate cAMP, and MEK to activate p42/p44 MAPK (reviewed in Ledeen, 1992; Arvanitis et al., 2002). Myelin also contains relatively high amounts of the signal transduction lipids phosphatidylinositol (PI), phosphatidic acid, phosphatidylinositol 4-phosphate (PIP), and phosphatidylinositol 4,5-bisphosphate (PIP_2) (Deshmukh et al., 1980). The lightest myelin fraction, which comes from compact myelin, was even more enriched in these signal transduction lipids. PIP_2 is 1.5% of the total phospholipid in myelin and 1.7% in the lightest fraction, compared to 1% reported for the plasma membrane of other cells (McLaughlin and Murray, 2005). MBP stimulates a PI-specific phospholipase C purified from human myelin, but deimination of MBP prevents this effect (Tompkins and Moscarello, 1993). Thus, the more highly charged isomers of MBP may be directly involved in a signaling system involving phosphoinositides in myelin. There is considerable turnover of phosphoinositides in myelin *in vivo* (Kahn and Morell, 1988), which may be regulated by communication with the axon (Witt and Brady, 2000). In addition to being a source of second messengers, PIP_2 is a potent regulator of many actin-binding proteins (Janmey et al., 1999). Thus, mechanisms are present in both OLs and myelin to regulate the interaction of MBP with the membrane and with the cytoskeleton by phosphorylation of MBP and by changes in lipid composition.

Trans interactions between GSLs in membrane signaling domains in apposed cells trigger signaling and have been proposed to result in formation of a glycosynapse between the cells (Hakomori, 2002). Glycosynapses could also form between the apposed extracellular surfaces in compact myelin through contact between GalC and sulfatide in signaling domains (Boggs et al., 2004; Boggs et al., 2008a). Ions and water released into the periaxonal space following the axonal action potential may be able to traverse throughout the extracellular space in compact myelin by passing through the tight junction pores of the radial component (Dyer, 2002) and could regulate this interaction. The signals may in turn be transmitted across the membrane to MBP and the cytoskeletal elements in myelin, as found in cultured OLs for anti-GalC antibody and GalC/sulfatide-containing liposomes, possibly by causing Ca^{2+} entry into the cytosolic domains. The accumulation of Ca^{2+} into the cytosolic domains of compact myelin, mediated by NMDA receptors, has recently been detected within compact myelin (Micu et al., 2005). Subsequent effects on the cytoskeleton may regulate opening and closing of the tight junction pores between the adjacent myelin layers (Dyer, 2002; Boggs et al., 2008a) as occurs for intestinal epithelial cells (van Itallie and Anderson, 2004). This process may allow transmission of signals from the axon throughout compact myelin.

Conclusion

In order to produce myelin, the OL must extend membranous processes and produce large membrane sheets which then must interact with and wrap around the axon. These events require the cytoskeleton for membrane extension, transport of membrane constituents, organization of membrane domains, and localization of signaling molecules. Movement of the leading edge of membrane processes is driven by actin polymerization, with the microtubules located proximal to the actin network. The actin filaments provide tracks for the microtubules to invade regions of new growth (Song et al., 2001). Many cytosolic proteins which bind cytoskeletal proteins to each other, and membrane peripheral proteins which connect the membrane to the cytoskeleton have been identified in various cells (Sonnenberg and Liem, 2007; Niggli, 1995); some are ubiquitous and others are specific to certain cells (Greengard et al., 1994; Worton, 1995). Since MBP has the ability to assemble actin and tubulin and to bind actin filaments to a membrane surface *in vitro,* since it is such an abundant protein in OLs and myelin (30% of the myelin protein and 10% of myelin by weight), and since it is the only myelin-specific structural protein shown so far to be essential for myelination, it is reasonable to suggest that it is a multifunctional protein, and that one of its functions *in vivo*, is to bind cytoskeletal proteins to the cytosolic membrane surface, and perhaps also to each other (Boggs, 2006). The OL receives extracellular signals from growth factors, extracellular matrix, and the axon, which regulate myelination and influence the mature myelin sheath throughout life. MBP is necessary for transmission of some of these signals to the cytoskeleton. Its interactions with the cytoskeleton and regulation of these interactions by Ca^{2+}-calmodulin, by post-translational modifications to MBP, and by membrane surface potential could be mechanisms to allow MBP to be involved in signaling in OLs and myelin (Boggs, 2006; Dyer et al., 1994).

Acknowledgements

The work from my laboratory on the interactions of MBP with actin was supported by the Canadian Institutes of Health Research and the Multiple Sclerosis Society of Canada. I thank Dr. George Harauz for a critical reading of the chapter, helpful discussions, and assistance with Fig. 4.

References

Arvanitis, D.N., Yang, Y., and Boggs, J.M. (2002). Myelin proteolipid protein, basic protein, the small isoform of myelin-associated glycoprotein, and p42 MAPK are associated in the triton X-100 extract of central nervous system myelin. *J. Neurosci. Res*, 70, 8-23.

Arvanitis, D.N., Min, W., Gong, Y., Meng, Y.M., and Boggs, J.M. (2005). Two types of detergent-insoluble, glycosphingolipid/cholesterol-rich membrane domains in isolated myelin. *J. Neurochem*, 94, 1696-1710.

Barylko B., and Dobrowolski Z. (1984). Ca^{2+}-calmodulin-dependent regulation of F-actin-myelin basic protein interaction. *Eur. J. Cell Biol*, 35, 327-335.

Bates, I.R., Libich, D.S., Wood, D.D., Moscarello, M.A., and Harauz, G. (2002). An Arg/Lys→Gln mutant of recombinant murine myelin basic protein as a mimic of the deiminated form implicated in multiple sclerosis. *Protein Expr. Purif*, 25, 330-341.

Behrisch, A., Dietrich, C., Noegel, A.A., Schleicher, M., and Sackmann, E. (1995). The actin-binding protein hisactophilin binds *in vitro* to partially charged membranes and mediates actin coupling to membranes. *Biochemistry*, 34, 15182-15190.

Benmerah, A., Scott, M., Poupon, V., and Marullo, S. (2003). Nuclear functions for plasma membrane-associated proteins? *Traffic*, 4, 503-511.

Boggs, J.M., and Wang, H. (2001). Effect of liposomes containing cerebroside and cerebroside sulfate on cytoskeleton of cultured oligodendrocytes. *J. Neurosci. Res*, 66, 242-253.

Boggs, J.M., and Wang, H. (2004). Co-clustering of galactosylceramide and membrane proteins in oligodendrocyte membranes on interaction with polyvalent carbohydrate and prevention by an intact cytoskeleton. *J. Neurosci. Res*, 76, 342-355.

Boggs, J.M., Rangaraj, G., and Koshy K.M. (1999). Analysis of the membrane-interacting domains of myelin basic protein by hydrophobic photolabeling. *Biochim. Biophys. Acta*, 1417, 254-266.

Boggs, J.M., and Rangaraj, G. (2000). Interaction of lipid-bound myelin basic protein with actin filaments and calmodulin. *Biochemistry*, 39, 7799-7806.

Boggs, J.M., Wang, H., Gao, W., Arvanitis, D.N., Gong, Y., and Min, W. (2004). A glycosynapse in myelin? *Glycoconj. J*, 21, 97-110.

Boggs, J.M., and Rangaraj, G. (2004). Effect of phosphorylation of myelin basic protein on its interaction with actin *in vitro*. *J. Neurochem*, 90 (Suppl. 1):73.

Boggs, J.M., Rangaraj, G., Hill, C.M.D., Bates, I.R., Heng, Y. M., and Harauz, G. (2005). Effect of arginine loss in myelin basic protein, as occurs in its deiminated charge isoform, on mediation of actin polymerization and actin binding to a lipid membrane *in vitro*. *Biochemistry*, 44, 3524-3534.

Boggs, J.M., Rangaraj, G., Gao, W., and Heng, Y.M. (2006). Effect of phosphorylation of myelin basic protein by MAPK on its interactions with actin and actin binding to a lipid membrane *in vitro*. *Biochemistry*, 45, 391-401.

Boggs, J.M., Gao, W., and Hirahara, Y. (2008a). Myelin glycosphingolipids, galactosylceramide and sulfatide, participate in carbohydrate-carbohydrate interactions between apposed membranes and may form glycosynapses between oligodendrocyte and/or myelin membranes. *Biochim. Biophys. Acta*, 1780, 445-455.

Boggs, J.M., Gao, W., and Hirahara, Y. (2008b). Signal transduction pathways involved in interaction of galactosylceramide/sulfatide-containing liposomes with cultured oligodendrocytes and requirement for myelin basic protein and glycosphingolipids. *J. Neurosci. Res*, 86, 1448-1458.

Bubb, M.R., Lenox, R.H., and Edison, A.S. (1999). Phosphorylation-dependent conformational changes induce a switch in the actin-binding function of MARCKS. *J. Biol. Chem*, 274, 36472-36478.

Chakraborty, G., Drivas, A. and Ledeen, R. (1999). The phosphoinositide signaling cycle in myelin requires cooperative interaction with the axon. *Neurochem. Res*, 24, 249-254.

Chan, K.F.J., Robb, N.D., and Chen, W.H. (1990). Myelin basic protein: interaction with calmodulin and gangliosides. *J. Neurosci. Res*, 25, 535-544.

Chou, F.C.H., Chou, C.H.J., Shapira, R., and Kibler, R.F. (1976). Basis of microheterogeneity of myelin basic protein. *J. Biol. Chem*, 251, 2671-2679.

Cook, W.J., and Bugg, C.E. (1975). Calcium-carbohydrate bridges composed of uncharged sugars. Structure of a hydrated calcium bromide complex of α-fucose. *Biochim. Biophys. Acta*, 389, 428-435.

DeBruin, L.S., Haines, J.D., Wellhauser, L.A., Radeva, G., Schonmann, V., Bienzle, D., and Harauz, G. (2005). Developmental partitioning of myelin basic protein into membrane microdomains. *J. Neurosci. Res*, 80, 211-225.

Deshmukh, D.S., Kuizon, S., Bear, W.D., and Brockerhoff, H. (1980). Distribution of phosphoinositides among subfractions of rat brain myelin. *Lipids,* 15, 14-18.

Dobrowolski, Z., Osinska, H., Mossakowska, M., and Barylko, B. (1986a). Ca^{2+}-calmodulin-dependent polymerization of actin by myelin basic protein. *Eur. J. Cell Biol*, 42, 17-26.

Dobrowolski, Z., Barylko, B., and Drabikowski, W. (1986b) Interaction of tropomyosin with myelin basic protein and its effect on the ATPase activity of actomyosin. *Eur. J. Cell Biol*, 41, 65–71.

Dominguez, R. (2004). Actin-binding proteins – a unifying hypothesis. *Trends Biochem. Sci*, 29, 572-5780.

Dyer, C.A. (1993). Novel oligodendrocyte transmembrane signaling systems. *Mol. Neurobiol*, 7, 1-22.

Dyer, C.A. (1997). *Myelin proteins as mediators of signal transduction.* In B.H.J. Juurlink, R.M. Devon, J.R. Doucette, A. J. Nazarali, D.J. Schreyer, V.M.K. Verge (Eds.), Cell Biology and Pathology of Myelin: Evolving Biological Concepts and Therapeutic Approaches (pp. 69-74). New York, NY, USA: Plenum Press.

Dyer, C.A. (2002). The structure and function of myelin: From inert membrane to perfusion pump. *Neurochem. Res*, 27, 1279-1292.

Dyer, C.A., and Benjamins, J.A. (1989). Organization of oligodendroglial membrane sheets. II. Galactocerebroside: antibody interactions signal changes in cytoskeleton and myelin basic protein. *J. Neurosci. Res*, 24, 212–221.

Dyer, C.A., Philibotte, T.M., Wolf, M.K., and Billings-Gagliardi, S. (1994). Myelin basic protein mediates extracellular signals that regulate microtubule stability in oligodendrocyte membrane sheets. *J. Neurosci. Res*, 39, 97-107.

Dyer, C.A., Philibotte, T.M., Billings-Gagliardi, S., and Wolf, M.K. (1995). Cytoskeleton in myelin-basic-protein-deficient shiverer oligodendrocytes. *Dev. Neurosci*, 17, 53-62.

Dyer, C.A., Phillbotte, T., Wolf, M.K., and Billings-Gagliardi, S. (1997). Regulation of cytoskeleton by myelin components: studies on shiverer oligodendrocytes carrying an MBP transgene. *Dev. Neurosci*, 19, 395-409.

Dyson, H.J., and Wright, P.E. (2002). Coupling of folding and binding for unstructured proteins. *Curr. Opin. Struct. Biol*, 12, 54-60.

Erickson, A.K., Payne, D.M., Martino, P.A., Rossomando, A.J., Shabanowitz, J., Weber, M.J., Hunt, D.F. and Sturgill, T.W. (1990). Identification by mass spectrometry of threonine 97 in bovine myelin basic protein as a specific phosphorylation site for mitogen-activated protein kinase. *J. Biol. Chem*, 265, 19728-19735.

Feijoo, C., Campbell, D.G., Jakes, R., Goedert, M., and Cuenda, A. (2005). Evidence that phosphorylation of the microtubule-associated protein Tau by SAPK4/p38delta at Thr50 promotes microtubule assembly. *J. Cell Sci*, 118, 397-408.

Fitzner, D., Schneider, A., Kippert, A., Mobius, W., Willig, K.I., Hell, S.W., Bunt, G., Gaus, K., and Simons, M. (2006). Myelin basic protein-dependent plasma membrane reorganization in the formation of myelin. *EMBO J*, 25, 5037-5048.

Fuxreiter, M., Simon, I., Friedrich, P., and Tompa, P. (2004). Preformed structural elements feature in partner recognition by intrinsically unstructured proteins. *J. Mol. Biol*, 338, 1015-1026.

Galiano, M.R., Andrieux, A., Deloulme, J.C., Bosc, C., Schweitzer, A., Job, D. and Hallak, M.E. (2006). Myelin basic protein functions as a microtubule stabilizing protein in differentiated oligodendrocytes. *J. Neurosci. Res*, 84, 534–541.

Gillespie, C.S., Wilson, R., Davidson, A., and Brophy, P.J. (1989). Characterization of a cytoskeletal matrix associated with myelin from rat brain. *Biochem. J*, 260, 689-696.

Gow, A., Southwood, C.M., Li, J.S., Pariali, M., Riordan, G.P., Brodie, S.E., Danias, J., Bronstein, J.M., Kachar, B., and Lazzarini, R.A. (1999). CNS myelin and sertoli cell tight junction strands are absent in Osp/claudin-11 null mice. *Cell,* 99, 649-659.

Grand, R.J.A. and Perry, S.V. (1980). The binding of calmodulin to myelin basic protein and histone H2B. *Biochem. J*, 189, 227-240.

Greengard, P., Benfenati, F., and Valtorta, F. (1994). Synapsin I, an actin-binding protein regulating synaptic vesicle traffic in the nerve terminal. *Adv. Second Messenger Phosphoprotein Res*, 29, 31-45.

Hakomori, S. (2002). The glycosynapse. *Proc. Natl. Acad. Sci*, USA, 99, 225-232.

Harauz, G., Ishiyama, N., and Bates, I.R. (2000). Analogous structural motifs in myelin basic protein and in MARCKS. *Mol. Cell. Biochem*, 209, 155-163.

Harauz, G., Ishiyama, N., Hill, C.M.D., Bates, I.R., Libich, D.S., and Farès, C. (2004). Myelin basic protein-diverse conformational states of an intrinsically unstructured protein and its roles in myelin assembly and multiple sclerosis. *Micron*, 35, 503-542.

Harauz, G., Musse, A., Boggs, J. M., and Gao, W. (2008). Myelin basic protein (MBP) as a "PIP_2-modulin": New biological function for an old CNS protein. *J. Neurochem*, 104 (Supplement 1), 44.

Haynes, C., and Iakoucheva, L.M. (2006). Serine/arginine-rich splicing factors belong to a class of intrinsically disordered proteins, *Nucl. Acids Res*, 34, 305-312.

Hill, C.M., Bates, I.R., White, G.F., Hallett, F.R., and Harauz, G. (2002). Effects of the osmolyte trimethylamine-N-oxide on conformation, self-association, and two-dimensional crystallization of myelin basic protein. *J. Struct. Biol*, 139, 3-26.

Hill, C.M.D., and Harauz, G. (2005). Charge effects modulate actin assembly by classic myelin basic protein isoforms. *Biochem. Biophys. Res. Commun*, 329, 362-369.

Hill, C.M.D., Libich, D.S., and Harauz, G. (2005). Assembly of tubulin by classic myelin basic protein isoforms and regulation by post-translational modification. *Biochemistry*, 44, 16672-16683.

Hirschberg, D., Radmark, O., Jornvall, H., and Bergman, T. (2003). Thr94 in bovine myelin basic protein is a second phosphorylation site for 42-kDa mitogen-activated protein kinase (ERK2). *J. Protein Chem*, 22, 177-181.

Isenberg, G., and Goldmann, W.H. (1992). Actin-membrane coupling: a role for talin. *J. Musc. Res. Cell Motility*, 13, 587-589.

Isenberg, G., and Goldmann, W.H. (1995). Actin-binding proteins - lipid interactions, in The Cytoskeleton, Structure and Assembly, Vol. 1, pp. 169-204. Greenwich, CT: JAI Press, Inc.

Janmey, P.A., Xian, W., and Flanagan, L.A. (1999). Controlling cytoskeleton structure by phosphoinositide-protein interactions: phosphoinositide binding protein domains and effects of lipid packing. *Chem. Phys. Lipids*, 101, 93-107.

Jo, E., and Boggs, J.M. (1995). Aggregation of acidic lipid vesicles by myelin basic protein: dependence on potassium concentration. *Biochemistry*, 34, 13705-13716.

Jones, L.S., Yazzie, B., and Middaugh, C.R. (2004). Polyanions and the proteome. *Mol. Cell. Proteomics*, 3, 746-769.

Kachar, B., Behar, T., and Dubois-Dalcq, M. (1986). Cell shape and motility of oligodendrocytes cultured without neurons. *Cell Tissue Res*, 244, 27-38.

Kahn, D.W., and Morell P. (1988). Phosphatidic acid and phosphoinositide turnover in myelin and its stimulation by acetylcholine. *J. Neurochem*, 50, 1542–1550.

Karthigasan, J., Kosaras, B., Nguyen, J., and Kirschner, D.A. (1994). Protein and lipid composition of radial component-enriched CNS myelin. *J. Neurochem*, 62, 1203-1213.

Laux, T., Fukami, K., Thelen, M., Golub, T., Frey, D., Caroni, P. (2000). GAP43, MARCKS, and CAP23 modulate PI(4,5)P(2) at plasmalemmal rafts, and regulate cell cortex actin dynamics through a common mechanism. *J. Cell Biol*, 149, 1455-1471.

Ledeen, R.W. (1992). *Enzymes and receptors of myelin.* In R. E. Martenson (Ed.), Myelin: Biology and Chemistry (pp. 531-570). Boca Raton, FL: CRC Press.

Li, X., Ionescu, A.V., Lynn, B.D., Lu, S., Kamasawa, N., Morita, M., Davidson, K.G., Yasumura, T., Rash, J.E., and Nagy, J.I. (2004). Connexin 47, connexin29 and connexin 32 co-expression in oligodendrocytes and Cx47 association with zonula occludens-1 (ZO-1) in mouse brain. *Neurosci*, 126, 611-630.

Libich, D.S., Hill, C.M.D., Bates, I.R., Hallett, F.R., Armstrong, S., Siemiarczuk, A., and Harauz G. (2003a). Interaction of the 18.5 kDa isoform of myelin basic protein with Ca^{2+}-calmodulin: effects of deimination assessed by intrinsic Trp fluorescence spectroscopy, dynamic light scattering, and circular dichroism. *Protein Sci*, 12, 1507-1521.

Libich, D.S., Hill, C.M., Haines, J.D., and Harauz, G. (2003b). Myelin basic protein has multiple calmodulin-binding sites. *Biochem. Biophys. Res. Commun*, 308, 313-319.

Lichtenberg, D., Goni, F.M., and Heerklotz, H. (2005). Detergent-resistant membranes should not be identified with membrane rafts. *Trends Biochem. Sci*, 30, 430-436.

LoPresti, P., Szuchet, S., Papasozomenos, S.C., Zinkowski, R.P., and Binder, L.I. (1995). Functional implications for the microtubule-associated protein tau: Localization in oligodendrocytes. *Proc. Natl. Acad. Sci*, USA, 92, 10369-10373.

Lunn, K.F., Baas, P.W., and Duncan, I.D. (1997). Microtubule organization and stability in the oligodendrocyte. *J. Neurosci*, 17, 4921-4932.

McMahon, H.T. and Gallop, J.L. (2005). Membrane curvature and mechanisms of dynamic cell membrane remodeling, *Nature*, 438, 590-596.

McLaughlin, S. and Murray, D. (2005). Plasma membrane phosphoinositide organization by protein electrostatics. *Nature*, 438, 605-611.

McLaughlin, S., Hangyas-Mihalyne, G., Zaitseva, I., and Golebiewska, U. (2005). Reversible – through calmodulin – electrostatic interactions between basic residues on proteins and acidic lipids in the plasma membrane. *Biochem. Soc. Symp*, 72, 189-198.

Martenson, R.E. (1980). *Myelin basic protein: What does it do?* In S. Kumar (Ed.), Biochemistry of Brain (pp. 49-79). Oxford: Pergamon Press.

Micu, I., Jiang, Q., Coderre, E., Ridsdale, A., Zhang, L., Woulfe, J., Yin, X., Trapp, B.D., McRory, J.E., Rehak, R., Zamponi, G.W., Wang, W., and Stys, P.K. (2005). NMDA receptors mediate calcium accumulation in myelin during chemical ischaemia. *Nature*, 439, 988-992.

Modesti, N.M. and Barra, H.S. (1986). The interaction of myelin basic protein with tubulin and the inhibition of tubulin carboxypeptidase activity. *Biochem. Biophys. Res. Commun*, 136, 482-489.

Moscarello, M.A. (1997). *Myelin basic protein, the 'executive' molecule of the myelin membrane*. In B.H.J. Juurlink, R.M. Devon, J.R. Doucette, A. J. Nazarali, D.J. Schreyer, V.M.K. Verge (Eds.), Cell Biology and Pathology of Myelin: Evolving Biological Concepts and Therapeutic Approaches (pp. 13-25). New York, NY, USA: Plenum Press.

Moscarello, M.A., Pang, H., Pace-Asciak, C.R., and Wood, D.D. (1992). The N terminus of human myelin basic protein consists of C2, C4, C6, and C8 alkyl carboxylic acids. *J. Biol. Chem*, 267, 9779-9782.

Muller, R., Heinrich, M., Heck, S., Blohm, D., and Richter-Landsberg, C. (1997). Expression of microtubule-associated proteins MAP2 and tau in cultured rat brain oligodendrocytes. *Cell Tissue Res*, 288, 239-249.

Murray, D., Ben-Tal, N., Honig, B., and McLaughlin, S. (1997). Electrostatic interaction of myristoylated proteins with membranes: simple physics, complicated biology. *Structure*, 5, 985-989.

Murray, D., Arbuzova, A., Hangyas-Mihalyne, G., Gambhir, A., Ben-Tal, N., Honig, B., and McLaughlin, S. (1999). Electrostatic properties of membranes containing acidic lipids and adsorbed basic peptides: theory and experiment. *Biophys. J*, 77, 3176-3188.

Niggli, V. (1995). Membrane Cytoskeleton, in The Cytoskeleton, Structure and Assembly, Vol.1, pp. 123-168, Greenwich, CT: JAI Press, Inc.

Pandey, N., Ganapath, M., Kumar, K., Dasgupta, D., Das Sutar, S.K., and Dash, D. (2004). Comparative analysis of protein unfoldedness in human housekeeping and non-housekeeping proteins. *Bioinformatics,* 20, 2904-2910.

Patil, A. and Nakamura, H. (2006). Disordered domains and high surface charge confer hubs with the ability to interact with multiple proteins in interactions networks. *FEBS Lett*, 580, 2041-2045.

Pedraza, L., Fidler, L., Staugaitis, S.M., and Colman, D.R. (1997). The active transport of myelin basic protein into the nucleus suggests a regulatory role in myelination. *Neuron*, 18, 579-589.

Pereyra, P.M., Horvath, E., and Braun, P.E. (1988). Triton X-100 extractions of central nervous system myelin indicate a possible role for the minor myelin proteins in the stability in lamellae. *Neurochem. Res*, 13, 583-595.

Pirollet, F., Derancourt, J., Haiech, J., Job, D., and Margolis, R.L. (1992). Ca^{2+}-calmodulin regulated effectors of microtubule stability in bovine brain. *Biochemistry*, 31, 8849-8855.

Polverini, E., Boggs, J.M., Bates, I.R., Harauz, G., and Cavatorta, P. (2004). Electron paramagnetic resonance spectroscopy and molecular modelling of the interaction of myelin basic protein (MBP) with calmodulin (CaM)-diversity and conformational adaptability of MBP CaM-targets. *J. Struct. Biol*, 148, 353-369.

Polverini, E., Rangaraj, G., Libich, D.S., Boggs, J.M., and Harauz, G. (2008). Binding of the proline-rich segment of myelin basic protein to SH3-domains - spectroscopic, microarray, and modelling studies of ligand conformation and effects of post-translational modifications. *Biochemistry*, 47, 267-282.

Prasad, K., Barouch, W., Martin, B.M., Greene, L.E., and Eisenberg, E. (1995). Purification of a new clathrin assembly protein from bovine brain coated vesicles and its identification as myelin basic protein. *J. Biol. Chem*, 270, 30551-30556

Richter-Landsberg, C. (2008). The Cytoskeleton in oligodendrocytes: Microtubule dynamics in health and disease. *J. Mol. Neurosci*, 35, 55-63.

Roth, G.A., Gonzalez, M.D., Monferran, C.G., De Santis, M.L., and Cumar, F.A. (1993). Myelin basic protein domains involved in the interaction with actin. *Neurochem. Int*, 23, 459-465.

Simons, M., and Trajkovic, K. (2006). Neuron-glia communication in the control of oligodendrocyte function and myelin biogenesis. *J. Cell Sci*, 119, 4381-4389.

Simons, M., and Trotter, J. (2007). Wrapping it up: the cell biology of myelination. *Curr. Opin. Neurobiol,* 17, 533-540.

Song, J., Goetz, B.D., Baas, P.W., and Duncan, I.D. (2001). Cytoskeletal reorganization during the formation of oligodendrocyte processes and branches. *Mol. Cell. Neurosci*, 17, 624-636.

Sonnenberg, A., and Liem, R.K. (2007). Plakins in development and disease. *Exp. Cell Res*, 313, 2189-203.

Stewart, R.J., and Boggs, J.M. (1993). The carbohydrate-carbohydrate interaction between galactosylceramide-containing liposomes and cerebroside sulfate-containing liposomes: dependence on the glycolipid ceramide composition. *Biochemistry*, 32, 10666-10674.

Taketomi, M., Kinoshita, N., Kimura, K., Kitada, M., Noda, T., Asou, H., Nakamura, T., and Ide, C. (2002). Nogo-A expression in mature oligodendrocytes of rat spinal cord in association with specific molecules. *Neurosci. Lett*, 332, 37-40.

Tompa, P., Szasz, C., and Buday, L. (2005). Structural disorder throws new light on moonlighting. *Trends Biochem. Sci*, 30, 484-489.

Tompa, P. (2005) The interplay between structure and function in intrinsically unstructured proteins. *FEBS Lett*, 579, 3346-3354.

Tompkins, T.A., and Moscarello, M.A. (1993). Stimulation of bovine brain phospholipase C activity by myelin basic protein requires arginyl residues in peptide linkage. *Arch. Biochem. Biophys*, 302, 476-483.

Uversky, V.N., Oldfield, C.J., and Dunker, A.K. (2005). Showing your ID: intrinsic disorder as an ID for recognition, regulation and cell signaling. *J. Mol. Recog*. 18, 343-384.

Van Itallie, C.M., and Anderson, J.M. (2004). The molecular physiology of tight junction pores. *Physiol*, 19, 331-338.

Wilson, R., and Brophy, P.J. (1989). Role for the oligodendrocyte cytoskeleton in myelination. *J. Neurosci. Res*, 22, 439-448.

Witt, A., and Brady, S.T. (2000). Unwrapping new layers of complexity in axon/glial relationships. *Glia,* 29, 112-117.

Wohnsland, F., Schmitz, A.A.P., Steinmetz, M.O., Aebi, U., and Vergères, G. (2000). Interaction between actin and the effector peptide of MARCKS-related protein. Identification of functional amino acid segments. *J. Biol. Chem,* 275, 20873-20879.

Wood, D.D., and Moscarello, M.A. (1989).The isolation, characterization, and lipid-aggregating properties of a citrulline containing myelin basic protein. *J. Biol. Chem*, 264, 5121-5127.

Worton, R. (1995). Muscular dystrophies: diseases of the dystrophin-glycoprotein complex. *Science*, 270, 755-756.

Yoshikawa, H. (2001). Myelin-associated oligodendrocytic basic protein modulates the arrangement of radial growth of the axon and the radial component of myelin. *Med. Electron Microsc*, 34, 160-164.

Zand, R., Li, M.X., Jin, X., and Lubman, D. (1998). Determination of the sites of posttranslational modifications in the charge isomers of bovine myelin basic protein by capillary electrophoresis-mass spectroscopy. *Biochemistry*, 37, 2441-2449.

In: Myelin Basic Protein
Editor: Joan M. Boggs
ISBN: 978-1-60456-699-4

Chapter IX

Molecular Modelling of the Interaction of Myelin Basic Protein Peptides with Signalling Proteins and Effects of Post-Translational Modifications

*Eugenia Polverini**

Abstract

Myelin basic protein (MBP) is a multifunctional protein involved in maintaining the stability and integrity of the myelin sheath by a variety of interactions with membranes, and with cytoskeletal and other proteins. In this chapter the interactions of MBP with two signalling proteins, calmodulin (CaM) and an SH3-domain containing protein, and their modulation by post-translational modifications (PTMs) were investigated at an atomic level by means of the docking simulation approach, to gain further structural insight into the putative signalling role of MBP. Based on previously obtained experimental results, three CaM and one SH3-domain putative targets were identified in the classic 18.5 kDa MBP isoform, and their interaction with the binding site of the receptor was modelled. The ability of CaM and of the SH3-domain to bind the MBP peptides was confirmed and the strength of the interaction seemed to be of the same order of magnitude. The docking results on CaM highlighted the conformational adaptability of MBP. The target peptides could adopt different binding modes, adapting the orientations of their side chains in such a way that the basic residues interacted with the negatively-charged clusters at the extremities of the CaM binding tunnel, and the hydrophobic ones anchored the MBP segments to its hydrophobic pockets. Thus it seems that CaM induces the binding mode that is most favourable for it, promoting the α-helical conformation of its targets. In contrast, the polyproline type II helical conformation that is a characteristic of SH3-domain ligands is already present on the MBP target in vitro under physiological

* Department of Physics, University of Parma, Viale G.P. Usberti, 7/A, 43100 Parma, Italy. Telephone: +39 0521 905254; Fax: +39 0521 905223; E-mail: eugenia.polverini@fis.unipr.it

conditions. The MBP - SH3-domain interaction occurs by means of salt bridges and cation-π interactions between the basic residues in the peptide and the negatively-charged and aromatic residues in the receptor, but the molecular recognition and association seems to be mediated by weak CH$\cdots\pi$ interactions of the ligand prolyl residues with the aromatic residues in the binding site. The effects of PTMs on MBP peptides show a lowering of the CaM-MBP affinity after deimination, while phosphorylation seems to have only a minor effect. Phosphorylation or methylation of the MBP ligand did not cause any major inhibition of binding with the SH3-domain, beyond a somewhat less favourable interaction for phosphorylated peptides. Although the conformation of the MBP peptide in the SH3-domain pocket changed significantly, new interactions were able to substitute for those lost, in order to stabilize the complex.

Abbreviations

ADM: asymmetrically-dimethylated; CaM: calmodulin; Cit: citrulline; EPR: electron paramagnetic resonance; Golli: genes of the oligodendrocyte lineage; MARCKS: myristoylated alanine-rich C-kinase substrate; MBP: myelin basic protein; MHC: major histocompatibility complex; MM: monomethylated; mMBP: murine MBP; NMR: nuclear magnetic resonance; NRTK: non-receptor tyrosine kinase; PDB: Protein Data Bank; PPII: polyproline-type-II; PS: phosphorylated serine; PSD95: presynaptic density protein 95; PT: phosphorylated threonine; PTM: post-translational modification; rmC1: recombinant murine MBP C1 component; rmC8: recombinant murine MBP C8 component; rmMBP: recombinant murine MBP; RMSD: root mean square deviation; SDSL: site-directed spin labelling; SDM: symmetrically-dimethylated; SH3: Src homology domain 3.

Introduction

Myelin basic protein (MBP), a family of developmentally regulated isoforms involved in formation and maintenance of the myelin sheath (Boggs, 2002; Campagnoni and Skoff, 2001; Harauz et al., 2004; Smith, 1992) is also known to interact with numerous cytoskeletal and signalling proteins, interactions that are modulated by post-translational modifications (PTM) of MBP (Baryłko and Dobrowolski, 1984; Boggs, 2006; Boggs and Rangaraj, 2000; Boggs and Rangaraj, 2004; Boggs and Wang, 2004; Boggs et al., 2003; Dobrowolski et al., 1986; Libich et al., 2003b; Polverini et al., 2004; Polverini et al., 2008). The various PTMs (see Chapters II and III) seem to be a characteristic of signalling proteins (Seet et al., 2006; Yang, 2005), and many of these proteins belong to the "intrinsically unstructured" or "conformationally adaptable" class (Dyson and Wright, 2005; Receveur-Bréchot et al., 2006; Tompa, 2005; Uversky et al., 2000). Three important modifications of classic MBP isoforms are deimination (arginyl residues are converted to citrulline), phosphorylation, and methylation, whose levels change during both myelin development and the pathogenesis of multiple sclerosis (Harauz and Musse, 2007; Kim et al., 2003; Mastronardi and Moscarello, 2005; Moscarello et al., 2007) (see Chapters II and III). They have been shown to affect the targeting of MBP to microdomains in myelin in vivo, as well as to modulate its interactions with lipid membranes and proteins such as actin, tubulin (see chapters VI and VIII), calmodulin (CaM), and the SH3-domain containing proteins such as the non-receptor protein-tyrosine kinases (NRTKs) (Boggs and Rangaraj, 2000; Boggs et al., 2006; Boggs et al., 2005; DeBruin et al., 2005; Deibler et al., 1990; Harauz and Musse, 2007; Harauz et al., 2004; Hill and Harauz, 2005; Hill et al., 2005; Libich et al., 2003b; Ramwani et al., 1989).

To better understand the new putative signalling role of MBP, its interactions with some signalling proteins and their modulation by PTMs have been investigated by means of both experimental and computational techniques. The complementarity of the experimental and theoretical approaches is essential and optimal to the structural/functional study of a protein system. The computational docking simulation technique, and its use to clarify the atomic details of the interactions of MBP with calmodulin and an SH3-domain, complementing and gaining insight into experimental studies, will be the subject of the present chapter.

What Is Experimentally Known?

MBP-CaM interaction

Calmodulin is a relatively small protein that acts as a primary intracellular calcium sensor protein, regulating target proteins through protein-protein interactions in a calcium-dependent manner, and ultimately transducing the initial Ca^{2+} signal into a wide range of cellular events and processes (Crivici and Ikura, 1995). Since CaM reverses the actin polymerisation and bundling effects of MBP, it might be a regulatory element, in concert with modifying enzymes such as kinases and deiminases, of the putative signalling role of MBP in vivo (Boggs and Rangaraj, 2000; Boggs et al., 2006; Boggs et al., 2005). This hypothesis has suggested the study of the MBP–CaM interaction, supported by the evidence in the MBP sequence of a CaM-binding motif, found using The Calmodulin Target Database (Yap et al., 2000) at the Cellular Calcium Information Server of the Ontario Cancer Institute (http://calcium.uhnres.utoronto.ca).

The 18.5 kDa MBP isoform's in vitro interaction with CaM has been investigated experimentally in detail (Kaur et al., 2003; Libich and Harauz, 2002a; Libich and Harauz, 2002b; Libich et al., 2003a; Libich et al., 2003b) using intrinsic Trp fluorescence spectroscopy, gel mobility assays, and dynamic light scattering. It had been demonstrated that CaM binds to 18.5 kDa MBP in a specific, Ca^{2+}-dependent manner, in a 1:1 stoichiometry, with affinities of the order of sub-micromolar, and with a significantly different interaction when deiminated. The deiminated protein appeared to have a possible second binding site in addition to the primary, predicted target at the C-terminus, which conjecture had been supported by further studies using N- and C-terminal deletion mutants of MBP (Libich et al., 2003a). Subsequently, site-directed spin labelling (SDSL) and electron paramagnetic resonance (EPR) spectroscopy were used to map more precisely the sites of interaction of recombinant murine 18.5 kDa MBP (rmMBP), both the unmodified rmC1 and the quasi-deiminated rmC8 forms, with CaM (Polverini et al., 2004). The results showed greatest immobilization of the nitroxide spin label on residue S44C in the N-terminal half in rmC1 and rmC8, and on residues H85C and V91C in the region mMBP(P82-P93) of rmC1. The spin label on residue S159C, at the end of the C-terminal domain mMBP(T147-D158), predicted earlier by Libich et al. (Libich et al., 2003b) to form an α-helical CaM-binding domain, was also significantly immobilised in rmC1, but not in rmC8.

In conjunction with the last experiments, the docking simulation approach was used to examine, at the atomic level, the interactions with CaM of three classic 18.5 kDa mMBP fragments and its post-translationally modified forms.

MBP - SH3-domain interaction

Whereas the interactions of MBP with the myelin membrane and with calmodulin, actin, and tubulin, may represent an indirect (passive) signalling role in terms of modifying a local milieu, a direct interaction of MBP with signalling molecules, such as the SH3-domain containing proteins, would indicate an active role in myelin signalling. The SH3-domains are found in a large number of intracellular signalling proteins, including the non-receptor protein-tyrosine kinases (NRTKs). The NRTK Fyn is of particular interest because it participates directly in MBP gene expression and in compact myelin sheath formation in the central nervous system (Osterhout et al., 1999; Seiwa et al., 2000; Seiwa et al., 2007; Sperber et al., 2001; Umemori et al., 1994; Umemori et al., 1999). In lipid rafts isolated from myelin, Fyn and Lyn colocalize with Golli and classic MBP isoforms (DeBruin et al., 2005; Krämer et al., 1999). In late myelination and in mature myelin, MBP and phospho-Thr95-MBP (murine 18.5 kDa sequence numbering, phosphorylated by a mitogen-activated protein kinase at Thr95) partition into lipid rafts (DeBruin et al., 2006; DeBruin et al., 2005).

As for CaM, the presence of a potential SH3-domain target (TPRTP, residues 92-96 in the murine 18.5 kDa sequence numbering) in the classic MBP protein sequence (DeBruin et al., 2002; Moscarello, 1997) strongly supported the study of this kind of interaction. Using circular dichroic spectroscopy, it was demonstrated that the MBP proline-rich segment that contains this potential SH3-ligand forms a polyproline-type-II helix in vitro (Polverini et al., 2008). Previous preliminary experimental investigations had shown that the 18.5 kDa MBP isoform interacted with SH3-domain containing proteins in vitro (DeBruin et al., 2002), and this association was later explored more comprehensively using SH3-domain microarrays (Polverini et al., 2008). It was shown that the minimally-modified rmC1 isoform bound to the SH3-domains of the Src tyrosine kinases Yes1, Fyn, and c-Src, and the kinases Abl, Itk and PSD95, and the actin-binding protein cortactin. Several Src kinases, including Fyn and Src, and cortactin are present in oligodendrocytes and/or myelin (Kim et al., 2006; Sperber et al., 2001). It is known that MBP binds to cytoskeletal proteins (Boggs, 2006; Boggs and Rangaraj, 2000; Boggs et al., 2006; Boggs et al., 2005; Harauz et al., 2004; Hill and Harauz, 2005; Hill et al., 2005) and is required to transmit extracellular signals, provided by anti-GalC antibody, to the oligodendrocyte cytoskeleton (Dyer et al., 1994). Binding of MBP to the SH3-domains of several proteins could support a role for MBP in signal transduction and membrane-cytoskeleton associations.

It was also demonstrated that the Fyn-SH3-domain can bind to membrane-associated MBP. Spin-labelled S99C, in membrane-associated rmC1 and rmC8, was exposed to the aqueous phase (Bates et al., 2003) suggesting that the entire target region of both MBP isomers should be accessible to enzymes and other proteins, including those with SH3-domains even when MBP is associated with the membrane. It was shown that this region of MBP binds Fyn to the vesicle surface, supporting the conclusion that the Pro-rich region is accessible on the bilayer surface, and suggesting that MBP could bind Fyn and other SH3-domain-containing proteins to the oligodendrocyte and/or myelin membrane (Polverini et al., 2008).

Again, experimental measurements were used in conjunction with computational methods to investigate the details of SH3-domain interactions with the putative target in MBP.

Molecular Docking Simulation Methods

The application of computational methods to study the formation of intermolecular complexes has been the subject of intensive research during the last decade. In their binding conformations, the molecules exhibit geometric and chemical complementarity, both of which are essential for successful activity. The computational process of searching for a ligand that is able to fit both geometrically and energetically the binding site of a protein is called molecular docking (Teodoro et al., 2003).

When the structure of the target protein is known, receptor-based computational methods can be employed. These involve explicit molecular docking of each ligand into the binding site of the target, producing a predicted binding mode together with a measure of the quality of the fit of the ligand in the target binding site. The docking process consists of sampling the coordinate space of the binding site and scoring each possible ligand pose, which is then taken as the predicted binding mode for that compound (Lyne, 2002).

The newest docking techniques model the docking of a ligand to a target in greater detail; the ligand begins randomly outside the protein, and the program explores translations, orientations, and conformations (i.e., torsion angle changes), until an ideal site is found. These techniques allow flexibility within the ligand to be modeled, and can utilize more detailed molecular mechanics to calculate the energy of the ligand in the context of the putative active site. Two major considerations must be taken into account: inaccuracies in the energy models used to score potential ligand/receptor complexes, and the inability of current methods to account for conformational changes in the receptor that occur during the binding process, and that could have a great effect on docking results. In particular, the last problem has been partially faced by incorporating the flexibility of selected side chains into search methods (Huey et al., 2007; Morris et al., 1998; Osterberg et al., 2002), or by docking the snapshot obtained during a molecular dynamics simulation of the receptor; still, the prediction of the receptor's structural rearrangements is very complex and has not yet been solved. A refinement of the results could be also obtained by performing a dynamics simulation on the more favoured complex, in order to correct the interactions found by the docking approach. Nevertheless, even if some proteins show significant motion upon binding to ligands, in other cases the protein target is relatively rigid and the crystallographic structure of the protein is an adequate representation of the conformation of the protein in the desired docked complex. In addition, if the crystallographic protein structure had already been solved in a complex with a given ligand, then the insertion of the selected target into the binding site being explored with a docking simulation inherently overcomes the "motion upon binding" because it has already happened.

Another important aspect is the influence of water molecules on the ligand binding affinity, and on the ligand binding orientation. Crystal water molecules and water molecules from the bulk solvent sometimes have to be considered during docking simulations, but in this field, quantitative studies are still lacking.

There are many docking programs available that differ in the sampling algorithms used, the handling of ligand and protein flexibility, the scoring functions that they employ, and the CPU time required to dock a molecule to a given target. An efficient automated docking method that predicts the bound conformations of flexible ligands to macromolecular targets is Autodock (Morris et al., 1998). This method applies a Lamarckian model of genetics, in which environmental adaptations of an individual's phenotype are reverse-transcribed into its genotype and become heritable traits. In this way, the program can handle ligands with several tens of degrees of freedom. Autodock has been applied with great success in the prediction of bound conformations of enzyme-inhibitor complexes (Lunney et al., 1994), peptide-antibody complexes (Friedman et al., 1994), and even protein-protein interactions (Stoddard and Koshland, 1992), giving results in good agreement with experiments. It is for these reasons that we have chosen Autodock to explore the binding modes of MBP segments to CaM and signalling protein receptors.

The key results of a docking simulation are the docked structures found at the end of each run, the energies of these docked structures, and their similarities to each other. A docking job is composed of several tens of docking runs. At the end of each docking run, AutoDock outputs a result which represents the lowest energy conformation of the ligand that it found during that run. This conformation is a combination of translation, rotation, and torsion that have led to the Cartesian coordinates of the docked ligand atoms. Each conformation is characterized by intermolecular energy, internal energy, and torsional energy. The ligand-protein interaction energy (score) is better defined by different scoring functions, that sometimes give also a binding affinity prediction. A docking experiment usually has several solutions. The 'best' docking result can be considered to be the conformation with the lowest docked (intermolecular plus internal) energy. Alternatively, the best result can be selected based on the similarity of docked structures, measured by computing the root-mean-square-deviation (rmsd) between the coordinates of the atoms. The reliability of a docking result depends on the similarity of its final docked conformations.

For quality assurance of the docking results, two evaluations are important (Huey and Morris, 2007): (i) The evaluation of the convergence, to determine the thoroughness of the search. If the conformational search is exhaustive enough, the results should be clustered, that is, a small number of 'best' results will be found. If the results do not show reasonable clustering, some docking parameters should be changed. (ii) The evaluation of the chemical reasonableness of the best results by examining the interactions between the receptor and the best docked conformation(s) of the ligand. For example, if the ligand is bound inside a pocket in the receptor; if the non-polar/polar atoms in the ligand are docked near non-polar/polar atoms in the receptor; if particular interactions that are known to exist are present in the docked results; if the ligand-receptor interactions seem reasonable in the context of what is known from other experimental results (Morris et al., 1998). A "best complex" structure could thus be taken into consideration as the most representative conformation of each binding mode. At the end, the interpretation of docking results is open-ended; to some extent it depends on the user's insight, but the conjunction with all the experimental data available is of great importance.

ANALYSIS OF THE STRUCTURAL FEATURES OF THE RECEPTOR AND THE LIGAND: MODEL BUILDING OF MBP FRAGMENTS

Target identification

The first step that supported the study of the interaction between MBP peptides and both the calmodulin protein and the Fyn-NRTK SH3-domain was the identification of characteristic sequence motifs in the primary structure of MBP.

```
mMBP     1   ASQKRPSQRS KYLATASTMD HARHGFLPRH RDTGILDSIG RFFSGDRGAP
rmC1     1   ASQKRPSQRS KYLATASTMD HARHGFLPRH RDTGILDSIG RFFSGDRGAP
rmC8     1   ASQKRPSQRS KYLATASTMD HAQHGFLPRH QDTGILDSIG RFFSGDRGAP

                                                        PT PT  PS
                                                        ↓  ↓   ↓
mMBP     51  KRGSGKDSHT RTTHYGSLPQ KSQHGRTQDE NPVVHFFKNI VTPRTPPPSQ
rmC1     51  KRGSGKDSHT RTTHYGSLPQ KSQHGRTQDE NPVVHFFKNI VTPRTPPPSQ
rmC8     51  KRGSGKDSHT RTTHYGSLPQ KSQHGRTQDE NPVVHFFKNI VTPRTPPPSQ

               MM
              ADM                                              E
              SDM                                              PS
               ↓                                               ↓
mMBP     101 GKGRGLSLSR FSWGAEGQKP GFGYGGRASD YKSAHKGFKG AYDAQGTLSK
rmC1     101 GKGRGLSLSR FSWGAEGQRP GFGYGGRASD YKSAHKGFKG AYDAQGTLSK
rmC8     101 GKGRGLSLSR FSWGAEGQQP GFGYGGQASD YKSAHKGFKG AYDAQGTLSK

                  Q
                 Cit
                  ↓
mMBP     151 IFKLGGRDSR SGSPMARR
rmC1     151 IFKLGGRDSR SGSPMARR LEHHHHHH
rmC8     151 IFKLGGQDSR SGSPMARQ LEHHHHHH
```

Figure 1. Amino acid sequences of the classic 18.5 kDa murine MBP isoform, the recombinant murine C1 isoform (rmC1) and quasi-deiminated C8 isoform (rmC8). The examined CaM target peptides are highlighted in gray, the SH3-domain target peptide is underlined. The Q substitutions in the rmC8 isoform are bolded. The examined PTMs are indicated with an arrow.

By means of The Calmodulin Target Database (Yap et al., 2000) at the Cellular Calcium Information Server of the Ontario Cancer Institute (http://calcium.uhnres.utoronto.ca), a primary CaM-binding motif was found, and two other ones were chosen on the basis of experimental results (Kaur et al., 2003; Polverini et al., 2004). Using murine MBP (mMBP) sequence numbering (168 residues for the classic 18.5 kDa isoform, see Figure 1), the first site was the C-terminal one T147-D158 (TLSKIFKLGGRD). The second one, the P82-P93 (PVVHFFKNIVTP) segment, lies in the central portion of the protein; it represents an immunodominant epitope and forms an amphipathic α-helix in reconstituted myelin-mimetic systems (Bates and Harauz, 2003; Bates et al., 2004), that becomes more highly exposed after deimination of membrane-bound MBP (Musse et al., 2006). The third is the R41-R52 segment (RFFSGDRGAPKR); it is quite close to the N-terminus and was chosen on the basis of the results obtained with SDSL/EPR experiments (Polverini et al., 2004). Such segments, whose lengths were more or less arbitrarily chosen, were in some cases shifted or extended, to compare similar binding characteristics and better delineate the protein target.

Subsequently, the presence of a potential SH3-domain target (TPRTP) situated in a proline-rich region of the classic MBP protein sequence was defined (DeBruin et al., 2002; Moscarello, 1997). It is flanked by various serine and threonine phosphorylation sites, and there is also an arginine methylation site, R104, in its vicinity. This putative SH3-target corresponds to the segment TPRTPPPSQGKGR, residues T92-R104 in the murine 18.5 kDa MBP sequence (Figure 1).

Structural features of the ligand and the receptor

The second step was the analysis of the structural features of the ligands and the receptor, to confirm the choice of the putative ligand peptides of MBP and to predict the possible binding modes.

The protein calmodulin consists of two structurally homologous Ca^{2+}-binding domains. In the calcium-free state, each domain is composed of a four-helix bundle and the two domains are bridged by a long helix that is dynamically disordered at its center (Babu et al., 1985; Kuboniwa et al., 1995). Upon binding calcium, each globular domain reorganizes its tertiary structure to form two helix-loop-helix or EF-hand motifs (Babu et al., 1985). This structural rearrangement exposes a large surface to solvent that will ultimately form extensive hydrophobic and ionic contacts with the target domain. Again, in the free but calcium-activated state, the two globular domains of calmodulin are tethered together by the bridging helix.

It has been shown (Afshar et al., 1994) that the main characteristics of CaM-binding peptides were the presence of at least two positive residues separated by nine or ten residues with a central hydrophobic core, and that an important binding characteristic was the pattern of positively-charged—hydrophobic—positively-charged residues. The positive residues were demonstrated to interact with both the negatively-charged extremities in the binding tunnel of CaM, in many cases forming salt-bridges. It was also demonstrated [ibid.] that the CaM-binding channel possesses a polarity created by a non-uniform distribution of acidic residues at its two margins. The second margin is slightly smaller than the first one, and also comprises a basic residue; thus, it is less negatively-charged overall. The electrostatic component is reported to play a major role both in the strong binding of a target peptide (Afshar et al., 1994; Imparl et al., 1995), and in its orientation (Afshar et al., 1994). In fact, the symmetry of this interaction, due to the positioning of positively-charged residues, permits a reverse-binding mode for the peptide, with roughly the same energy. The insertion of the target peptide into the binding tunnel usually occurs spanning the biggest first margin. Hydrophobic interactions are also an important force for driving the protein-peptide complex formation. In each CaM domain there is a deep, hydrophobic pocket that can accommodate the bulkiest residues (such as Trp, Phe, and Leu) of the target peptide. These pockets contain nine methionyl residues involved in peptide binding, the presence of which is a distinguishing feature of CaM (Meador et al., 1992; Vogel and Zhang, 1995; Yang et al., 2004). It is the balance of electrostatic and hydrophobic contributions, due to the different kinds and placement of residues of the target peptide, which determines the final binding mode.

The SH3-domain fold is highly conserved and is composed of five β–strands organised into two β-sheets packed at right angles to form a sandwich (Musacchio et al., 1994; Noble et al., 1993). One side of the sandwich is rather hydrophobic and constitutes the ligand-binding surface. It is delimited on one side by an RT-loop between the βA and βB strands and on the other side by a small loop between the βB and βC strands (called the "n-Src insertion point loop") and by a 5-residue long 3_{10} helix, which separates the βD and βE strands. The effect of this particular fold is to bring most of the conserved residues in the SH3-domain close to each other, resulting in the formation of an aromatic rich patch on the protein surface that forms the ligand binding site. One end of this site is lined by negatively-charged residues in the RT-loop and in the n-SRC loop.

Canonical SH3-ligands usually contain two identical XP dipeptides, separated by a scaffolding residue (often a proline) (Cesareni et al., 2002; Dalgarno et al., 1997; Jia et al., 2005; Li, 2005; Macias et al., 2002; Mayer, 2001). The two XP moieties in the core motif (XP-x-XP) occupy two hydrophobic pockets at the binding surface, and a stacking interaction between one of the Pro residues and the Trp of the binding pocket is well-conserved. The proline residues of the ligand seem to be implicated in having a role in molecular association employing C-H···π interactions with aromatic residues at the binding site. The third SH3-domain binding pocket, lined by negatively-charged residues, can host a positively-charged side chain flanking the core motif.

Modeling of the target

The third step for this kind of theoretical approach is to use all the known structural characteristics of the ligand of interest to model the potential ligand peptides extracted from the MBP protein sequence. In fact, a docking simulation needs a starting structure to which are assigned a number of torsional degrees of freedom that can vary during the simulation, to find the best conformation for the peptide with respect to the receptor, namely for the complex. If this number of degrees of freedom is too high, the results of the simulation are not reliable, from the point of view of the conformational searching algorithm, due to the high computational cost. Therefore, it is necessary to keep some of the active torsion angles rigid to diminish this number. It is clear that this is one of the limitations of the docking simulations, regarding the ligand flexibility as well as that of the receptor. A test simulation run, using the receptor and ligand coordinates (both extracted from an analogous complex) could be useful to check if it is possible to reach the correct results by the docking algorithm, and therefore to set the best simulation parameters for obtaining such a result. This initial step permits the use of such parameters for a similar ligand (in our case the MBP peptide) with the same high number of degrees of freedom, docked into an identical binding site.

To choose which angles to keep rigid, we can assign a particular conformation to a backbone region (or to the side chains) that is known to be present in the ligand complex. It is emphasized here that a docking simulation cannot investigate the binding process but only the binding mode, that is, the more favoured conformation of the ligand in the binding site.

The canonical mode of 1:1 binding of CaM ligands is an α–helix clamped by the N-terminal (residues 1–75) and C-terminal (residues 82–148) domains of CaM; therefore, the backbones of the MBP peptides were built in such a way that their torsion angles were kept rigid. The side chain torsion angles were left free to rotate.

Regarding the SH3-domain binding mode, the 7-9 amino acid long core regions of the ligands bind to SH3-receptors in a left-handed poly-proline II (PPII) helical conformation. Therefore, the backbone of the segment comprising residues T92 to P98 was built in a PPII conformation, leaving free to rotate the torsion angles of the backbone of the S99-R104 region, and the sidechains of the whole peptide.

Choice of the receptor structure

A careful choice of the receptor structure is equally important in a docking simulation due to the fact that it is kept completely rigid. After the necessary biological considerations regarding, e.g., the organism studied and the protein function, other important characteristics to check for the choice of a resolved structure are the completeness, i.e., the number of amino acids that are actually resolved with respect to the whole; the resolution; the experimental technique used, i.e., X-ray crystallography (higher resolution but more rigid structure), NMR spectroscopy (lower resolution, but done on samples in solution, thus giving information on the flexible regions), theoretical modeling (usually very low resolution but useful if nothing else is available). The comparison of the different available structures, better if complexed with a ligand, is very informative, in particular regarding the orientation of some conserved side chains or ligand contacts. Of course, for practical use in docking simulations, the coordinates of any co-resolved ligand are excised and the "clean" structure of the receptor is used to dock the putative built target.

The availability of complexes is useful also to examine the motion upon binding of the receptor active site, comparing the bound and the unbound resolved structures. For both our receptors (CaM and SH3-domain), no difference was noticeable between the bound and the unbound receptor conformations. Furthermore, no water molecule is known to be involved in the binding; this fact makes reasonable the approximation of ignoring it in the docking simulations.

CaM can assume at least three different conformations, depending on the binding of Ca^{2+} ions and on the peptide target. The CaM structure that we used was extracted from the recently derived entry 1CKK (Osawa et al., 1999) of the Protein Data Bank (http://www.rcsb.org/pdb/) (Berman et al., 2000). It was determined by solution NMR and represents the bent, complexed conformation of CaM with four bound Ca^{2+} ions and a segment of a protein kinase kinase (Osawa et al., 1999). Comparing all the available structures, the orientation of the acidic residues, which are involved in the peptide binding, was verified to be the same. The CaM from the 1CKK structure is from *Xenopus laevis* (African clawed frog), but it has the same amino acid sequence and identical structure as bovine CaM, which was used in previous experimental investigations (Libich et al., 2003b).

The choice of the SH3-domain structure to use for the docking simulations was made on the basis of several considerations. Because the Fyn NRTK plays a major role in central nervous system myelination and contains an SH3-domain, we sought a convenient structure for the SH3-domain amongst the crystal structures available for Fyn in the Protein Data Bank

(PDB) (Berman et al., 2000). After a careful analysis and comparison, we chose the unbound receptor crystal structure 1shf (PDB entry) of human Fyn, chain A, with a resolution of 1.9 Å (Musacchio et al., 1994; Noble et al., 1993). The torsion angle of the sidechain of Tyr91 (a residue that is usually involved in the interaction with the ligand) of 1shf was modified according to the conformation that it has in the other Fyn crystal structures.

In silico post-translational modifications of the MBP target peptides

As we have indicated above, MBP is highly post-translationally modified, and we have examined in silico the effects of deimination, phosphorylation and methylation of specific residues that are known to be modified in vivo and that are present in the studied MBP peptides.

The modified amino acids are quite easy to build in silico, but particular attention has to be paid to the correct attribution of the partial charges on each atom, to use in the docking simulation. In some cases it could be useful to utilize the Gasteiger charge set (Gasteiger and Marsili, 1980), that is usually used for a generic ligand, rather than the Kollmann charge set (Weiner et al., 1984), that should be better for a peptidic one. The factors that could influence the binding mode of a modified peptidic ligand, both in vitro and in silico, are a different distribution of charges, as well as different instances of steric hindrance. Nonetheless, some test simulations performed using both the Gasteiger and the Kollmann charge sets to dock a ligand peptide into its binding site, led to the same results (Polverini et al., 2008).

On the C-terminal, predicted CaM target mMBP(T147-D158), two specific residues are known to be modified in vivo. Therefore, the following modifications were introduced: (i) quasi-deimination R157Q, that is the emulation of deimination by an R $\rightarrow$ Q substitution; (ii) deimination R157Cit; (iii) quasi-phosphorylation S149E, that is the emulation of phosphorylation by an S $\rightarrow$ E substitution; (iv) phosphorylation S149[PS].

On the putative proline-rich SH3 target mMBP(T92-R104), both various serine and threonine phosphorylation sites and the arginine methylation site are present, and the following modifications were applied: (i) phosphorylation S99[PS], (ii) phosphorylation T92[PT] and T95[PT] (both individually and together), (iii) monomethylation (MM), asymmetrical-dimethylation (ADM), and symmetrical-dimethylation (SDM) of R104.

CaM - MBP Complexes: The Docking Results

As we have previously considered, mMBP presents two amphipathic and one non-amphipathic α-helical targets that show a symmetrical distribution of both basic and hydrophobic residues, in this way equally favouring different binding modes.

Due to the amphipathicity of the α-helices, the basic residues in the peptide are spatially quite close and can interact with the same negatively-charged cluster on CaM in almost all cases. In addition, the loss for MBP of the canonical spacing of 12/14 residues between two hydrophobic anchor residues in the target peptides is a characteristic that has been found in some other CaM-binding proteins (Yap et al., 2000), such as MARCKS (Yamauchi et al.,

2003); as a consequence, in most cases, only one hydrophobic pocket of CaM at a time is involved in the binding.

In the following sections, the structural characteristics of the binding complexes for the three putative MBP targets of CaM will be described in detail.

The C-terminal T147-D158 target

This peptide has the two most distant positively-charged residues (K150 and R157) separated by only six positions, whereas K153 is flanked by several hydrophobic residues (Figure 1). Constructing the peptide in an α-helical conformation shows a clear amphipathic pattern, with the three positively-charged residues on the same side of the helix, thus forming a cluster that interacts strongly with the negatively-charged residues that are present at the extremities of the binding tunnel in CaM. As predicted by the symmetrical distribution of the basic residues, the docking simulation finds both orientations of the peptide in the binding tunnel as energetically favourable, with the positive cluster interacting with either the first or with the second margin, forming salt bridges with acidic residues, while the hydrophobic ones are accommodated into the hydrophobic pockets of CaM. These results are entirely consistent with the work of (Afshar et al., 1994), and support the theoretical prediction (Yap et al., 2000) and experimental demonstrations (Libich and Harauz, 2002a; Libich and Harauz, 2002b; Libich et al., 2003a; Libich et al., 2003b) that this region of MBP is a CaM-binding site.

The possibility of not only a reverse binding mode, but also a peptide displacement along the helix axis, was studied by (Afshar et al., 1994) who showed that the CaM interaction surface could accommodate large peptide translation/rotation with local sidechain optimisation. To investigate the effect of the presence in the target of two positively-charged residues at the canonical distance of nine residues (ibid.) a shifted peptide mMBP(S149-R160) was investigated. It presents the first positively-charged K150 and the last R160 separated by nine residues, although there are two other basic residues (K153 and R157) in between (Figure 1). With the presence of four positively-charged residues in total, the interaction involves two negatively-charged clusters of CaM at the same time at the opposite extremities of the binding channel, so that there is an extensive electrostatic interaction. Despite the symmetrical positioning of the basic residues in the shifted peptide, the orientation with the C-terminus inserted into the CaM binding tunnel is favoured, probably because of the accommodation of the two nearby, bulky hydrophobic residues (F152 and L154) in the largest pocket of the C-terminal domain of CaM (Figure 2A). This orientation is in agreement with the experimentally found strong immobilization of spin-labeled residue S159C of rmC1 (Polverini et al., 2004), which would be buried in the CaM-binding tunnel.

Altogether, from an energetic point of view, the shift and the presence of the "canonically spaced" basic residues does not enhance the strength of the binding with respect to the original mMBP(T147-D158) segment.

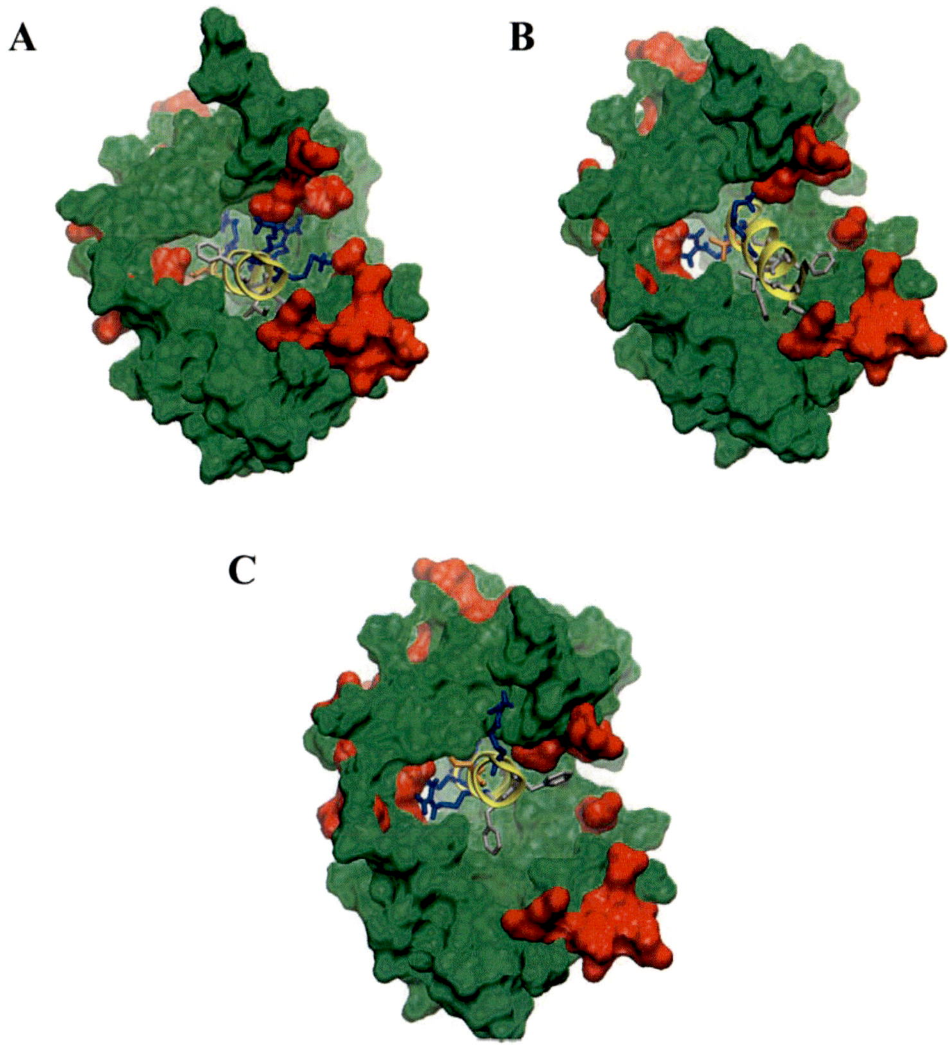

Figure 2. (A) Complex of the CaM with the mMBP(S149-R160) C-terminal shifted peptide in the more favoured conformation. The peptide C-terminus is inserted into the binding channel. The orange stick is the buried S159 residue, found strongly immobilized by the SDSL/EPR experiments. (B) Complex of the CaM with the mMBP(P82-R94) central extended peptide in the more favoured conformation. The peptide C-terminus is inserted into the binding channel. The orange stick is the buried V91 residue, found strongly immobilized by the SDSL/EPR experiments. (C) Complex of the CaM with the mMBP (R41-R52) N-terminal peptide in the more favoured conformation. The peptide C-terminus is inserted into the binding channel. The orange stick is the S44 residue, found strongly immobilized by the SDSL/EPR experiments. The CaM is represented with its accessible surface (in green); the acidic residues at the two extremities of the binding channel are in red. The MBP peptide helix is in a yellow cartoon ribbon structure; its basic residues are represented with blue sticks, the bulky hydrophobic ones with gray sticks.

The central P82-P93 target

The conformation of the putative secondary CaM-target mMBP(P82-P93) has been reported to be variable in solution, depending on the length of the peptide being analyzed (Whitaker et al., 1990). Several experimental studies have shown that this segment is an amphipathic α-helix when bound to lipid (Bates et al., 2003; Bates et al., 2004; Harauz et al., 2004; Libich et al., 2004), which probably reflects its in vivo conformation. Thus, the α-helical conformation appears to be physiologically relevant and it is for this reason that this fragment was examined, although it does not have the characteristic pattern of other CaM-binding peptides, and is not predicted to be a CaM-target by database searches.

The SDSL/EPR experimental results obtained for eight spin-labeled residues examined in this domain showed that all of them, in particular V91C, were strongly immobilised by CaM (Polverini et al., 2004). In silico, the peptide mMBP(P82-P93), modelled as an α–helix, binds CaM with a less favourable energy with respect to the C-terminal CaM target mMBP(T147-D158), even though it interacts with the same negatively-charged residue clusters on CaM. This peptide has only one central, positively-charged residue (K153) and it can bind equally well in both orientations and in rotated positions, nicely accommodating the hydrophobic residues into the hydrophobic pockets of CaM.

To confirm the importance of the α-helical conformation, the same peptide in an extended conformation (that which it has in the major histocompatibility complex (MHC) protein binding site (Li et al., 2000)) was investigated, resulting in a very unfavourable energy for CaM-binding.

Because the residue immediately following the sequence of mMBP(P82-P93) is R94, the interaction in the presence of an extra positive charge in the target peptide was also investigated. Although the interaction is almost as favourable as that of the mMBP(P82-P93) peptide, the two positively-charged residues interact in the same complex with two different clusters at the opposite ends of the binding tunnel. The orientation that inserts the target C-terminus into the binding channel is greatly favoured over the reverse one, because the terminal arginyl residue on this segment of mMBP proffers a significant electrostatic contribution to the interaction with CaM, whereas the extension of the chain favours the hydrophobic accommodation of the two consecutive, bulky mMBP peptide residues F86 and F87 into the C-terminal hydrophobic pocket of CaM (Figure 2B). Such an orientation is also supported by the SDSL/EPR experimental results obtained for eight spin-labeled residues examined in this domain, that showed for all of them, in particular V91C a strong immobilization by CaM (Polverini et al., 2004). The last hydrophobic residue is oriented in silico in such a way that it is buried in the hydrophobic pocket of CaM.

The N-terminal R41-R52 target

The peptide mMBP(R41-R52) was chosen on the basis of the results obtained with SDSL/EPR experiments showing that the spin-labeled residue S44C was immobilised (Polverini et al., 2004). Two positively-charged residues (R41 and K51) are separated by nine residues, although there is another positively-charged residue (R47) in between. At the C-terminus, there is a cluster of two positively-charged residues, K51 and R52 (Figure 1). The peptide can interact in both orientations with the same three clusters of acidic residues in CaM through the entire binding tunnel, and with a very favourable docked energy. The orientation with the C-terminus inserted is the prevalent one, probably due to the presence of the two bulky F42 and F43 residues at the N-terminus of the peptide that interact with the deep and largest hydrophobic pocket in the C-terminus domain of CaM, at the largest entrance of the binding tunnel (Figure 2C). Probably, this hydrophobic interaction gives a greater contribution to the free energy of binding than the electrostatic interaction of the bi-basic cluster at the peptide C-terminus. This result is in general agreement with those obtained with the mMBP(P82-P93) and mMBP(P82-R94) peptides, which also contain a Phe-Phe pair. The position of these two residues seems to influence the orientation of the peptide in the CaM binding site, probably due to the capability to form strong hydrophobic interactions.

SH3-Domain - MBP Complex: The Docking Results

The T92-R104 proline-rich segment

The results of the docking simulation performed on the mMBP(T92-R104) peptide, led to a best complex conformation with a very favourable docked energy, slightly better than the one calculated for the MBP-calmodulin interaction (Polverini et al., 2004). The dissociation constant (K_d) of MBP with calmodulin has been experimentally measured to be in the μM range (Libich et al., 2003a; Libich et al., 2003b). Generally, the interactions of SH3-domains with proteins containing proline-rich sequences are also weak, with experimentally measured dissociation constants in this range, enabling rapid association and dissociation (Mayer, 2001); therefore, it could be surmised that they are of a similar order of magnitude also for the interaction of SH3-domains with MBP.

The PPII helix segment (T92-P98) of the MBP peptide lies in the canonical SH3-binding pocket, and the flexible backbone region at the C-terminus (S99-R104) is bent close to the SH3-domain surface (Figure 3). Analysis of the contact residues between the two molecules showed that the positively-charged amino acids in the ligand "canonically" interact with the acidic residues that line the binding pocket. In particular, R104 makes a salt bridge with Asp118 in the n-Src loop and interacts with the residues of the 3_{10} helix; the other arginine R94, which lies just inside the core motif, interacts with the negatively-charged residues in the RT-loop, making a salt bridge with Asp99. The third positive residue in the MBP peptide, K102, makes a cation–π interaction with W119, and interacts also with the 3_{10} helix region. Residue P96 in MBP-peptide is stacked with the aromatic ring of W119 in the SH3-domain, forming a C-H···π interaction, that is usually present in one of the characteristic SH3-domain hydrophobic binding pockets (Cesareni et al., 2002). The proline residues of the ligand in SH3 complexes are, in fact, implicated in molecular association, employing these interactions with conserved aromatic residues at the binding site (Bhattacharyya and Chakrabarti, 2003). Here, P98 also forms a C-H···π interaction with W119, and P96 with Y132, as was similarly found for the Abl tyrosine kinase (Bhattacharyya and Chakrabarti, 2003). Such aromatic residues (along with N136) form canonical and non-canonical (i.e., by means of their CH group) hydrogen bonds with the backbone oxygens of the ligand.

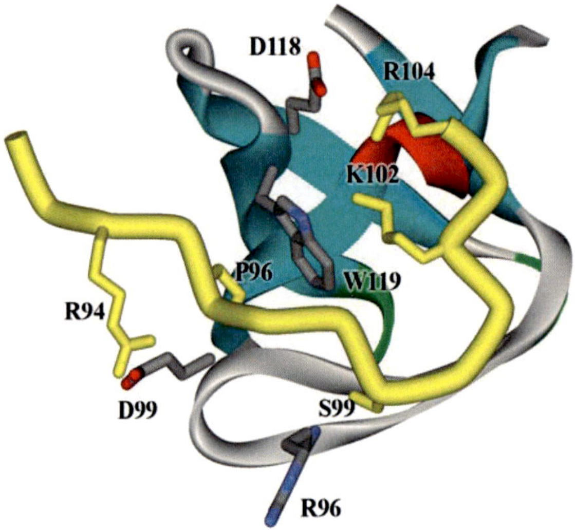

Figure 3. Complex of the Fyn-SH3-domain with the mMBP(T92-R104) peptide (in yellow) in the more favoured conformation. The SH3-domain is represented with its secondary structure elements (the β-sheets in cyan, the 3_{10}-helix in red, and the turns in green). The relevant contact residues are in a stick representation, coloured by atom type in the SH3-domain, and in yellow in the MBP peptide. Reprinted from Biochemistry 47 (2008) 267-282, DOI 10.1021/bi701336n with permission from the American Chemical Society. Copyright 2007 American Chemical Society.

Effects of Post-Translational Modifications

CaM binding target mMBP(T147-D158)

The effects of deimination and phosphorylation of two specific residues in the C-terminus of MBP that are known to be modified in vivo (Kim et al., 2003) were examined by docking simulations on the mutated mMBP(T147-D158) target. First, the peptides with the quasi-deiminated and deiminated residues R157Q and R157Cit display very similar binding behaviour, with roughly equal docked energy, but less favourable than the unmodified peptide. Due to the lack of the positive charge at the modified peptide's C-terminus, thus creating a non-symmetrical distribution of the basic residues, the penetration of the C-terminus into the CaM channel is still favoured, with the accommodation of the aromatic or long-chained hydrophobic residues, that are mainly present at the N-terminus of the peptide, in the deep hydrophobic pocket in the C-terminal domain of CaM. In this case, the addition of a terminal arginyl residue as in the shifted segment mMBP(S149-R160, R157Q) leads to a more favourable docked energy for the segment oriented in the reverse direction (Figure 4A). This observation is consistent with the experimental SDSL/EPR data showing increased mobility of spin-labeled S159C on rmC8 relative to rmC1 (Polverini et al., 2004), that is at the C-terminal end of the peptide that is located at the external end of the binding tunnel. Overall, the deimination seems to perturb the MBP–CaM interaction, as has been observed experimentally (Libich et al., 2003b).

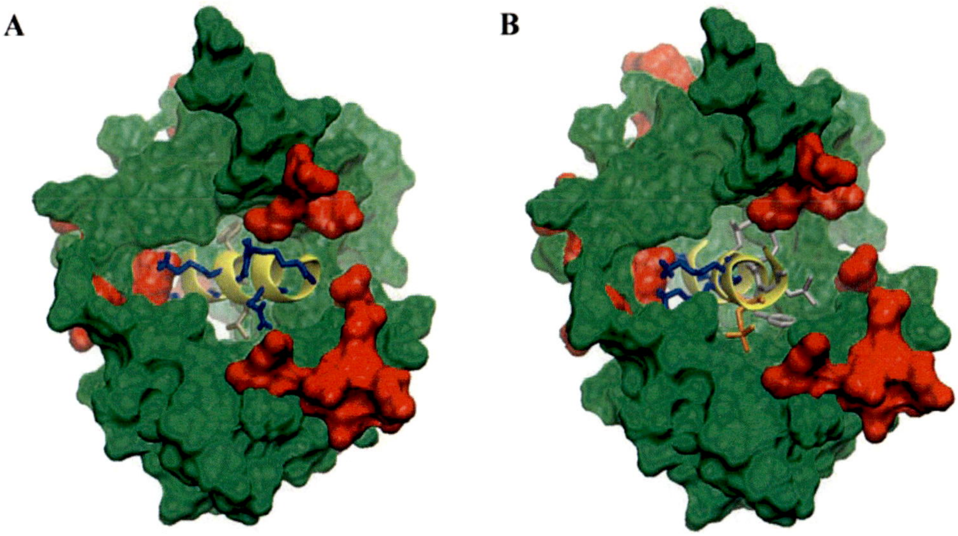

Figure 4. (A) Complex of the CaM with the quasi-deiminated mMBP(S149-R160, R157Q) peptide. The peptide N-terminus is inserted into the binding channel. (B) Complex of the CaM with the phosphorylated mMBP(T147-D158, S149[PS]) peptide. The peptide N-terminus is inserted into the binding channel. The orange stick is the S149[PS] residue, that remains oriented outside the binding tunnel. The CaM and MBP peptide representations are as in Figure 2.

The pseudophosphorylated or phosphorylated mMBP(T147-D158) peptides containing substitutions S149E and S149[PS] have essentially the same docked energy as the unmodified mMBP(T147-D158) segment. However, as for the unmodified target, the insertion of the C-terminus into the binding channel is favoured also in this case, presumably because, even if a positive charge is not lacking here, they possess an N-terminal negatively-charged group, whose sidechain remains oriented outside the binding tunnel with no contacts with its negatively-charged extremity (Figure 4B). These results are consistent with those of (Yamauchi et al., 2003) who indicated that phosphorylation of a fragment of MARCKS (myristoylated alanine-rich C-kinase substrate) towards one end of the binding domain, where the phosphate would be relatively exposed to the solvent, had only a minor effect on the strength of its interaction with CaM.

It is interesting to note that the interacting clusters of CaM are the same for all of the peptides (native, quasi-deiminated, deiminated, quasi-phosphorylated, and phosphorylated), resulting in very similar bound complexes. In addition, the positively-charged residues in the peptides form salt bridges with the same acidic residues in CaM, namely E11, D80, E84, and E87.

One further practical conclusion that arises from this study is that quasi-deimination (as emulated by R $\rightarrow$ Q substitution) and quasi-phosphorylation (as emulated by S $\rightarrow$ E substitution) are appropriate mimics for natural deimination and phosphorylation reactions, respectively. The results justify the use of R/K $\rightarrow$ Q substituted mutants of rmMBP (Bates et al., 2003; Bates et al., 2002; Hill et al., 2002; Libich et al., 2003b) to emulate the naturally deiminated form. The mimicking of phosphorylation by S $\rightarrow$ E substitution has been utilized in studies of tau protein (Eidenmüller et al., 2000; Eidenmüller et al., 2001), and is suggested here to be worthwhile for rmMBP. Thus, the computational comparisons performed here suggest that experimental studies using site-directed mutants of MBP are good mimics of the naturally modified forms, which are more difficult to purify in large quantities.

SH3-domain binding target

Three docking simulations were performed to investigate the effects of three kinds of phosphorylation on the MBP peptide T92-R104 : on residue S99, on residue T95, and on both residues T92 and T95. The best complex conformations resulting from the docking of the phosphorylated peptides superimposed well with each other and with that of the unmodified peptide, indicating a very similar binding mode. The PPII region of the ligand lies in the same site and, although the C-termini show some differences in their backbone and sidechain torsion angles, they remain oriented in the same direction forming a bundle of structures occupying the same space (Figure 5A).

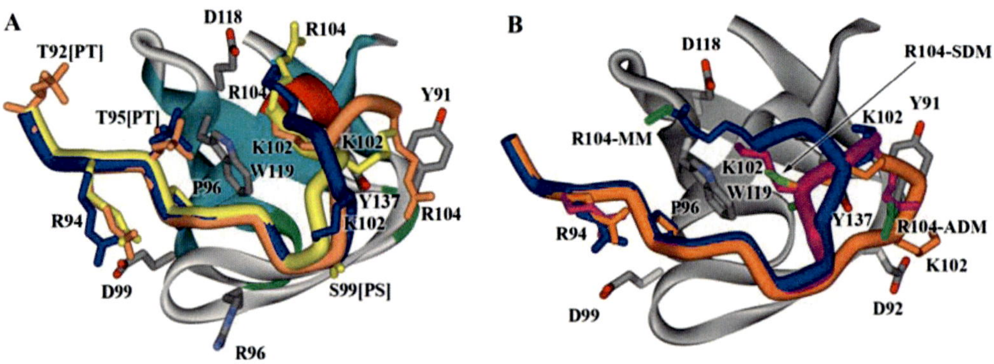

Figure 5. (A) Complexes of the Fyn-SH3-domain with the mMBP(T92-R104) phosphorylated peptides, superimposed for comparison. The MBP peptide with the post-translational modification S99[PS] is in yellow; the one with the post-translational modification T95[PT] alone is in blue; the one with the post-translational modifications T92[PT] plus T95[PT] is in orange. The cartoon ribbon representation of the SH3-domain is coloured as in Figure 3. (B) Complexes of the Fyn-SH3-domain with the mMBP(T92-R104) R104-methylated peptides, superimposed for comparison. The MBP peptide with R104-MM is coloured dark blue; the one with R104-ADM is coloured magenta; the one with R104-SDM is coloured orange. The methylated atoms are highlighted in green. The cartoon ribbon representation of the SH3-domain is totally grey for the sake of clarity. The relevant contact residues are in a stick representation, coloured by atom type in the SH3-domain, and by the ligand colour in the docked peptides. Reprinted from Biochemistry 47 (2008) 267-282, DOI 10.1021/bi701336n with permission from the American Chemical Society. Copyright 2007 American Chemical Society.

In comparison with the unmodified peptide, the main preserved interactions in all the complexes are the R94-D99 salt bridge, the C-H···π Pro-aromatic interactions, in particular the stacking of P96 with W119, and the characteristic H-bonds between the ligand backbone oxygens and the residues N136 and Y137. The R104 residue maintains the salt bridge with D118 in all the complexes except that with the double-phosphorylated peptide, in which it is substituted with a cation-π interaction with Y91 and Y137 in the RT-loop. The K102 residue is the more variable one, changing the orientation of its long sidechain away from the 3_{10} helix towards the RT-loop, but forming in most cases cation-π interactions with the conserved aromatic residues of the binding site (Y91 and Y137 for the S99[PS] and W119 for the T92[PT] + T95[PT] peptides,).

The S99[PS] sidechain shows a different orientation compared to the unmodified S99 residue, due to the phosphate steric hindrance, and the complexes of the SH3-domain with the S99[PS]-peptide have a less favourable docked energy than the unmodified MBP peptide complex. The docked energies of the complexes of the SH3-domain with the T95[PT] and the double-phosphorylated T92[PT] and T95[PT] peptides are also less favourable than for the unmodified MBP peptide (slightly less for the first and comparable to those of the S99[PS] peptide for the second). It is possible that the rigidity (real and built-in) of the PPII backbone segment that flanks the phosphorylated residues prevents the modified sidechain from better adapting to the receptor surface, giving a less favourable complex interaction energy.

With the methylation of residue R104, which remained charged, the steric hindrance caused by the mono- or dimethyl (symmetrical and asymmetrical) groups results in the ligand peptide adjusting the position of this long sidechain to optimise the interaction and its conformation in the binding site. Thus, the best methylated complexes have good docked energies, of the same value as the unmodified MBP peptide. Also in this case, the best complexes obtained for the three methylated ligands superimpose very well with each other, and with the unmodified one in the PPII region (Figure 5B). The stacking of P96 with W119, mediated by the weak CH···π interactions, is still present, underlying the importance of pairing these residues for molecular recognition and association. Canonical and non-canonical hydrogen bonds are preserved, in particular with N136.

The C-terminus (G101-R104) shows a larger shift than for the phosphorylated ligands, in order to make more favourable interactions with the R104 methylated residue. In the monomethylated ligand, the R104-MM residue is near the n-Src loop, but nevertheless forms a salt bridge with D118. Instead, R104-ADM shifts towards the first residues of the RT-loop, interacting with the aromatic rings of Y91 and Y137 by cation-π interactions. The post-translational modification R104-SDM is peculiar to MBP and only a few other proteins. The C-terminal backbone (from G101 to R104-SDM) is affected the most by this modification, being strongly bent to the center of the binding pocket in a way that causes the R104 sidechain, with its symmetric methyl groups, to interact with W119, making more hydrophobic contacts than polar ones. Nonetheless, a cation-π interaction with Y137 is still present. Due to the shift of the C-terminus, the charged K102 also has different orientations (towards the RT-loop or the n-Src loop), but almost always forming salt-bridges or cation-π interactions.

CONCLUSION

Myelin basic protein is a multifunctional protein based on its variety of splice isoforms and myriad post-translational modifications, and interactions with membranes and proteins (Boggs, 2006; Harauz et al., 2004). In order to gain further structural insight, we have investigated via molecular docking simulations the interaction of putative CaM and SH3-

domain ligands of the classic 18.5 kDa MBP isoform with CaM and an SH3-domain-containing proteins.

The results on CaM support the previous conjecture that it can potentially bind different amphipathic α-helical segments of MBP, an especially important consideration for such a flexible protein (Libich et al., 2003a). Even if the simulations were to keep the calmodulin rigid, it was shown how the target peptides can adapt the orientations of their sidechains, and adopt different binding modes by translating, rotating, or reversing, with the basic residues interacting with the negatively-charged clusters at the extremities of the CaM binding tunnel and the hydrophobic ones anchored to its hydrophobic pockets. This conformational adaptability of MBP is consistent with its intrinsically unstructured nature (Harauz et al., 2004; Hill et al., 2002; Hill et al., 2003). Thus, it seems that CaM induces the binding mode that is most favourable for it, promoting the α-helical conformation on its targets.

On the other hand, it was shown by CD spectroscopy that the proline-rich segment that contains the potential SH3-ligand forms a PPII helix already in vitro under physiological conditions. Also in this case the ability of the SH3-domain to bind the MBP target was confirmed. The molecular recognition and association by the SH3-domain seems to be mediated by the weak CH···π interactions of the ligand prolyl residues with the aromatic residues in the binding site, in particular by the stacking of a proline with the tryptophanyl residue 119 in the n-Src loop, that is a characteristic of this kind of interaction and was identified in particular for SH3-domain binding (Bhattacharyya and Chakrabarti, 2003). Basic residues in the peptide interacted via salt bridges and cation-π interactions with negatively-charged and aromatic residues in the SH3-domain.

From an energetic point of view, the docking results, coupled with the experimental ones (Polverini et al., 2004), suggest that the strength of the interaction for the SH3-domain-MBP complex would be of the same order of magnitude as with calmodulin, from which we can hypothesize that the dissociation constants (K_d) are in the same μM range. Thus, there would be rapid association and dissociation, an important feature for signalling hubs.

Regarding the effects of PTMs, the docking results on CaM with deiminated and quasi-deiminated MBP suggest a means by which deimination perturbs the MBP–CaM interaction at the atomic level, as has been observed experimentally (Libich et al., 2003a), lowering the protein's affinity for CaM. The phosphorylation of the CaM target seems to have only a minor effect on the strength of its interaction with CaM.

Concerning the formation of the MBP-SH3-domain complex, post-translational modification (phosphorylation or methylation) of the ligand did not cause any major inhibition of the binding beyond a somewhat less favourable interaction for phosphorylated peptides. The methylation of R104 may not be important in modulating the interaction with SH3-domains, however, in agreement with experimental results (Polverini et al., 2008). Although the conformation of the MBP peptide in the SH3-domain pocket changed significantly with Arg methylation or with phosphorylation, new interactions were able to substitute for those lost, in order to stabilize the complex. These multiple potential interaction sites on the SH3-domain thus allow an SH3-domain to bind promiscuously to a variety of ligands (Li, 2005). Although R104 methylation in 18.5 kDa MBP is associated with healthy myelin (Kim et al., 2003), its exact biological significance remains unknown.

An important consideration is that the whole protein, especially when membrane-associated, should be able to interact more strongly with the signalling proteins (cf., (Boggs and Rangaraj, 2000; Boggs et al., 2006; Boggs et al., 2005)). Moreover, the interaction could involve also other structured regions in addition to the MBP target segments, as is the case for the SH3-Nef protein complex (Arold et al., 1997; Lee et al., 1996) and as is confirmed by the observations that the classic MBP isoforms interact strongly with calmodulin, but Golli-MBP isoforms (with almost identical predicted calmodulin targets on them) interact weakly (Bamm et al., 2007; Kaur et al., 2003; Libich et al., 2003a; Libich et al., 2003b). Also the post-translational modifications at sites other than the receptor-binding segment might have an indirect effect on the interaction by altering MBP's conformation and exposing new sites.

Acknowledgements

A part of this work has been published in Journal of Structural Biology, vol. 148, Polverini, E., Boggs, J. M., Bates, I. R., Harauz, G., and Cavatorta, P., Electron paramagnetic resonance spectroscopy and molecular modelling of the interaction of myelin basic protein (MBP) with calmodulin (CaM) - diversity and conformational adaptability of MBP CaM-targets, page nos. 353-369, Copyright Elsevier (2004). Another part has been published in Biochemistry, vol. 47, Polverini, E., Rangaraj G., Libich D.S., Boggs J.M., and Harauz G., Binding of the proline-rich segment of myelin basic protein to SH3-domains - spectroscopic, microarray, and modelling studies of ligand conformation and effects of post-translational modifications, page nos. 267-282, Copyright American Chemical Society (2008).

I am grateful to Prof. George Harauz for his kind assistance and suggestions.

References

Afshar, M., Caves, L.S., Guimard, L., Hubbard, R.E., Calas, B., Grassy, G., and Haiech, J. (1994). Investigating the high affinity and low sequence specificity of calmodulin binding to its targets. *Journal of Molecular Biology*, 244, 554-571.

Arold, S., Franken, P., Strub, M.P., Hoh, F., Benichou, S., Benarous, R., and Dumas, C. (1997). The crystal structure of HIV-1 Nef protein bound to the Fyn kinase SH3 domain suggests a role for this complex in altered T cell receptor signaling. *Structure (London, England)*, 5, 1361-1372.

Babu, Y.S., Sack, J.S., Greenhough, T.J., Bugg, C.E., Means, A.R., and Cook, W.J. (1985). Three-dimensional structure of calmodulin. *Nature*, 315, 37-40.

Bamm, V.V., Ahmed, M.A.M., Ladizhansky, V., and Harauz, G. (2007). Purification and spectroscopic characterization of the recombinant BG21 isoform of murine golli myelin basic protein. *Journal of Neuroscience Research*, 85, 272-284.

Baryłko, B., and Dobrowolski, Z. (1984). Ca^{2+}-calmodulin-dependent regulation of F-actin-myelin basic protein interaction. *European Journal of Cell Biology*, 35, 327-335.

Bates, I.R., and G. Harauz, 2003. Molecular dynamics exposes alpha-helices in myelin basic protein. *Journal of Molecular Modeling*, 9, 290-297.

Bates, I.R., Boggs, J.M., Feix, J.B., and Harauz, G. (2003). Membrane-anchoring and charge effects in the interaction of myelin basic protein with lipid bilayers studied by site-directed spin labeling. *The Journal of Biological Chemistry*, 278, 29041-29047.

Bates, I.R., Feix, J.B., Boggs, J.M., and Harauz, G. (2004). An immunodominant epitope of myelin basic protein is an amphipathic alpha-helix. *The Journal of Biological Chemistry*, 279, 5757-5764.

Bates, I.R., Libich, D.S., Wood, D.D., Moscarello, M.A., and Harauz, G. (2002). An Arg/Lys-->Gln mutant of recombinant murine myelin basic protein as a mimic of the deiminated form implicated in multiple sclerosis. *Protein Expression and Purification*, 25, 330-341.

Berman, H.M., Westbrook, J., Feng, Z., Gilliland, G., Bhat, T.N., Weissig, H., Shindyalov, I.N., and Bourne, P.E. (2000). The Protein Data Bank. *Nucleic Acids Research*, 28, 235-242.

Bhattacharyya, R., and Chakrabarti, P. (2003). Stereospecific interactions of proline residues in protein structures and complexes. *Journal of Molecular Biology*, 331, 925-940.

Boggs, J.M. (2002). *Myelin basic protein*. Encyclopedia of Molecular Medicine. pp. 2174-2177, Wiley, New York, USA.

Boggs, J.M. (2006). Myelin basic protein: a multifunctional protein. *Cellular and Molecular Life Sciences*, 63, 1945-1961.

Boggs, J.M., and Rangaraj, G. (2000). Interaction of lipid-bound myelin basic protein with actin filaments and calmodulin. *Biochemistry*, 39, 7799-7806.

Boggs, J.M., and Rangaraj, G. (2004). Effect of phosphorylation of myelin basic protein on its interaction with actin in vitro. *Journal of Neurochemistry,* 90 (Suppl. 1), 73.

Boggs, J.M., and Wang, H. (2004). Co-clustering of galactosylceramide and membrane proteins in oligodendrocyte membranes on interaction with polyvalent carbohydrate and prevention by an intact cytoskeleton. *Journal of Neuroscience Research*, 76, 342-355.

Boggs, J.M., Rangaraj, G., Gao, W., and Heng, Y.M. (2006). Effect of phosphorylation of myelin basic protein by MAPK on its interactions with actin and actin binding to a lipid membrane in vitro. *Biochemistry*, 45, 391-401.

Boggs, J.M., Rangaraj, G., Heng, Y.M., Bates, I., and Harauz, G. (2003). Interaction of membrane-bound myelin basic protein with actin filaments in vitro. *Journal of Neurochemistry,* 85 (suppl. 1), 48.

Boggs, J.M., Rangaraj, G., Hill, C.M.D., Bates, I.R., Heng, Y.M., and Harauz, G. (2005). Effect of arginine loss in myelin basic protein, as occurs in its deiminated charge isoform, on mediation of actin polymerization and actin binding to a lipid membrane in vitro. *Biochemistry*, 44, 3524-3534.

Campagnoni, A.T., and Skoff, R.P. (2001). The pathobiology of myelin mutants reveal novel biological functions of the MBP and PLP genes. *Brain Pathology (Zurich, Switzerland)*, 11, 74-91.

Cesareni, G., Panni, S., Nardelli, G., and Castagnoli, L. (2002). Can we infer peptide recognition specificity mediated by SH3 domains? *FEBS Letters*, 513, 38-44.

Crivici, A., and Ikura, M. (1995). Molecular and structural basis of target recognition by calmodulin. *Annual Review of Biophysics and Biomolecular Structure*, 24, 85-116.

Dalgarno, D.C., Botfield, M.C., and Rickles, R.J. (1997). SH3 domains and drug design: ligands, structure, and biological function. *Biopolymers*, 43, 383-400.

DeBruin, L.S., Wood, D.D., Moscarello, M.A., and Harauz, G. (2002). *Interaction of myelin basic protein with SH3-domain containing proteins*. European Science Foundation Exploratory Workshop on Myelin Structure and its Role in Autoimmunity, Potenza, Italy.

DeBruin, L.S., Haines, J.D., Bienzle, D., and Harauz, G. (2006). Partitioning of myelin basic protein into membrane microdomains in a spontaneously demyelinating mouse model for multiple sclerosis. *Biochemistry and Cell Biology = Biochimie et Biologie Cellulaire*, 84, 993-1005.

DeBruin, L.S., Haines, J.D., Wellhauser, L.A., Radeva, G., Schonmann, V., Bienzle, D., and Harauz, G. (2005). Developmental partitioning of myelin basic protein into membrane microdomains. *Journal of Neuroscience Research*, 80, 211-225.

Deibler, G.E., Stone, A.L., and Kies, M.W. (1990). Role of phosphorylation in conformational adaptability of bovine myelin basic protein. *Proteins*, 7, 32-40.

Dobrowolski, Z., Osińska, H., Mossakowska, M., and Baryłko, B. (1986). Ca^{2+}-calmodulin-dependent polymerization of actin by myelin basic protein. *European Journal of Cell Biology*, 42, 17-26.

Dyer, C.A., Philibotte, T.M., Wolf, M.K., and Billings-Gagliardi, S. (1994). Myelin basic protein mediates extracellular signals that regulate microtubule stability in oligodendrocyte membrane sheets. *Journal of Neuroscience Research*, 39, 97-107.

Dyson, H.J., and Wright, P.E. (2005). Intrinsically unstructured proteins and their functions. *Nature Reviews. Molecular Cell Biology*, 6, 197-208.

Eidenmüller, J., Fath, T., Hellwig, A., Reed, J., Sontag, E., and Brandt, R. (2000). Structural and functional implications of tau hyperphosphorylation: information from phosphorylation-mimicking mutated tau proteins. *Biochemistry*, 39, 13166-13175.

Eidenmüller, J., Fath, T., Maas, T., Pool, M., Sontag, E., and Brandt, R. (2001). Phosphorylation-mimicking glutamate clusters in the proline-rich region are sufficient to simulate the functional deficiencies of hyperphosphorylated tau protein. *The Biochemical Journal*, 357, 759-767.

Friedman, A.R., Roberts, V.A., and Tainer, J.A. (1994). Predicting molecular interactions and inducible complementarity: fragment docking of Fab-peptide complexes. *Proteins*, 20, 15-24.

Gasteiger, J., and Marsili, M. (1980). Iterative partial equalisation of orbital electronegativity - a rapid access to atomic charges. *Tetrahedron*, 36, 3219-3228.

Harauz, G., and Musse, A.A. (2007). A tale of two citrullines--structural and functional aspects of myelin basic protein deimination in health and disease. *Neurochemical Research*, 32, 137-158.

Harauz, G., Ishiyama, N., Hill, C.M.D., Bates, I.R., Libich, D.S., and Farès, C. (2004). Myelin basic protein-diverse conformational states of an intrinsically unstructured protein and its roles in myelin assembly and multiple sclerosis. *Micron (Oxford, England)*, 35, 503-542.

Hill, C.M.D., Bates, I.R., White, G.F., Hallett, F.R., and Harauz, G. (2002). Effects of the osmolyte trimethylamine-*N*-oxide on conformation, self-association, and two-dimensional crystallization of myelin basic protein. *Journal of Structural Biology*, 139, 13-26.

Hill, C.M.D., and Harauz, G. (2005). Charge effects modulate actin assembly by classic myelin basic protein isoforms. *Biochemical and Biophysical Research Communications*, 329, 362-369.

Hill, C.M.D., Libich, D.S., and Harauz, G. (2005). Assembly of tubulin by classic myelin basic protein isoforms and regulation by post-translational modification. *Biochemistry*, 44, 16672-16683.

Hill, C.M.D., Haines, J.D., Antler, C.E., Bates, I.R., Libich, D.S., and Harauz, G. (2003). Terminal deletion mutants of myelin basic protein: new insights into self-association and phospholipid interactions. *Micron (Oxford, England)*, 34, 25-37.

Huey, R., and Morris, G.M. (2007). *Using AutoDock4 with AutoDockTools: A Tutorial*. The Scripps Research Institute, La Jolla, California.

Huey, R., Morris, G.M., Olson, A.J., and Goodsell, D.S. (2007). A semiempirical free energy force field with charge-based desolvation. *Journal of Computational Chemistry*, 28, 1145-1152.

Imparl, J.M., Senshu, T., and Graves, D.J. (1995). Studies of calcineurin-calmodulin interaction: probing the role of arginine residues using peptidylarginine deiminase. *Archives of Biochemistry and Biophysics*, 318, 370-377.

Jia, C.Y.H., Nie, J., Wu, C., Li, C., and Li, S.S.C. (2005). Novel Src homology 3 domain-binding motifs identified from proteomic screen of a Pro-rich region. *Molecular and Cellular Proteomics*, 4, 1155-1166.

Kaur, J., Libich, D.S., Campagnoni, C.W., Wood, D.D., Moscarello, M.A., Campagnoni, A.T., and Harauz, G. (2003). Expression and properties of the recombinant murine Golli-myelin basic protein isoform J37. *Journal of Neuroscience Research*, 71, 777-784.

Kim, H.J., DiBernardo, A.B., Sloane, J.A., Rasband, M.N., Solomon, D., Kosaras, B., Kwak, S.P., and Vartanian, T.K. (2006). WAVE1 is required for oligodendrocyte morphogenesis and normal CNS myelination. *The Journal of Neuroscience*, 26, 5849-5859.

Kim, J.K., Mastronardi, F.G., Wood, D.D., Lubman, D.M., Zand, R., and Moscarello, M.A. (2003). Multiple sclerosis: an important role for post-translational modifications of myelin basic protein in pathogenesis. *Molecular and Cellular Proteomics*, 2, 453-462.

Krämer, E.M., Klein, C., Koch, T., Boytinck, M., and Trotter, J. (1999). Compartmentation of Fyn kinase with glycosylphosphatidylinositol-anchored molecules in oligodendrocytes facilitates kinase activation during myelination. *The Journal of Biological Chemistry*, 274, 29042-29049.

Kuboniwa, H., Tjandra, N., Grzesiek, S., Ren, H., Klee, C.B., and Bax, A. (1995). Solution structure of calcium-free calmodulin. *Nature Structural Biology*, 2, 768-776.

Lee, C.H., Saksela, K., Mirza, U.A., Chait, B.T., and Kuriyan, J. (1996). Crystal structure of the conserved core of HIV-1 Nef complexed with a Src family SH3 domain. *Cell*, 85, 931-942.

Li, S.S.C. (2005). Specificity and versatility of SH3 and other proline-recognition domains: structural basis and implications for cellular signal transduction. *The Biochemical Journal*, 390, 641-653.

Li, Y.L., Li, H.M., Martin, R., and Mariuzza, R.A. (2000). Structural basis for the binding of an immunodominant peptide from myelin basic protein in different registers by two HLA-DR2 proteins. *Journal of Molecular Biology*, 304, 177-188.

Libich, D.S., and Harauz, G. (2002a). Interactions of the 18.5 kDa isoform of myelin basic protein with Ca^{2+}-calmodulin: in vitro studies using gel shift assays. *Molecular and Cellular Biochemistry*, 241, 45-52.

Libich, D.S., and Harauz, G. (2002b). Interactions of the 18.5-kDa isoform of myelin basic protein with Ca(2+)-calmodulin: in vitro studies using fluorescence microscopy and spectroscopy. *Biochemistry and Cell Biology = Biochimie et Biologie Cellulaire*, 80, 395-406.

Libich, D.S., Hill, C.M.D., Haines, J.D., and Harauz, G. (2003a). Myelin basic protein has multiple calmodulin-binding sites. *Biochemical and Biophysical Research Communications*, 308, 313-319.

Libich, D.S., Robertson, V.J., Monette, M.M., and Harauz, G. (2004). Backbone resonance assignments of the 18.5 kDa isoform of murine myelin basic protein (MBP). *Journal of Biomolecular NMR*, 29, 545-546.

Libich, D.S., Hill, C.M.D., Bates, I.R., Hallett, F.R., Armstrong, S., Siemiarczuk, A., and Harauz, G. (2003b). Interaction of the 18.5-kD isoform of myelin basic protein with Ca^{2+}-calmodulin: effects of deimination assessed by intrinsic Trp fluorescence spectroscopy, dynamic light scattering, and circular dichroism. *Protein Science*, 12, 1507-1521.

Lunney, E.A., Hagen, S.E., Domagala, J.M., Humblet, C., Kosinski, J., Tait, B.D., Warmus, J.S., Wilson, M., Ferguson, D., and Hupe, D. (1994). A novel nonpeptide HIV-1 protease inhibitor: elucidation of the binding mode and its application in the design of related analogs. *Journal of Medicinal Chemistry*, 37, 2664-2677.

Lyne, P.D. (2002). Structure-based virtual screening: an overview. *Drug Discovery Today*, 7, 1047-1055.

Macias, M.J., Wiesner, S., and Sudol, M. (2002). WW and SH3 domains, two different scaffolds to recognize proline-rich ligands. *FEBS Letters*, 513, 30-37.

Mastronardi, F.G., and Moscarello, M.A. (2005). Molecules affecting myelin stability: a novel hypothesis regarding the pathogenesis of multiple sclerosis. *Journal of Neuroscience Research*, 80, 301-308.

Mayer, B.J. (2001). SH3 domains: complexity in moderation. *Journal of Cell Science*, 114, 1253-1263.

Meador, W.E., Means, A.R., and Quiocho, F.A. (1992). Target enzyme recognition by calmodulin: 2.4 A structure of a calmodulin-peptide complex. *Science (New York, N.Y.)*, 257, 1251-1255.

Morris, G.M., Goodsell, D.S., Halliday, R.S., Huey, R., Hart, W.E., Belew, R.K., and Olson, A.J. (1998). Automated docking using a Lamarckian genetic algorithm and an empirical binding free energy function. *Journal of Computational Chemistry*, 19, 1639-1662.

Moscarello, M.A. (1997). Myelin basic protein, the "executive" molecule of the myelin membrane, p. 13-25, in: B. H. J. Juurlink, et al., Eds.). *Cell Biology and Pathology of Myelin: Evolving Biological Concepts and Therapeutic Approaches*, Plenum Press, New York.

Moscarello, M.A., Mastronardi, F.G., and Wood, D.D. (2007). The role of citrullinated proteins suggests a novel mechanism in the pathogenesis of multiple sclerosis. *Neurochemical Research*, 32, 251-256.

Musacchio, A., Saraste, M., and Wilmanns, M. (1994). High-resolution crystal-structures of tyrosine kinase SH3 domains complexed with proline-rich peptides. *Nature Structural Biology*, 1, 546-551.

Musse, A.A., Boggs, J.M., and Harauz, G. (2006). Deimination of membrane-bound myelin basic protein in multiple sclerosis exposes an immunodominant epitope. *Proceedings of the National Academy of Sciences of the United States of America*, 103, 4422-4427.

Noble, M.E., Musacchio, A., Saraste, M., Courtneidge, S.A., and Wierenga, R.K. (1993). Crystal structure of the SH3 domain in human Fyn; comparison of the three-dimensional structures of SH3 domains in tyrosine kinases and spectrin. *The EMBO Journal*, 12, 2617-2624.

Osawa, M., Tokumitsu, H., Swindells, M.B., Kurihara, H., Orita, M., Shibanuma, T., Furuya, T., and Ikura, M. (1999). A novel target recognition revealed by calmodulin in complex with Ca^{2+}-calmodulin-dependent kinase kinase. *Nature Structural Biology*, 6, 819-824.

Osterberg, F., Morris, G.M., Sanner, M.F., Olson, A.J., and Goodsell, D.S. (2002). Automated docking to multiple target structures: incorporation of protein mobility and structural water heterogeneity in AutoDock. *Proteins*, 46, 34-40.

Osterhout, D.J., Wolven, A., Wolf, R.M., Resh, M.D., and Chao, M.V. (1999). Morphological differentiation of oligodendrocytes requires activation of Fyn tyrosine kinase. *The Journal of Cell Biology*, 145, 1209-1218.

Polverini, E., Rangaraj, G., Libich, D.S., Boggs, J.M., and Harauz, G. (2008). Binding of the proline-rich segment of myelin basic protein to SH3-domains - spectroscopic, microarray, and modelling studies of ligand conformation and effects of posttranslational modifications. *Biochemistry*, 47, 267-282.

Polverini, E., Boggs, J.M., Bates, I.R., Harauz, G., and Cavatorta, P. (2004). Electron paramagnetic resonance spectroscopy and molecular modelling of the interaction of myelin basic protein (MBP) with calmodulin (CaM) - diversity and conformational adaptability of MBP CaM-targets. *Journal of Structural Biology*, 148, 353-369.

Ramwani, J.J., Epand, R.M., and Moscarello, M.A. (1989). Secondary structure of charge isomers of myelin basic protein before and after phosphorylation. *Biochemistry*, 28, 6538-6543.

Receveur-Bréchot, V., Bourhis, J.M., Uversky, V.N., Canard, B., and Longhi, S. (2006). Assessing protein disorder and induced folding. *Proteins*, 62, 24-45.

Seet, B.T., Dikic, I., Zhou, M.M., and Pawson, T. (2006). Reading protein modifications with interaction domains. *Nature Reviews. Molecular Cell Biology*, 7, 473-483.

Seiwa, C., Sugiyama, I., Yagi, T., Iguchi, T., and Asou, H. (2000). Fyn tyrosine kinase participates in the compact myelin sheath formation in the central nervous system. *Neuroscience Research*, 37, 21-31.

Seiwa, C., amamoto, M. Y, Tanaka, K., Fukutake, M., Ueki, T., Takeda, S., Sakai, R., Ishige, A., Watanabe, K., Akita, M., Yagi, T., Tanaka, K., and Asou, H. (2007). Restoration of FcRgamma/Fyn signaling repairs central nervous system demyelination. *Journal of Neuroscience Research*, 85, 954-966.

Smith, R. (1992). The basic protein of CNS myelin: its structure and ligand binding. *Journal of Neurochemistry*, 59, 1589-1608.

Sperber, B.R., Boyle-Walsh, E.A., Engleka, M.J., Gadue, P., Peterson, A.C., Stein, P.L., Scherer, S.S., and McMorris, F.A. (2001). A unique role for Fyn in CNS myelination. *The Journal of Neuroscience*, 21, 2039-2047.

Stoddard, B.L., and Koshland, D.E. Jr. (1992). Prediction of the structure of a receptor-protein complex using a binary docking method. *Nature*, 358, 774-776.

Teodoro, M.L.,. Phillips, G.N.Jr., and Kavraki, L.E. (2003). Understanding protein flexibility through dimensionality reduction. Journal of Computational Biology, *A Journal of Computational Molecular Cell Biology*, 10, 617-634.

Tompa, P. (2005). The interplay between structure and function in intrinsically unstructured proteins. *FEBS Letters*, 579, 3346-3354.

Umemori, H., Sato, S., Yagi, T., Aizawa, S., and Yamamoto, T. (1994). Initial events of myelination involve Fyn tyrosine kinase signalling. *Nature*, 367, 572-576.

Umemori, H., Kadowaki, Y., Hirosawa, K., Yoshida, Y., Hironaka, K., Okano, H., and Yamamoto, T. (1999). Stimulation of myelin basic protein gene transcription by Fyn tyrosine kinase for myelination. *The Journal of Neuroscience*, 19, 1393-1397.

Uversky, V.N., Gillespie, J.R., and Fink, A.L. (2000). Why are "natively unfolded" proteins unstructured under physiologic conditions? *Proteins*, 41, 415-427.

Vogel, H.J., and Zhang, M. (1995). Protein engineering and NMR studies of calmodulin. *Molecular and Cellular Biochemistry*, 149-150, 3-15.

Weiner, S.J., Kollman, P.A., Case, D.A., Chandra Singh, U., Ghio, C., Alagona, G., Profeta, S.Jr., and Weiner, P. (1984). A new force field for molecular mechanical simulation of nucleic acids and proteins. *Journal of the American Chemical Society*, 106, 765-784.

Whitaker, J.N., Moscarello, M.A., Herman, P.K., Epand, R.M., and Surewicz, W.K. (1990). Conformational correlates of the epitopes of human myelin basic protein peptide 80-89. *Journal of Neurochemistry*, 55, 568-576.

Yamauchi, E., Nakatsu, T., Matsubara, M., Kato, H., and Taniguchi, H. (2003). Crystal structure of a MARCKS peptide containing the calmodulin-binding domain in complex with Ca^{2+}-calmodulin. *Nature Structural Biology*, 10, 226-231.

Yang, C., Jas, G.S., and Kuczera, K. (2004). Structure, dynamics and interaction with kinase targets: computer simulations of calmodulin. *Biochimica Et Biophysica Acta*, 1697, 289-300.

Yang, X.J. (2005). Multisite protein modification and intramolecular signaling. *Oncogene*, 24, 1653-1662.

Yap, K.L., Kim, J., Truong, K., Sherman, M., Yuan, T., and Ikura, M. (2000). Calmodulin target database. *Journal of Structural and Functional Genomics*, 1, 8-14.

In: Myelin Basic Protein
Editor: Joan M. Boggs

ISBN: 978-1-60456-699-4

Chapter X

STRUCTURE AND DYNAMICS OF THE MYELIN BASIC PROTEIN FAMILY BY SOLUTION AND SOLID-STATE NMR

***George Harauz*[1,3*] *and Vladimir Ladizhansky*[2,3]**

ABSTRACT

The myelin basic protein (MBP) family has a rich conformational and functional repertoire defined by its environment, degree of post-translational modification, and association with myriad other proteins and ligands. Nuclear magnetic resonance (NMR) spectroscopy is a powerful tool for deriving local and global structural detail, and dynamics, of biological macromolecules. We review here the basic principles of the methodology, and the prior application of NMR spectroscopy to studies of the classic 18.5 kDa MBP isoform. The current work from our group, using emerging solution and solid-state NMR techniques to probe the conformational choreography and multifunctionality of both Golli- and classic MBP isoforms, is described in detail. A fundamental challenge in these investigations is reconstituting the MBP isoform of

[1] Department of Molecular and Cellular Biology, University of Guelph, Ontario, N1G 2W1, Canada.
[2] Department of Physics, University of Guelph, Ontario, N1G 2W1, Canada.
[3] Biophysics Interdepartmental Group, University of Guelph, Guelph, Ontario, Canada.
[*] Telephone: 1-519-824-4120, Ext. 52535; Fax: 1-519-837-1802; E-mail: gharauz@uoguelph.ca

interest into an environment that is both physiologically relevant, and experimentally tractable to NMR spectroscopy.

ABBREVIATIONS

C1-C8 - MBP charge isomers, or components, 1 to 8; CPMAS – cross-polarization magic angle spinning; DPC-d_{38} – perdeuterated dodecylphosphocholine; Golli - genes of oligodendrocyte lineage; GIP - Golli-interacting protein; HSQC - heteronuclear single quantum coherence; IDP - intrinsically disordered protein; LIM - Lin-11/Isl-1/Mec-3; MAS - magic angle spinning; MBP - myelin basic protein; NMR - nuclear magnetic resonance; *NOE* - nuclear Overhauser effect; NOESY - nuclear Overhauser effect spectroscopy; ppm - parts per million; rmBG21 or rmMBP - recombinant murine Golli-MBP isoform BG21, or 18.5 kDa classic MBP isoform, respectively; SH3 - Src homology 3; TFE-d_2 - perdeuterated trifluoroethanol;

INTRODUCTION

Myelin Basic Protein – an intrinsically disordered protein of the central nervous system

The myelin basic protein family is produced by differential splicing of a single mRNA transcript arising from one of three transcription start sites of the gene complex called Golli (Genes of OLigodendrocyte LIneage) [1-4] (Figure 1; see also Chapter I).

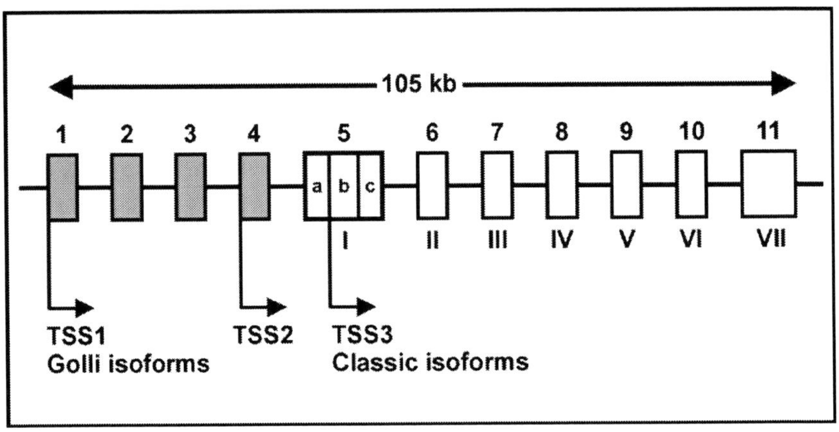

Figure 1. Genetic origin of myelin basic protein splice isoforms. The murine Golli (genes of oligodendrocyte lineage) genetic unit encompasses 11 exons and 3 transcription start sites (labelled TSS). The "classic" MBP exons are labelled with Roman numerals as per popular (albeit dated) convention, and are derived primarily from TSS3 in mature myelin; the Golli exon numbering is shown with Sanskrit/Arabic numerals. The Golli-MBP isoforms are derived primarily from TSS1 in early development, and comprise some classic protein segments in addition to the Golli-specific ones (*viz.*, the first 133 amino acids). The human Golli genetic unit has a similar structure. Figure courtesy of Dr. Christopher Hill, based on reference [3].

Various Golli proteins are produced in developing myelin from transcription start site 1, and are called BG21, J37, and TP8 in the mouse [1, 3-5]. The Golli-MBP isoforms translocate between the nucleus and cellular processes, suggesting multiple roles in modulating the activity of signal transduction pathways in myelin formation or in T-cell activation, or in gene regulation [3-12]. In mammals, the classic MBP isoforms arise from transcription start site 3, and represent splice variants ranging in molecular mass from 14 kDa to 21.5 kDa. The major classic MBP isoform is 18.5 kDa in adult central nervous system myelin, which is peripherally membrane-associated and found in cell processes and compact myelin [13-16], where it functions as a membrane adhesive and microdomain modeller [17-21]. This isoform is extremely positively-charged (+19 at neutral pH), and exists as a number of charge isomers due to myriad, combinatorial post-translational modifications, representing further diversification [13-16, 22] (see Chapters II and III). Whereas genetic modifications in other myelin proteins are associated with a variety of dysmyelinating diseases [23], post-translational modifications of MBP may be involved in the pathophysiology of multiple sclerosis, a human demyelinating disease [22, 24-26].

A fundamental tenet of structural molecular biology has been that a protein's three-dimensional structure determines its function, a concept derived from the successes of X-ray crystallography since the 1950s, and the correlation between the activity and (relatively) rigid structure of enzymes, with their lock-and-key mechanisms of substrate recognition and modification [27]. Despite many attempts, it has not been possible to crystallise MBP with a high degree of order [15, 28-32], except for peptide fragments bound to larger immune system complexes [33-39] (see Chapter V). The reason is that the eukaryotic cellular proteome comprises many intrinsically disordered proteins (IDPs), with properties similar to MBP, whose lack of globular (folded) structure may be directly part of their inherent function (e.g., springs or linkers – entropic chains), or is required for recognition and binding [40-61]. These proteins are multifunctional, and their conformations are highly dependent on their environment or association with some ligand. It appears that IDPs often function as hubs or linkers in protein interaction networks [62, 63], or as molecular switches in transcriptional and translational control [64-67]. Many IDPs are important in signal transduction, where their binding affinities can be regulated by post-translational modifications such as phosphorylation [68-71]; *cf.*, [66]. Their intermolecular associations may involve pre-formed structural elements such as α-helices, and/or an induced fit, i.e., coupled folding and binding [53, 64-66, 69, 72-85].

Whereas some of the classic MBP isoforms are peripherally membrane-associated, all forms of MBP are IDPs. The classic isoforms, the ones that have been most widely studied so far, have long been known to be highly flexible [86, 87] and are extended, extreme polycations whose net charge is reduced by combinatorial post-translational modifications representing a molecular barcode. We may posit these proteins' various functions to be in both the entropic chain category (as an adhesive of the cytoplasmic leaflets of the oligodendrocyte membranes, potentially defining various microdomains of distinct lipid composition, and linked also to proteolipid and other proteins in compact myelin, and the underlying cytoskeleton in the paranodal regions of myelin), and in molecular recognition and networking in signalling pathways (potentially with Src homology 3 (SH3)-domain containing proteins) [16, 88, 89] (see Chapter IX). Thus, we must consider the idea of "structure" in the MBP family of proteins in a variety of environments, and/or when they are interacting with other biological molecules or cofactors: lipids, other proteins, small molecules, and ions (e.g., Zn^{2+}). In addition, the various post-translational modifications may also have significant structural effects that could modulate its diverse interactions. Perturbations of the structure-function networks of IDPs have been correlated with human illnesses such as Parkinson's and cardiovascular diseases [61, 90, 91], and have been hypothesised by us to be operative in multiple sclerosis [88, 92-94]. In order to assess these relationships fully for the many MBP isoforms and modified components, complementary biophysical (non-crystallographic) strategies are required [15, 53, 58].

Principles of nuclear magnetic resonance (NMR) spectroscopy

NMR spectroscopy is a powerful tool that can provide information on a protein's structure and dynamics in a site-specific manner [95-99]. The technique relies on the fact that many nuclei have a non-zero magnetic moment associated with spin quantum number, and can interact with one another and with their surroundings, yielding many correlations that can be measured [100]. In biomolecular NMR, the primary nuclei of interest are ^{1}H (natural abundance 99.98%), ^{13}C (natural abundance 1.11%), ^{15}N (natural abundance 0.37%), and ^{31}P (natural abundance 100%). More exotic nuclei such as ^{19}F (natural abundance 100%) are also occasionally of interest [101], and deuterated (^{2}H, natural abundance 0.016%) molecules are essential in many experiments to "silence" specific proton resonances [102]. Since the ^{15}N and ^{13}C isotopes are of low natural abundance, it is almost a routine matter in modern structural NMR to use isotopic enrichment to incorporate ^{13}C and ^{15}N labels into the biomolecules. This labelling is accomplished most efficiently by expressing a protein in *E. coli* grown in a minimal medium with $^{15}NH_4Cl$ as the only nitrogen source, and with ^{13}C-glucose as the only carbon source [98, 103]. Such an expression strategy results in a sample with uniform incorporation of ^{13}C and ^{15}N nuclei. Cell-free expression systems are currently costly, but will be of increasing interest in the future [104].

Each of the widely-used stable isotopes ^{1}H, ^{13}C, and ^{15}N can be thought of as a magnetic dipole aligned along the field when placed in a strong external magnetic field of a superconducting magnet. These dipoles can absorb energy at specific frequencies in the radio-frequency spectrum, and change their orientation with respect to the magnetic field. Thus, the NMR spectrum can be thought of as a collection of individual spectral lines, each with its own frequency. The frequency positions of the lines, or isotropic chemical shifts, are defined by the local chemical environment (e.g., in CH and CH_2 groups, both protons and carbons would have very distinct chemical shifts), by the secondary structure, and by the amino acid type. Isotropic chemical shifts (referred to simply as chemical shifts in the following discussion) constitute primary NMR observables, and allow *site-specific* studies of a molecule.

Two typical solution ^{1}H NMR spectra are shown in Figure 2, one of an IDP (the Golli-MBP isoform BG21 in Figure 2A), and one of a canonically-folded globular protein (the B1 domain of protein L [105] in Figure 2B). In proteins with a well-defined tertiary fold, and with both α-helical and β-sheet secondary structural elements, the amide proton resonances cover a wide range of chemical shifts from 6.5-9.5 ppm, whereas in α-helical and especially

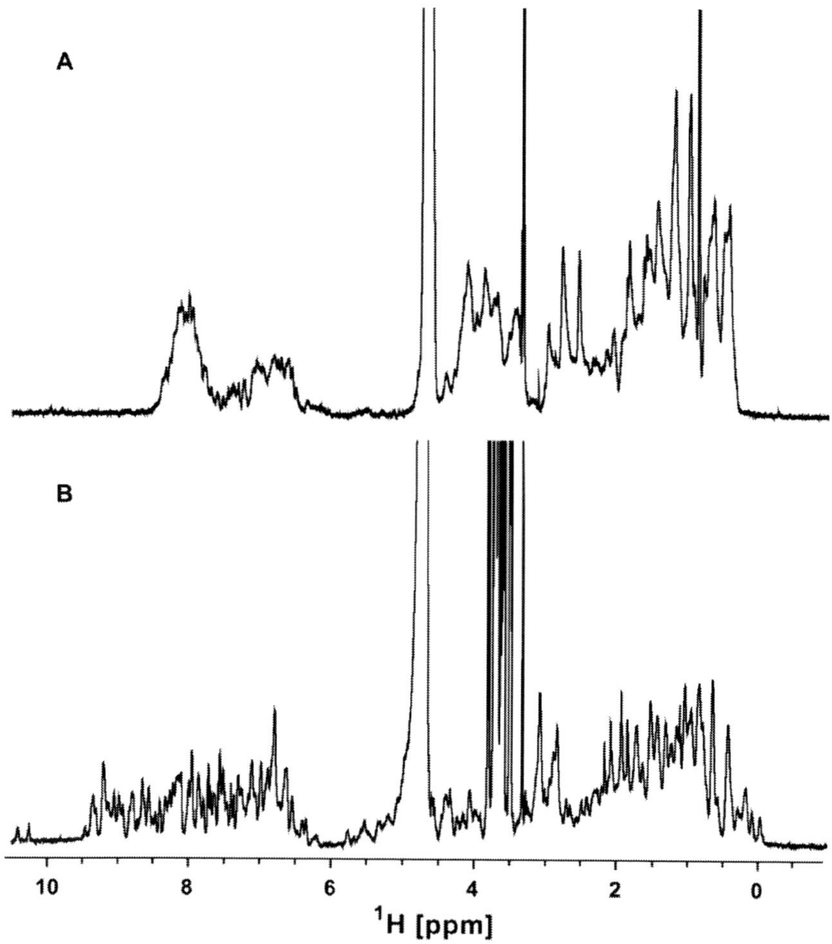

Figure 2. One-dimensional proton spectra of (A) recombinant murine Golli-MBP BG21 (rmBG21), an intrinsically disordered protein, and (B) the B1 domain of protein L, a globular protein (105). Figure courtesy of Mr. Mumdooh Ahmed.

unstructured proteins, the dispersion is much smaller. This pattern is further highlighted in Figure 3, where a ^{15}N-^{1}H heteronuclear single quantum coherence (HSQC) spectrum for the intrinsically unstructured Golli-isoform BG21, is shown.

For comparison, the HSQC spectrum for the B1 domain of protein L is shown in Figure 4.

These ^{15}N-^{1}H HSQC spectra show the correlations between amide protons and their nitrogens. For intrinsically unstructured BG21, the chemical shifts of amide protons are primarily in the range of 7.8-8.5 ppm, whereas for protein L-B1 they span a larger range of 7.2-9.4 ppm. Nuclear spins can also interact with each other, either through-bond (scalar or J-coupling), or through-space (dipolar interaction).

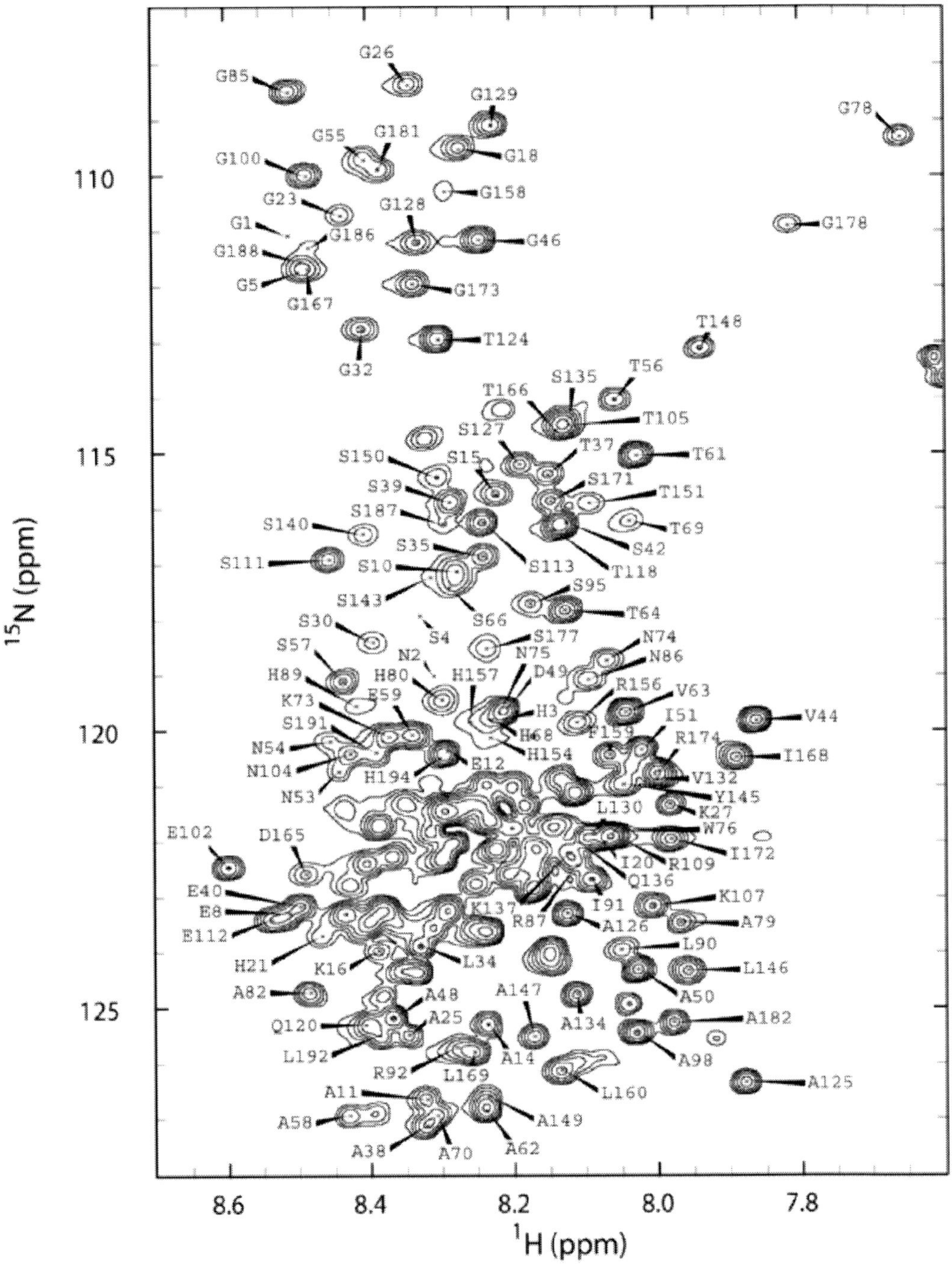

Figure 3. The ^{1}H-^{15}N heteronuclear single quantum coherence (HSQC) spectrum of the recombinant murine Golli-MBP isoform rmBG21, with backbone assignments of most resolved non-prolyl resonances. Figure adapted with permission from reference (233).

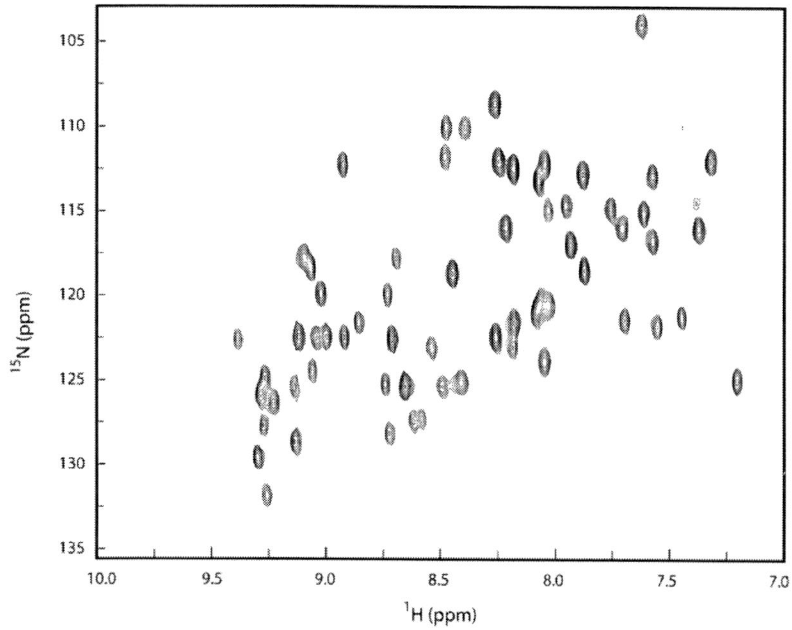

Figure 4. The ^{1}H-^{15}N heteronuclear single quantum coherence (HSQC) spectrum of the the B1 domain of protein L. Figure courtesy of Mr. Mumdooh Ahmed.

Solution versus solid-state NMR spectroscopy

NMR spectroscopy can be applied to proteins either in solution or in the solid state. One fundamental difference between solution and solid-state samples, which has a direct impact on the complexity of NMR spectra and on the ease of NMR measurements, is the time scale and the extent of molecular motions. In solution, molecules tumble rapidly and isotropically, with overall tumbling times on the order of 10^{-9}-10^{-8} s for proteins up to 20-25 kDa. This motion eliminates line-broadening mechanisms associated with the effects of anisotropic interactions such as chemical shift anisotropy and dipolar couplings, and results in high-resolution spectra of all nuclei: ^{1}H, ^{15}N, or ^{13}C. Proteins of larger size (or protein complexes) have slower tumbling times, and shorter relaxation times, as a consequence.

In contrast, the molecular motions are either slow and/or anisotropic, or absent in the solid state, and the magic angle spinning (MAS) technique [106, 107] is typically employed to get high-resolution spectra of ^{13}C or ^{15}N, but not of protons. The lack of the proton dimension for detection in solid-state NMR is associated with a significant loss in sensitivity, which must be overcome before solid-state NMR can become a widely available and versatile structural tool. Despite this current sensitivity obstacle, solid-state NMR has no *fundamental* size limitation, and can be used to study biological molecules and complexes of any size.

In this last regard, solid-state NMR has a tremendous advantage over solution NMR, and can provide unique structural information for membrane proteins, large protein assemblies, and noncrystalline amyloid fibrils [108-122[. Solid-state NMR of oriented samples has been applied to obtaining structures of small membrane-associated proteins (< 10 kDa), particularly the orientations of α-helices with respect to the membrane. Recent landmark studies [123-126] have demonstrated that magic angle spinning (MAS) solid-state NMR of non-oriented samples can also be used to determine complete three-dimensional arrangements of insoluble and non-crystalline peptides and proteins. In particular, the technique has yielded assignments for solid-state samples of uniformly $^{13}C^{15}N$-labelled human ubiquitin [127-130], bovine pancreatic trypsin inhibitor (BPTI) [131], for the α-spectrin SH3-domain [132, 133], and for a number of other globular proteins. Complete three-dimensional structures have been determined for a tripeptide Met-Leu-Phe [125], an amyloid-forming fragment of transthyretin [126], and for the α-spectrin SH3-domain [123]. Various groups are applying magic angle spinning solid-state NMR to studies of larger membrane proteins [101, 109, 134-141]. The purpose of these investigations is to achieve complete three-dimensional structure determinations in physiologically relevant environments.

Although solid-state NMR measurements are currently limited by low sensitivity, as indicated above, this limitation can be overcome in the future through (i) the introduction of proton detection in the solid state, (ii) improvement of the instrumentation (e.g., cryogenic magic angle spinning probes are being developed by a number of academic and industrial research groups across the world), and (iii) through the introduction of signal enhancement techniques, such as Dynamic Nuclear Polarisation [142].

Fundamental measurements and experiments in NMR spectroscopy

The identification of individual resonances (spectroscopic assignments) is the essential prerequisite for a biomolecular NMR study of a protein (Figure 2-4). In solution NMR, the assignments are usually noted on an HSQC spectrum, introduced above in Figures 3 and 4, which shows the correlations between amide protons and their nitrogens. Then, a multitude of further analyses can proceed. Chemical shifts are sensitive to the microenvironment of the particular nucleus, and have been correlated with secondary structure types [143, 144]. Conformational constraints can be used to generate an ensemble of three-dimensional structures that match them [96, 145, 146]; in solution NMR, these are usually *NOE* data, whereas in solids, dipolar recoupling approaches provide internuclear distances [147]. For practical reasons, structures of soluble proteins up to 20-25 kDa in molecular mass, that tumble quickly in solution, can be almost routinely obtained in this way. So far, structure determination by solid-state NMR has been demonstrated on small (less than 10 kDa) microcrystalline proteins [123]. The NMR relaxation parameters are also sensitive to molecular motions on various time scales, and NMR spectroscopy can shed insight into local and global motions due to bond rearrangements, folding processes, and other conformational exchanges [148-154]. This ability is particularly useful for IDPs [58, 69, 73, 85, 155-164].

Literature review – NMR of MBP in the 20th century

We review here some selected themes in the considerable literature in which NMR has been applied to study MBP (as opposed to myelin imaging by NMR e.g., [165, 166]. First of all, it was realised from the outset, in the 1970s, that MBP was conformationally flexible, and not only sensitive to the type of environment, but that changes in environment such as pH or hydrophobicity (organic solvents) could induce some degree of intramolecular folding [167-172]. Subsequent studies of the protein or its specific peptide fragments in aqueous solution were focussed on searching for local intramolecular interactions representative of a particular structure type, particularly reverse turns or bends [173-196]. The conformational changes in the protein or peptide fragments, due to interaction with detergents or lipids, were also monitored, as was the nature of the association with these amphiphiles [169, 197-224].

Almost all of these earlier investigations were performed on heterogeneous mixtures of classic 18.5 kDa isoform charge components, extracted from brain of various species (human, bovine, rabbit, porcine, guinea pig), and only rarely on purified charge components (denoted C1, C2, C3, C4, C5, or C8) [194, 208, 224-226]. These components are derived, of course, from combinatorial post-translational modifications of the protein [15, 22]. The methylated arginyl residue was probed early on by ^{1}H spectroscopy to preclude an intramolecular interaction with neighbouring phenylalanyl residues, and to demonstrate a conformational change upon lipid-association [174, 198], whereas ^{31}P NMR spectroscopy was used to confirm the stoichiometry of one phosphorylation event on residue Thr98 (human 18.5 kDa numbering, one of the two mitogen-activated protein kinase sites) in the 17.2 and 18.5 kDa human MBP classic isoforms [194]. Persaud et al. performed an *in vitro* glycosylation of threonyl residues 95 and 98 (the two mitogen-activated protein kinase phosphorylation sites), and showed by one-dimensional ^{1}H NMR spectroscopy that the modification was sequential, which they interpreted as indicating that the protein was highly structured in this region [227]. One-dimensional ^{1}H NMR spectroscopy was also used to probe glutaminyl deamidation, methionyl oxidation, and arginyl methylation in purified charge components C1-C3 of bovine MBP [225, 226]. The only protein-protein interactions hitherto investigated by NMR have been monoclonal antibody binding to MBP, primarily to measure the dissociation constant, and to show that similar results were obtained with the whole protein as with a peptide fragment comprising the epitope [228].

The Devil is in the details! NMR spectroscopy provides highly-detailed, site-specific structural information that for IDPs is highly-dependent on the protein sample and its preparation conditions. It is devilish, then, to glean what the literature is telling us. In summary, the 20th century NMR studies established that MBP was conformationally flexible, being highly extended in aqueous solution, and that it interacted strongly (by electrostatic and hydrophobic interactions) with membranes, upon which association it gained a considerable proportion of α-helical structure. These investigations were performed most often on unlabelled protein, or on protein that was isotopically-labelled only at specific sites [200, 204], using generally low magnetic fields (100-400 MHz or so, the levels available at the time). The one-dimensional experiments were the most accessible, with later multi-dimensional studies performed on peptide fragments. The intrinsically disordered nature of MBP was thus characterised spectroscopically well before it was identified as part of the intrinsically disordered proteome (as well as being a peripheral membrane protein) at the dawn of the new millenium [31, 43, 44].

Since NMR spectroscopy is a rapidly developing tool, radically different NMR approaches are now available in this, the 21st century. Primarily, there have been substantial recent advances in instrumentation and in methodology of multi-dimensional NMR techniques, both in solution and in the solid state. Ultra-high magnetic fields have enabled studies of large proteins and protein complexes. Most importantly, NMR methodology can also be applied to unfolded (including intrinsically disordered) proteins [155, 159, 160], unlike X-ray or electron crystallography. In this chapter, we shall focus on our own groups' initiated NMR investigations of structure and dynamics of proteins from the myelin family [15, 229-234]. We have, to begin, introduced the use of recombinant murine 18.5 kDa MBP (rmMBP) [235, 236], as well as recombinant forms of early developmental Golli isoforms [231, 237], enabling their uniform labelling by ^{13}C and ^{15}N for structural NMR studies.

Solution NMR spectroscopy of Golli-MBP isoform BG21 – spectral assignment and dynamics

The murine Golli-MBP isoforms J37 and BG21 have an amino-terminal, Golli-specific sequence of 133 residues (the first 3 exons of the Golli gene derived from transcription start site 1), and a carboxy-terminal sequence comprising classic MBP exons [1, 2] (Figure 1). We have chosen to study the BG21 isoform first by solution NMR spectroscopy since it is considerably smaller than the J37 isoform (21.6 kDa *vs*. 27.6 kDa for the recombinant forms, respectively) [231, 237], and spectral analyses are thus proportionally easier. The eventual goal is to study the interaction (and hypothesised induced-folding) of BG21 with Golli-interacting protein (GIP) [11]. As described in Chapter I by Campagnoni and Campagnoni, the GIP is found in the nuclei of neuronal cells; it is a phosphatase that acts on the carboxy-terminal domain of the largest subunit of RNA polymerase II, and may function in silencing neuronal genes. It has been suggested that GIP may be part of a transcriptional regulatory complex, modulating genes regulated by LIM family members (Lin-11/Isl-1/Mec-3, homeodomains with a common Zn^{2+}-binding motif). Thus, BG21 appears to be a linker protein that functions as part of a larger interaction network, representing an important biological process and a challenging structural biological system.

The Golli-MBP isoforms are not as highly-charged as the classic ones, but are still IDPs [15, 31]. In order to be able to study BG21 and its interactions with GIP by NMR, we first have had to optimise protein overexpression and purification, and characterise these proteins using optical spectroscopic approaches [231]. The recombinant form of murine BG21 (rmBG21) was expressed in *E. coli*, and isolated to 96% purity via metal chelation chromatography, with yields of 6-8 mg protein per liter of culture in minimal M9 media. Circular dichroic spectroscopy showed that this isoform gained α-helical structure in the presence of lipids, like classic 18.5 kDa MBP; these segments may represent putative recognition motifs for binding other proteins, many of which are probably yet to be discovered. One- and two-dimensional proton NMR spectroscopy confirmed this protein to be intrinsically disordered in aqueous buffer, in contrast to a "nicely-folded" globular protein (Figure 2A *vs*. Figure 2B, and Figure 3 *vs*. Figure 4, respectively).

We used triple resonance multi-dimensional heteronuclear NMR spectroscopy to achieve sequence-specific resonance assignments of rmBG21 in physiologically relevant buffer, to analyse its secondary structure using chemical shift indexing, and to investigate its backbone dynamics using ^{15}N spin relaxation measurements [233]. The ^{1}H-^{15}N HSQC spectrum of this protein is shown in Figure 3. The sample used here for the multidimensional NMR studies was uniformly $^{13}C^{15}N$-labelled rmBG21 dissolved in water with 10% D_2O, 100 mM KCl, and 0.005% sodium azide at pH 6.9. This preparation was found to be very stable (as judged by criteria such as the undiminished NMR signal intensity) for at least 1 year in these solvent conditions, and in a wide range of temperatures. Whereas the classic MBP isoforms are primarily membrane-associated *in vivo*, justifying the use of membrane-mimetic solvents and other such conditions, the Golli protein is not necessarily lipid-associated in the cell. Thus, the solution condition used here for rmBG21 was chosen as being simple yet still physiologically relevant, especially for future protein-protein interaction studies.

Because of the low degree of ordered secondary structure, the chemical shifts in this intrinsically disordered protein are poorly dispersed compared to canonically-folded ones (Figure 3 *vs*. Figure 4; *cf.*, Figure 2). Spin system overlap and degeneracy make it difficult to use the standard methodology for assignment. Our sequential assignment strategy relied heavily on the use of CO-based triple-resonance experiments. Out of a total of 199 residues, 184 residues were assigned unambiguously. The assignments of eight other residues, N2, N86, G5, G188, K6, K189, E24, and E47, were ambiguous due to sequence similarity and the disordered nature of the protein, which resulted in identical chemical shifts for corresponding spins. The first two residues, N2 and N86, are parts of the sequence-similar fragments G1-N2-H3 and G85-N86-R87, respectively. Likewise, the glutamyl residues E24 and E47 are found in the middle of similar triplet fragments: G23-E24-A24 and G46-E47-A48. Finally, the G5-K6 and G188-K189 pairs are found in the similar fragments S4-G5-K6-R7 and S187-G188-K189-V190. The remaining seven unassigned residues included R164, D170, and five histidyl residues belonging to the His_6 tag (H195-H199), which could not be assigned because of the spin system overlap.

After residue assignment, the chemical shift index analysis revealed little ordered secondary structure under these conditions, with only some small fragments having a slight tendency towards α-helicity, which again may represent putative recognition motifs (Figure 5).

These conclusions were further supported by the results of ^{15}N relaxation measurements. Negative heteronuclear ^{1}H-$\{^{15}N\}$ *NOE* values and low order parameters for the N-terminal Golli-specific portion (residues S5-T69) indicated a high degree of flexibility and disorder, whereas the rest of the protein, although not as flexible, exhibited relaxation parameters typical of proteins in the extended conformation (Figure 6). The highly dynamic nature of this N-terminal region may be to provide additional plasticity, or conformational adaptability, in protein-protein interactions. Another highly mobile segment, A126-S127-G128-G129, may function as a hinge. These studies provide the basis for further NMR investigations of the interaction between BG21 and other proteins such as GIP.

Solution NMR spectroscopy of the full-length 18.5 kDa classic MBP isoform – assignments under aqueous and membrane-mimetic conditions

As indicated above, our group has introduced the use of recombinant murine 18.5 kDa MBP (rmMBP) to structural studies of the protein [235, 236], enabling its uniform labelling by ^{13}C and ^{15}N, and subsequently multidimensional NMR spectroscopy. Clearly, as can be appreciated by the preceding description of rmBG21, a major concern in studying IDPs in solution is their inherent flexibility, and their extreme dependence on the global environment, necessitating novel NMR strategies. A condition that creates a homogeneous population in solution allows for a 'snapshot' of the protein to be taken using solution NMR techniques. The direct application of solution NMR to the membrane-associated classic 18.5 kDa MBP

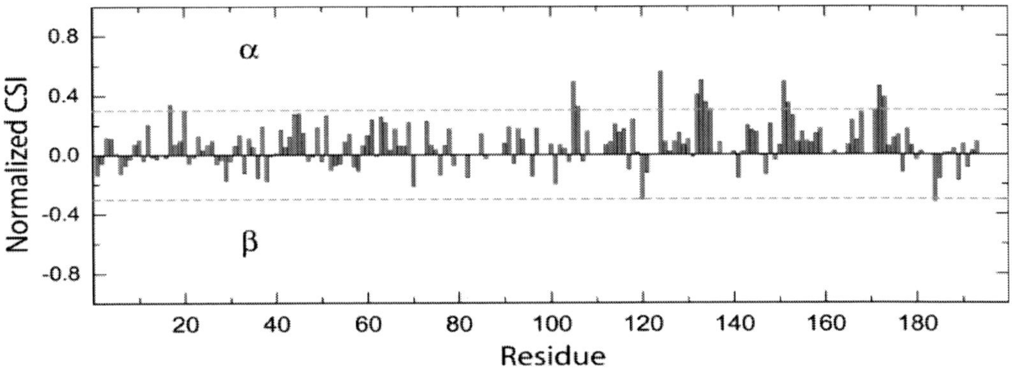

Figure 5. Overall normalised chemical shift difference index for rmBG21. The dashed reference line corresponds to an approximate threshold value at which the chemical shift indices start to reflect regions of ordered secondary structure. This protein is extremely disordered, even compared to other IDPs. Figure adapted with permission from reference (233).

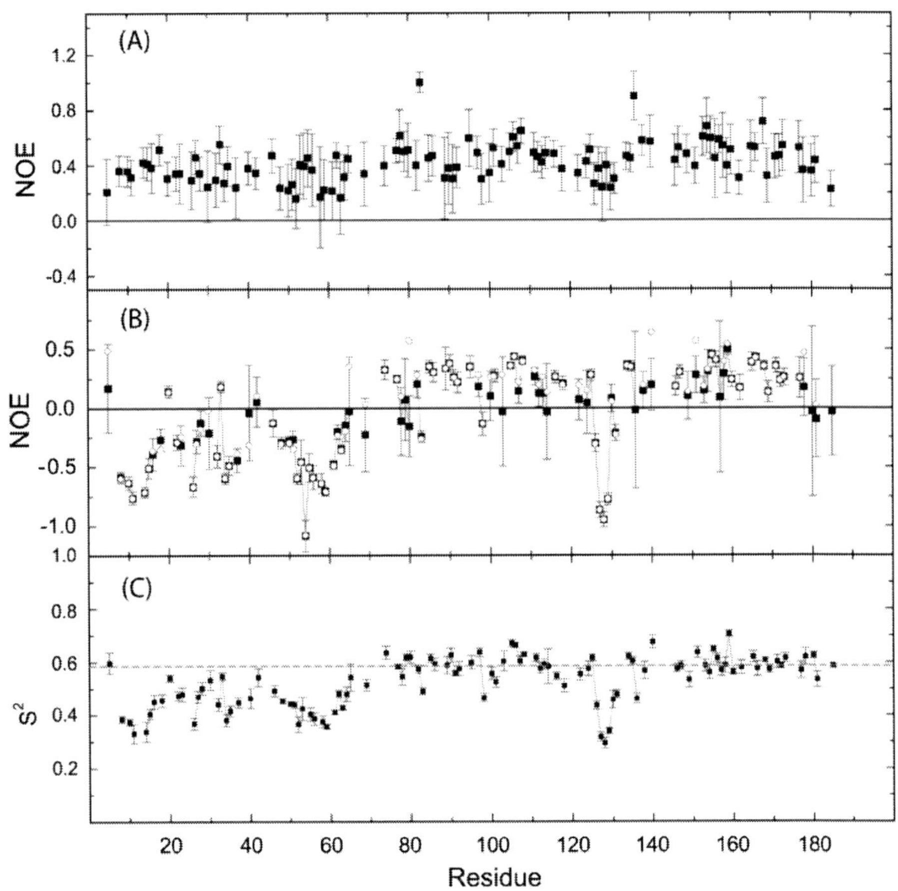

Figure 6. Dynamics and flexibility of the Golli-MBP isoform rmBG21. (A) Relaxation measurements of rmBG21 at 7°C and at 600 MHz field strength - heteronuclear $\{^1H\}$-^{15}N *NOE* values. (B) Experimental relaxation measurements of rmBG21 at 27°C and at 600 MHz field strength – heteronuclear $\{^1H\}$-^{15}N *NOE* values. (C) Order parameter S^2 for rmBG21 determined by a Lipari-Szabo model-free analysis as described [233]. Figure adapted with permission from reference [233].

isoform is problematic due to the reduced mobility of the protein in a reconstituted protein-lipid assembly. The challenge is to find sample preparation conditions that would allow high-resolution NMR studies of uniformly $^{13}C^{15}N$-labelled 18.5 kDa MBP in an environment most closely mimicking the native myelin sheath.

A comparative study of an immunodominant epitope of 18.5 kDa MBP has hitherto shed the most detailed insight into the dependence of conformation on environment [230]. Using solution NMR spectroscopy, three-dimensional structures were obtained under three conditions for an 18-residue synthetic polypeptide fragment comprising this highly conserved region of MBP. The peptide fragment formed a stable, amphipathic, α-helix under organic (30% vol TFE-d_2) and membrane-mimetic (100 mM dodecylphosphocholine, DPC-d_{38}, micelles) conditions, but had only a partially helical conformation in aqueous solution. These results were consistent with electron paramagnetic resonance experiments that suggested an α-helical structure when bound to a membrane [238] (see Chapter VI). Although there had been several previous NMR studies of other MBP-derived polypeptides (*vide supra*), they could not, at the time, be compared with other structural analyses in environments representative of the *in vivo* situation. In addition to providing a complete characterisation of the peptide *per se*, this work represented a step towards establishing and optimising physiologically relevant and experimentally tractable solution NMR conditions that could eventually be applied to structural studies of the intact protein. This last theme recurs throughout this chapter.

With regard now to the full-length 18.5 kDa MBP, we first performed NMR studies of uniformly $^{13}C^{15}N$-labelled rmMBP in 30% TFE-d_2 (perdeuterated trifluoroethanol) [229], which mimics the membrane environment of the myelin sheath, and later in aqueous solution [232]. There are advantages and drawbacks from either the physiological or experimental point of view: preparations of $^{13}C^{15}N$-rmMBP in aqueous solutions serve as a general reference point upon which the effect of various "membrane-mimetic" conditions such as TFE can be evaluated, and also provide a suitable framework for studying interactions between MBP and other proteins. The same difficulties were encountered in spectral assignment for these classic 18.5 kDa MBP preparations as described above for BG21.

The fluorinated alcohol TFE is a convenient solvent for studying MBP's conformation in a membrane-mimetic environment, but it is known to induce potentially spurious α-helical structures, as well as stabilise real ones [239-243]. Although one must be cautious in interpreting the results of NMR studies of MBP dissolved in TFE, there are many benefits compared to other solvents, and the usage of fluorinated alcohols is widespread for membrane-associated proteins. We have evaluated many solvents and found the intact protein to be insufficiently soluble in methanol or in dimethylsulphoxide, and thus it could not be concentrated to the extent required for NMR spectroscopy. In contrast, MBP can be dissolved in TFE in high concentration, and results in well-resolved NMR peaks, despite poor dispersion typical in unstructured proteins (Figure 7A) [229].

Almost complete resonance assignments obtained in TFE have facilitated further chemical shift analyses of backbone resonances, which have revealed some propensity towards formation of α-helical structure. Interestingly, the regions where α-helicity is observed match those expected and predicted by other experimental approaches, especially site-directed spin-labelling and electron paramagnetic resonance spectroscopy of lipid-associated MBP [238, 244, 245] (see Chapter VI). These results are suggestive that the

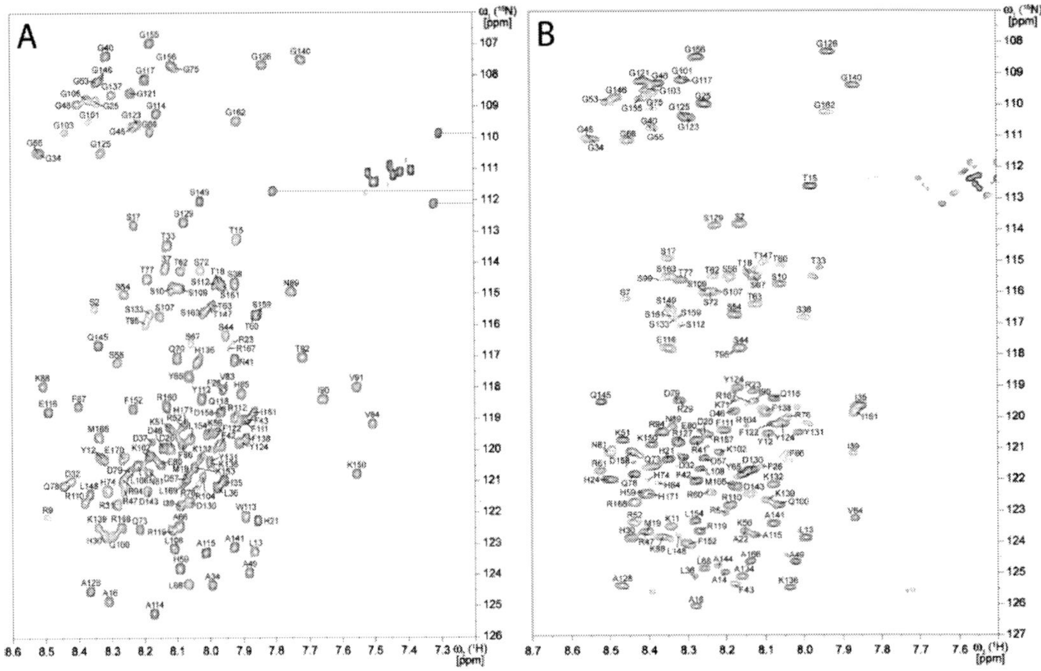

Figure 7. The ^{1}H-^{15}N HSQC spectrum of uniformly ^{15}N-labelled rmMBP (176 residues, including a C-terminal LEH_6 tag) dissolved in (A) 30% TFE-d_2, pH 6.5, 27°C at a concentration of 1.72 mM or (B) in 100 mM KCl, pH 6.5, 10% D_2O, 27°C at a concentration of 1.75 mM. A total of 164 of 176 peaks are observable (the N-terminus and 11 prolyl residues are excluded). In panel A (TFE-d_2), 158 resonances were observed and assigned or 96% of the observable peaks. For rmMBP dissolved in KCl (panel B), a total of 142 resonances were observed and assigned, or 87% of the protein. Six other residues were assigned but appear below the shown contour levels: Q8, R9, G105, G114, H135, and L169. Although both spectra display typical IDP peak dispersion in both the proton and nitrogen dimensions, peaks that were assigned to the central immunodominant epitope display the upfield shifts associated with α-helix formation (panel A, 7.4-7.8 ppm for ^{1}H, and 114-122 ppm for ^{15}N). Secondary structure analysis confirms the increase of α-helicity, stabilised in rmMBP dissolved in TFE, over rmMBP dissolved in KCl. Figure courtesy of Dr. David Libich.

TFE/water system is stabilising real but transient α-helical structural elements *cf.*, [246], and not inducing them. Moreover, these data provide a useful complement to solid-state NMR studies of MBP reconstituted with membranes, which will be discussed in the following section.

Although the membrane-mimetic conditions represent a workable starting point, the TFE solvent is unsuitable for studies of MBP-protein (e.g., CaM) interactions. Therefore, 100 mM KCl was also evaluated to emulate the intracellular cytosolic environment and facilitate future studies of rmMBP-CaM and rmMBP-SH3-domain complexes, to further our understanding of MBP's putative role as a signalling linker protein during myelin development and remodelling [16, 89]. Nearly complete ^{1}H, ^{15}N, and ^{13}C backbone resonance assignments (93%) were achieved in 100 mM KCl (Figure 7B) [232]. The chemical shifts of rmMBP dissolved in 100 mM KCl are significantly different than those obtained for the protein under membrane-mimetic conditions. Notably, the amphipathic α-helix (V83-T92) present in the protein in the presence of TFE, and also found in reconstituted myelin [238, 245], appears disordered under these solution conditions. This conformational lability would have consequences on 18.5 kDa MBP's interactions with calmodulin or SH3-domain-containing proteins, because this segment is either a putative target or immediately adjacent to one [89, 247].

Solid-state NMR spectroscopy of 18.5 kDa MBP reconstituted with lipid vesicles – conformation and dynamics of the protein

Valuable as these solution studies are for studying MBP-protein interactions, a major goal is the determination of the detailed structure of 18.5 kDa MBP *in situ* in the myelin membrane. The proposed structural role of MBP *in vivo* is to maintain compactness of myelin sheaths by appropriately binding and spacing adjacent membranes. Reconstitution of the classic 18.5 kDa MBP isoform with lipid vesicles under physiological buffer conditions yields large assemblies mimicking the natural environment of the myelin sheath (248), which cannot be probed by solution NMR. Thus, a new approach is required. Biomolecular solid-state NMR is a rapidly developing field, and methods of this class can be applied to membrane proteins (such as MBP) reconstituted in lipid bilayers, while best mimicking lipid-protein electrostatic and hydrophobic interactions.

In this respect, we have initiated solid-state NMR studies of MBP reconstituted with lipids [234]. The mobility of MBP in such a system is variable, depends on the local strength of the protein-lipid interaction, and in general, is on such a time scale that the dipolar interactions are averaged out, as illustrated by Figure 8.

Solid-state NMR spectroscopy provides ways to differentiate between highly mobile and immobilised portions of the protein. The signal excitation techniques that are used in solution NMR experiments work best if relaxation times are long, and thus allow one to filter out rapidly relaxing signals from immobilised residues, whereas the solid-state NMR excitation methods that rely on dipolar couplings will filter out signals from mobile residues, in which the dipolar interactions are missing.

Solid-state NMR spectroscopy of lipid-reconstituted MBP is in its infancy. We have so far focussed on the identification and analysis of the mobile residues in rmMBP reconstituted with membranes. Using a combination of three-dimensional chemical shift correlation methods (Figure 9), we have assigned many of the backbone resonances for mobile fragments of the protein [234]. Chemical shift indexing, or a comparison between the observed chemical shifts and the random coil values shown in Fig. 10, indicates a mostly unstructured character of the mobile parts of the protein, perhaps with a slight propensity towards α-helix.

Additional NOESY-based experiments revealed strong dipolar interactions between water and nearly all residues found in the mobile part, indicating that the mobile fragments were exposed to solvent, and were likely located outside the lipid bilayer, or in its hydrophilic portion.

It is also noteworthy that, although we can only report and analyse the chemical shifts for the residues that we observe in our measurements, the "invisibility" of approximately two thirds of the protein is the direct consequence of its reduced mobility, perhaps indicative of increased interactions of these residues with the lipid bilayer.

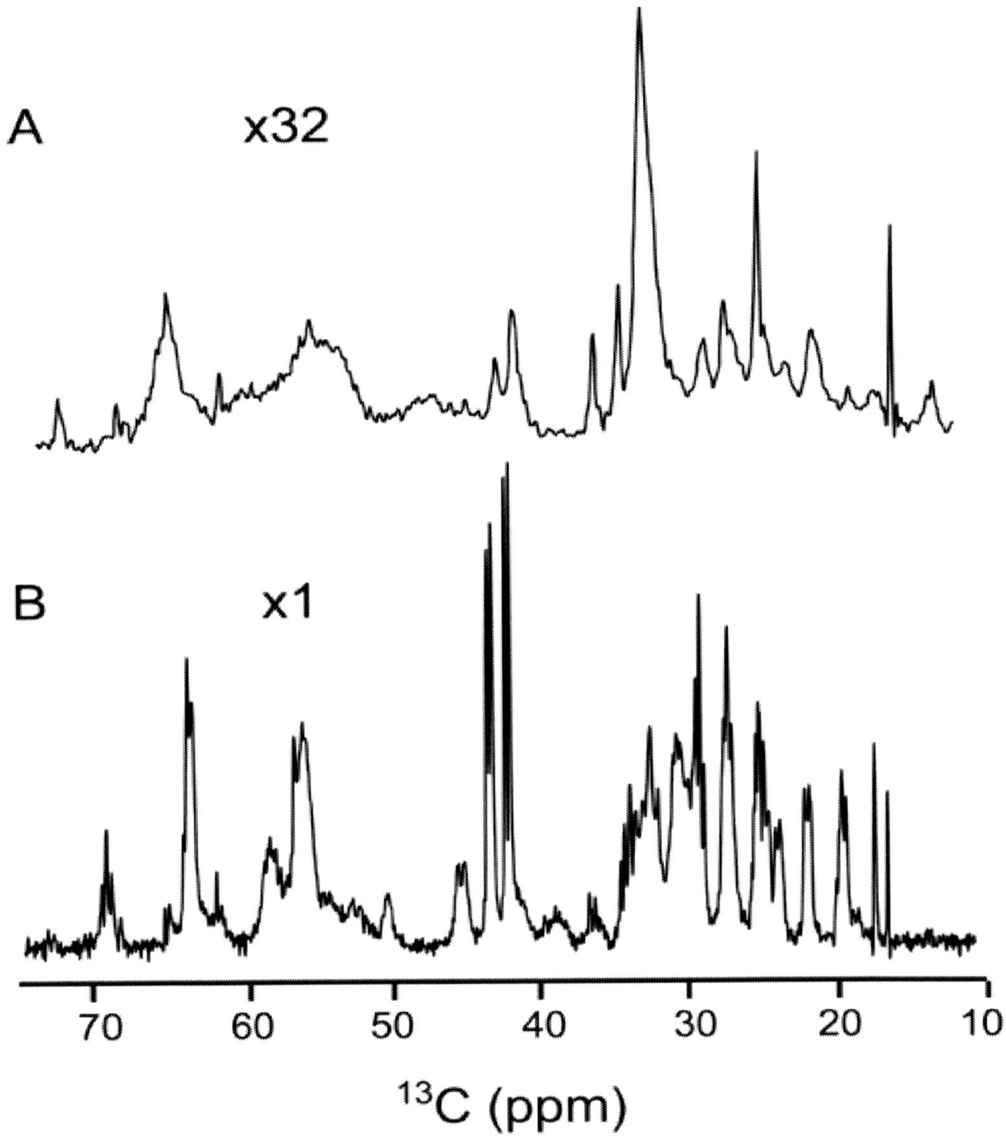

Figure 8. The cross-polarisation (CPMAS) carbon spectrum (A) and insensitive nuclear enhancement by polarisation transfer (INEPT) carbon spectrum (B) of fully $^{13}C^{15}N$ labelled 18.5 kDa rmMBP samples reconstituted with lipids. Two-pulse phase modulation (TPPM) decoupling of 71.4 kHz was applied during acquisition. Both spectra were taken at 600 MHz, and at a temperature of 32°C. The CPMAS spectrum was collected with a contact time of 3 ms, and with 4180 scans, and at a spinning frequency of 20 kHz. Experiments with shorter cross-polarisation mixing times and different cross-polarisation power levels resulted in spectra of similar intensities. The INEPT spectrum was collected with 32 scans, at a spinning frequency of 10 kHz. The acquisition lengths were 21 ms and 40 ms in the CPMAS and INEPT experiments, respectively. All spectra were processed with exponential function apodisation of 10 Hz. Figure adapted with permission from reference [234].

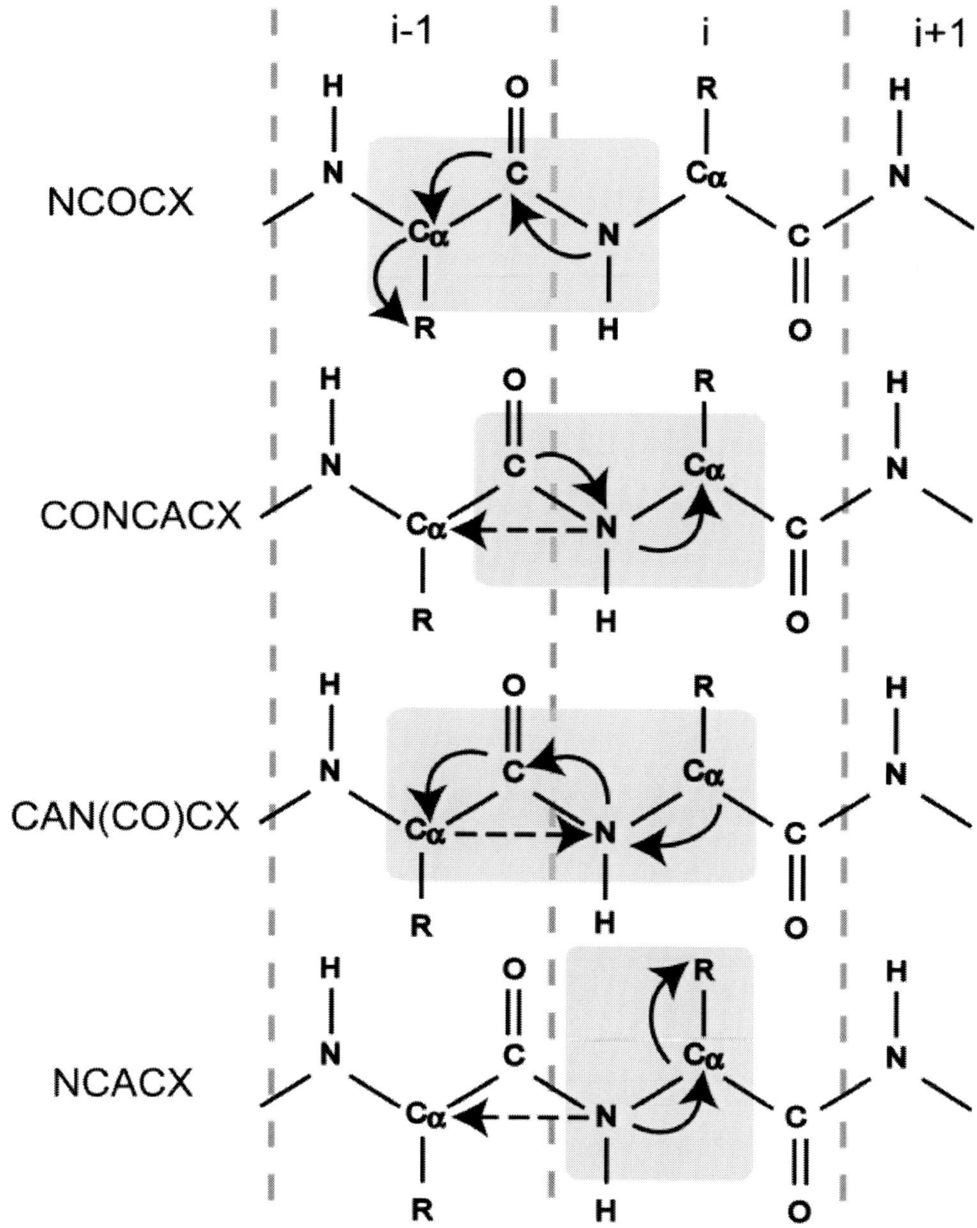

Figure 9. Schematic representation of the four triple-resonance experiments used to obtain sequential assignments. Correlated residues are indicated by light gray boxes. Solid arrows show dominating one-bond polarisation transfers, whereas dashed arrows show two-bond N[*i*]-C^{α} [*i*-1] transfers, which could not be observed in our spectra. Figure adapted with permission from reference [234].

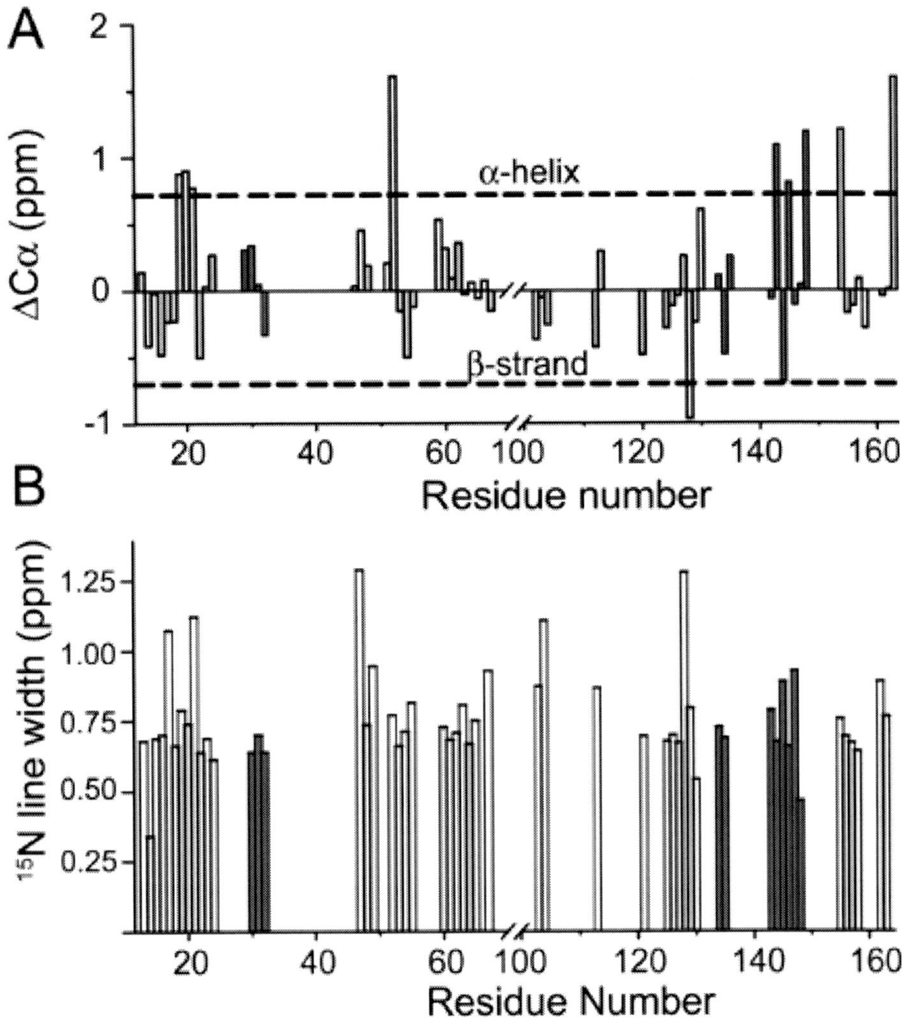

Figure 10. (A) Chemical shift index analysis of C^{α} of the assignments of the rmMBP backbone resonances in lipids. Secondary chemical shifts for tentatively assigned residues are shown in gray. Only 62 residues could be observed in the INEPT-based experiments, which selected only highly mobile parts of the protein. The remaining spectroscopically "invisible" residues were subjected to restricted motions, perhaps because of their stronger interactions with the lipid bilayer. Most notably, the hydrophobic fragments in the N- and C-termini of the protein, known to interact strongly with the membrane [244, 249], do not show up in these NMR spectra. (B) The ^{15}N linewidths for assigned residues. Linewidths for tentatively assigned residues are shown in gray. Figure adapted with permission from reference [234].

Additional information on the degree of order and mobility of the protein can be derived from the analysis of the line widths for the observed resonances. The detected line widths are relatively narrow and uniform, on the order of 0.4-0.7 ppm, and are indicative of the presence of local motions that produce time-averaged narrow lines. On the other hand, large-scale undulating motions of the entire extra-membranous fragments must be present to result in dynamic averaging of the dipolar interactions. In summary, our solid-state NMR measurements support the notion that MBP is primarily a peripheral membrane protein, with some fragments embedded in the membrane and immobilised by the strong interactions with lipids. A significant portion of rmMBP (approximately 35%) was found to be highly mobile and experiencing strong interactions with water, indicating that these fragments were located outside the lipid bilayer or in the interface region. Interestingly, the hydrophobic fragments in the N- and C-termini of the protein, known by site-directed spin labelling and electron paramagnetic resonance spectroscopy to interact strongly with the membrane [244, 249], were not visible in our experiments. Our future studies will be focussed on probing these immobilised segments.

Solid-state NMR spectroscopy of 18.5 kDa MBP reconstituted with lipid vesicles – Effects on the lipid bilayer

Biological membranes are also dynamic entities, and can also result in the averaging of dipolar interactions. Deuterium (^{2}H) and phosphorus (^{31}P) NMR spectroscopy has long been a powerful tool to study phospholipid headgroup conformations, and bilayer phase transitions and dynamics (250-252), particularly upon association with a protein or peptide [253, 254]. The analyses of ^{31}P lineshapes and relaxation parameters have indicated, in general, that MBP interacts with lipids by a combination of electrostatic and hydrophobic interactions [165, 198, 201, 203, 207-209, 212, 213, 215-219, 221-224, 234]. Most of these studies suggested that MBP stabilised lipid bilayers [208, 209, 213, 217, 220, 222]. However, Bloom and colleagues [218, 221] used solid-state NMR to demonstrate the fragmentation by MBP of lipid bilayers comprising phosphatidylcholine, but not of lipid bilayers comprising also negatively-charged lipids. These results were called into question by Pointer-Keenan et al. [224], who suggested that they were due simply to the freeze-thaw procedure used in the sample preparation, and that gentler protocols were required cf., [255]. It is noteworthy that Pointer-Keenan et al. [224] also used MBP that had been fractionated into charge isomers, not heterogeneous mixtures of all modified variants as had been used in almost all previous studies.

In our own solid-state NMR studies of unmodified 18.5 kDa rmMBP reconstituted with synthetic lipids (dimyristoylphosphatidylcholine and dimyristoylphosphatidylglycerol, DMPC and DMPG, respectively, mixed in a 1,1 mass ratio), the observed ^{31}P chemical shift anisotropy strengths were close to those reported for lipid vesicles composed of equal molar amounts of DMPC and DMPG [234]. The approximate axial symmetry of the ^{31}P chemical shift anisotropy tensors indicated fast axial diffusion of lipid molecules, not significantly disturbed by the presence of the protein. Future studies involving more complex mixtures of myelin-specific lipids may indicate a tighter association of the protein with the membrane, but are presently difficult because of the sensitivity of natural lipid preparations to the high power conditions encountered during the prolonged solid-state NMR experiments.

Conclusions and Future Directions

Knowledge of the tertiary structural repertoire of the various isoforms and modified variants of MBP is essential to understanding the mechanism by which these proteins stabilise compacted myelin multilayers and interact with other proteins, potentially in events coupling cellular signals to cytoskeletal organisation during myelination or remyelination. Only NMR spectroscopy can provide detailed structural information at both the local and global levels of conformationally adaptive proteins such as MBP. Modern high-resolution NMR techniques, both solution and solid-state, are being developed to probe this protein's overall folds and conformations in various environments, like those it encounters *in vivo*. Complementary application of NMR and other techniques is required, with numerous stages of cross-validation and refinement. A central segment representing an immunodominant epitope has been confirmed to be α-helical, and other segments shown potentially to adopt this secondary structure, even if transiently. In membrane-associated form, a significant portion of the protein is mobile. The development of new techniques for structure determination of uniformly-labelled proteins by solid-state NMR is an active area of research. This work will significantly enhance our understanding of myelin architecture, and may help elucidate the molecular mechanisms that transpire in demyelinating diseases such as multiple sclerosis. Moreover, some of the methodological advances developed for MBP, and described in this chapter, will likely have more general impact on the field of peripheral membrane proteins and peptides.

Acknowledgements

Our NMR spectroscopic work on myelin proteins has been supported by the Canadian Institutes of Health Research (CIHR, Operating Grants MOP 43982 to GH, and MOP 74468 to GH and VL), the Natural Sciences and Engineering Research Council of Canada (NSERC, Discovery Grants RG298480-04 to VL, and RG121541 to GH), the Canada Foundation for Innovation, and the Ontario Innovation Trust. VL is a Canada Research Chair in Biophysics at the University of Guelph, and is a recipient of an Early Researcher Award from the Ontario Ministry of Research and Innovation. We are grateful to Dr. Lewis Kay (University of Toronto) for the gift of the protein L-B1 domain sample, to Dr. Vladimir Bamm for his pivotal role in protein and sample preparation, and to Mr. Mumdooh Ahmed, Dr. David Libich, and Mr. Ligang Zhong for their thesis data and assistance with the figures.

References

[1] Campagnoni, A.T., Pribyl, T.M., Campagnoni, C.W., Kampf, K., Amur-Umarjee, S., Landry, C.F., Handley, V.W., Newman, S.L., Garbay, B., and Kitamura, K. (1993). Structure and developmental regulation of Golli-mbp, a 105-kilobase gene that encompasses the myelin basic protein gene and is expressed in cells in the oligodendrocyte lineage in the brain. *J. Biol. Chem*, 268, 4930-4938.

[2] Pribyl, T.M., Campagnoni, C.W., Kampf, K., Kashima, T., Handley, V.W., McMahon, J., and Campagnoni, A.T. (1993). The human myelin basic protein gene is included within a 179-kilobase transcription unit: expression in the immune and central nervous systems. *Proc. Natl. Acad. Sci*, USA, 90, 10695-10699.

[3] Givogri, M.I., Bongarzone, E.R., and Campagnoni, A.T. (2000). New insights on the biology of myelin basic protein gene: the neural-immune connection. *J. Neurosci. Res*, 59, 153-159.

[4] Givogri, M.I., Bongarzone, E.R., Schonmann, V., and Campagnoni, A.T. (2001). Expression and regulation of golli products of myelin basic protein gene during *in vitro* development of oligodendrocytes. *J. Neurosci. Res*, 66, 679-690.

[5] Campagnoni, A.T., and Skoff, R.P. (2001). The pathobiology of myelin mutants reveal novel biological functions of the MBP and PLP genes. *Brain Pathol*, 11, 74-91.

[6] Landry, C.F., Ellison, J.A., Pribyl, T.M., Campagnoni, C., Kampf, K., and Campagnoni, A.T. (1996). Myelin basic protein gene expression in neurons: developmental and regional changes in protein targeting within neuronal nuclei, cell bodies, and processes. *J. Neurosci*, 16, 2452-2462.

[7] Landry, C.F., Ellison, J., Skinner, E., and Campagnoni, A.T. (1997). Golli-MBP proteins mark the earliest stages of fiber extension and terminal arboration in the mouse peripheral nervous system. *J. Neurosci. Res*, 50, 265-271.

[8] Landry, C.F., Pribyl, T.M., Ellison, J.A., Givogri, M.I., Kampf, K., Campagnoni, C.W., and Campagnoni, A.T. (1998). Embryonic expression of the myelin basic protein gene: identification of a promoter region that targets transgene expression to pioneer neurons. *J. Neurosci*, 18, 7315-7327.

[9] Reyes, S.D., and Campagnoni, A.T. (2002). Two separate domains in the golli myelin basic proteins are responsible for nuclear targeting and process extension in transfected cells. *J. Neurosci. Res*, 69, 587-596.

[10] Feng, J.M., Fernandes, A.O., Campagnoni, C.W., Hu, Y.H., and Campagnoni, A.T. (2004). The golli-myelin basic protein negatively regulates signal transduction in T lymphocytes. *J. Neuroimmunol*, 152, 57-66.

[11] Fernandes, A.O., Campagnoni, C.W., Kampf, K., Feng, J.M., Handley, V.W., Schonmann, V., Bongarzone, E.R., Reyes, S., and Campagnoni, A.T. (2004). Identification of a protein that interacts with the golli-myelin basic protein and with nuclear LIM interactor in the nervous system. *J. Neurosci. Res*, 75, 461-471.

[12] Feng, J.M., Hu, Y.K., Xie, L.H., Colwell, C.S., Shao, X.M., Sun, X.P., Chen, B., Tang, H., and Campagnoni, A.T. (2006). Golli protein negatively regulates store depletion-induced calcium influx in T cells. *Immunity*, 24, 717-727.

[13] Moscarello, M.A. (1997). *Myelin basic protein, the "executive" molecule of the myelin membrane*. Pages 13-25, in Juurlink, B.H.J., Devon, R.M., Doucette, J.R., Nazarali, A.J., Schreyer, D.J., and Verge, V.M.K. (eds.), Cell Biology and Pathology of Myelin: Evolving Biological Concepts and Therapeutic Approaches, Plenum Press, New York, USA.

[14] Wood, D.D., and Moscarello, M.A. (1997). *Molecular Biology of the Glia: components of myelin - myelin basic protein - the implication of post-translational changes for demyelinating disease*. Pages 37-54, in Russell, W.C. (ed.), Molecular biology of multiple sclerosis, John Wiley and Sons, Chichester.

[15] Harauz, G., Ishiyama, N., Hill, C.M.D., Bates, I.R., Libich, D.S., and Farès, C. (2004). Myelin basic protein-diverse conformational states of an intrinsically unstructured protein and its roles in myelin assembly and multiple sclerosis. *Micron*, 35, 503-542.

[16] Boggs, J.M. (2006). Myelin basic protein: a multifunctional protein. *Cell. Mol. Life Sci*, 63, 1945-1961.

[17] Sedzik, J., Blaurock, A.E., and Höchli, M. (1984). Lipid/myelin basic protein multilayers. A model for the cytoplasmic space in central nervous system myelin. *J. Mol. Biol,* 174, 385-409.

[18] Riccio, P., Fasano, A., Borenshtein, N., Bleve-Zacheo, T., and Kirschner, D.A. (2000). Multilamellar packing of myelin modeled by lipid-bound MBP. *J. Neurosci. Res*, 59, 13-521.

[19] Fitzner, D., Schneider, A., Kippert, A., Mobius, W., Willig, K.I., Hell, S.W., Bunt, G., Gaus, K., and Simons, M. (2006). Myelin basic protein-dependent plasma membrane reorganization in the formation of myelin. *EMBO J*, 25, 5037-5048.

[20] Rosetti, C.M., and Maggio, B. (2007). Protein-induced surface structuring in myelin membrane monolayers. *Biophys. J*, 93, 4254-4267.

[21] Hu, Y., and Israelachvili, J. (2007). Lateral reorganization of myelin lipid domains by myelin basic protein studied at the air-water interface. *Colloids Surf. B Biointerfaces*, 62, 22-30.

[22] Kim, J.K., Mastronardi, F.G., Wood, D.D., Lubman, D.M., Zand, R., and Moscarello, M.A. (2003). Multiple sclerosis: an important role for post-translational modifications of myelin basic protein in pathogenesis. *Mol. Cell. Proteomics,* 2, 453-462.

[23] Jacobs, E.C. (2005). Genetic alterations in the mouse myelin basic proteins result in a range of dysmyelinating disorders. *J. Neurol. Sci*, 228, 195-197.

[24] Moscarello, M.A., Wood, D.D., Ackerley, C., and Boulias, C. (1994). Myelin in multiple sclerosis is developmentally immature. *J. Clin. Invest*, 94, 146-154.

[25] Wood, D.D., Bilbao, J.M., O'Connors, P., and Moscarello, M.A. (1996). Acute multiple sclerosis (Marburg type) is associated with developmentally immature myelin basic protein. *Ann. Neurol*, 40, 18-24.

[26] Whitaker, J.N., and Mitchell, G.W. (1996). A possible role for altered myelin basic protein in multiple sclerosis. *Ann. Neurol*, 40, 3-4.

[27] Fersht, A. (1999). *Structure and Mechanism in Protein Science - A Guide to Enzyme Catalysis and Protein Folding*. Freeman, New York, USA.

[28] Sedzik, J., and Kirschner, D.A. (1992). Is myelin basic protein crystallizable? *Neurochem. Res*, 17, 157-166.

[29] Ishiyama, N., Bates, I.R., Hill, C.M.D., Wood, D.D., Matharu, P., Viner, N.J., Moscarello, M.A., and Harauz, G. (2001). The effects of deimination of myelin basic protein on structures formed by its interaction with phosphoinositide-containing lipid monolayers. *J. Struct. Biol*, 136, 30-45.

[30] Kursula, P. (2001). The current status of structural studies on proteins of the myelin sheath (Review). *Int. J. Mol. Med*, 8, 475-479.

[31] Hill, C.M.D., Bates, I.R., White, G.F., Hallett, F.R., and Harauz, G. (2002). Effects of the osmolyte trimethylamine-*N*-oxide on conformation, self-association, and two-dimensional crystallization of myelin basic protein. *J. Struct. Biol*, 139, 13-26.

[32] Tzakos, A.G., Troganis, A., Theodorou, V., Tselios, T., Svarnas, C., Matsoukas, J., Apostolopoulos, V., and Gerothanassis, I.P. (2005). Structure and function of the myelin proteins: current status and perspectives in relation to multiple sclerosis. *Curr. Med. Chem*, 12, 1569-1587.

[33] Smith, K.J., Pyrdol, J., Gauthier, L., Wiley, D.C., and Wucherpfennig, K.W. (1998). Crystal structure of HLA-DR2 (DRA*0101, DRB1*1501) complexed with a peptide from human myelin basic protein. *J. Exp. Med*, 188, 1511-1520.

[34] Li, Y., Li, H., Martin, R., and Mariuzza, R.A. (2000). Structural basis for the binding of an immunodominant peptide from myelin basic protein in different registers by two HLA-DR2 proteins. *J. Mol. Biol*, 304, 177-188.

[35] Li, Y., Li, H., Dimasi, N., McCormick, J.K., Martin, R., Schuck, P., Schlievert, P.M., and Mariuzza, R.A. (2001). Crystal structure of a superantigen bound to the high-affinity, zinc-dependent site on MHC class II. *Immunity,* 14, 93-104.

[36] Wucherpfennig, K.W. (2001). Structural basis of molecular mimicry. *J. Autoimmun*, 16, 293-302.

[37] Wucherpfennig, K.W. (2001). Insights into autoimmunity gained from structural analysis of MHC-peptide complexes. *Curr. Opin. Immunol*, 13, 650-656.

[38] Hahn, M., Nicholson, M.J., Pyrdol, J., and Wucherpfennig, K.W. (2005). Unconventional topology of self peptide-major histocompatibility complex binding by a human autoimmune T cell receptor. *Nat. Immunol*, 6, 490-496.

[39] Li, Y., Huang, Y., Lue, J., Quandt, J.A., Martin, R., and Mariuzza, R.A. (2005). Structure of a human autoimmune TCR bound to a myelin basic protein self-peptide and a multiple sclerosis-associated MHC class II molecule. *EMBO J*, 24, 2968-2979.

[40] Romero, P., Obradovic, Z., Kissinger, C.R., Villafranca, J.E., Garner, E., Guilliot, S., and Dunker, A.K. (1998). Thousands of proteins likely to have long disordered regions. *Pac. Symp. Biocomput*, 437-448.

[41] Wright, P.E., and Dyson, H.J. (1999). Intrinsically unstructured proteins: re-assessing the protein structure-function paradigm. *J. Mol. Biol*, 293, 321-331.

[42] Dunker, A.K., Obradovic, Z., Romero, P., Garner, E.C., and Brown, C.J. (2000). Intrinsic protein disorder in complete genomes. Genome Inform. *Ser. Workshop Genome Inform*, 11, 161-171.

[43] Uversky, V.N., Gillespie, J.R., and Fink, A.L. (2000). Why are "natively unfolded" proteins unstructured under physiologic conditions? *Proteins*, 41, 415-427.

[44] Uversky, V.N. (2002). What does it mean to be natively unfolded? *Eur. J. Biochem*, 269, 2-12.

[45] Vucetic, S., Brown, C.J., Dunker, A.K., and Obradovic, Z. (2003). Flavors of protein disorder. *Proteins*, 52, 573-584.

[46] Radivojac, P., Obradovic, Z., Smith, D.K., Zhu, G., Vucetic, S., Brown, C.J., Lawson, J.D., and Dunker, A.K. (2004). Protein flexibility and intrinsic disorder. *Protein Sci,* 13, 71-80.

[47] Ward, J.J., Sodhi, J.S., McGuffin, L.J., Buxton, B.F., and Jones, D.T. (2004). Prediction and functional analysis of native disorder in proteins from the three kingdoms of life. *J. Mol. Biol*, 337, 635-645.

[48] Dyson, H.J., and Wright, P.E. (2005). Intrinsically unstructured proteins and their functions. *Nat. Rev. Mol. Cell Biol*, 6, 197-208.

[49] Tompa, P. (2005). The interplay between structure and function in intrinsically unstructured proteins. *FEBS Lett*, 579, 3346-3354.

[50] Tompa, P., Szasz, C., and Buday, L. (2005). Structural disorder throws new light on moonlighting. *Trends Biochem. Sci*, 30, 4-489.

[51] Tompa, P., Dosztanyi, Z., and Simon, I. (2006). Prevalent structural disorder in *E. coli* and *S. cerevisiae* proteomes. *J. Proteome Res*, 5, 1996-2000.

[52] Dyson, H.J., and Wright, P.E. (2006). According to current textbooks, a well-defined three-dimensional structure is a prerequisite for the function of a protein. Is this correct? *IUBMB Life*, 58, 107-109.

[53] Receveur-Bréchot, V., Bourhis, J.M., Uversky, V.N., Canard, B., and Longhi, S. (2006). Assessing protein disorder and induced folding. *Proteins*, 62, 24-45.

[54] Csizmok, V., Dosztanyi, Z., Simon, I., and Tompa, P. (2007). Towards proteomic approaches for the identification of structural disorder. *Curr. Protein Pept. Sci*, 8, 173-179.

[55] Dosztanyi, Z., Sandor, M., Tompa, P., and Simon, I. (2007). Prediction of protein disorder at the domain level. *Curr. Protein Pept. Sci*, 8, 161-171.

[56] Sickmeier, M., Hamilton, J.A., LeGall, T., Vacic, V., Cortese, M.S., Tantos, A., Szabo, B., Tompa, P., Chen, J., Uversky, V.N., Obradovic, Z., and Dunker, A.K. (2007). DisProt: the Database of Disordered Proteins. *Nucleic Acids Res*, 35, D786-D793.

[57] Le Gall, T., Romero, P.R., Cortese, M.S., Uversky, V.N., and Dunker, A.K. (2007). Intrinsic disorder in the Protein Data Bank. *J. Biomol. Struct. Dyn*, 24, 325-342.

[58] Radivojac, P., Iakoucheva, L.M., Oldfield, C.J., Obradovic, Z., Uversky, V.N., and Dunker, A.K. (2007). Intrinsic disorder and functional proteomics. *Biophys. J*, 92, 1439-1456.

[59] Xie, H., Vucetic, S., Iakoucheva, L.M., Oldfield, C.J., Dunker, A.K., Uversky, V.N., and Obradovic, Z. (2007). Functional anthology of intrinsic disorder. 1. Biological processes and functions of proteins with long disordered regions. *J. Proteome Res*, 6, 1882-1898.

[60] Vucetic, S., Xie, H., Iakoucheva, L.M., Oldfield, C.J., Dunker, A.K., Obradovic, Z., and Uversky, V.N. (2007). Functional anthology of intrinsic disorder. 2. Cellular components, domains, technical terms, developmental processes, and coding sequence diversities correlated with long disordered regions. *J. Proteome Res*, 6, 1899-1916.

[61] Xie, H., Vucetic, S., Iakoucheva, L.M., Oldfield, C.J., Dunker, A.K., Obradovic, Z., and Uversky, V.N. (2007). Functional anthology of intrinsic disorder. 3. Ligands, post-translational modifications, and diseases associated with intrinsically disordered proteins. *J. Proteome Res*, 6, 1917-1932.

[62] Dosztanyi, Z., Chen, J., Dunker, A.K., Simon, I., and Tompa, P. (2006). Disorder and sequence repeats in hub proteins and their implications for network evolution. *J. Proteome Res*, 5, 2985-2995.

[63] Patil, A., and Nakamura, H. (2006). Disordered domains and high surface charge confer hubs with the ability to interact with multiple proteins in interaction networks. *FEBS Lett*, 580, 2041-2045.

[64] Hershey, P.E., McWhirter, S.M., Gross, J.D., Wagner, G., Alber, T., and Sachs, A.B. (1999). The Cap-binding protein eIF4E promotes folding of a functional domain of yeast translation initiation factor eIF4G1. *J. Biol. Chem*, 274, 21297-21304.

[65] Chi, S.W., Lee, S.H., Kim, D.H., Ahn, M.J., Kim, J.S., Woo, J.Y., Torizawa, T., Kainosho, M., and Han, K.H. (2005). Structural details on mdm2-p53 interaction. *J. Biol. Chem*, 280, 38795-38802.

[66] Solt, I., Magyar, C., Simon, I., Tompa, P., and Fuxreiter, M. (2006). Phosphorylation-induced transient intrinsic structure in the kinase-inducible domain of CREB facilitates its recognition by the KIX domain of CBP. *Proteins*, 64, 749-757.

[67] Takeuchi, K., and Wagner, G. (2006). NMR studies of protein interactions. *Curr. Opin. Struct. Biol*, 16, 109-117.

[68] Iakoucheva, L.M., Radivojac, P., Brown, C.J., O'Connor, T.R., Sikes, J.G., Obradovic, Z., and Dunker, A.K. (2004). The importance of intrinsic disorder for protein phosphorylation. *Nucleic Acids Res*, 32, 1037-1049.

[69] Finerty, P.J., Jr., Mittermaier, A.K., Muhandiram, R., Kay, L.E., and Forman-Kay, J.D. (2005). NMR dynamics-derived insights into the binding properties of a peptide interacting with an SH2 domain. *Biochemistry*, 44, 694-703.

[70] Fuxreiter, M., Tompa, P., and Simon, I. (2007). Local structural disorder imparts plasticity on linear motifs. *Bioinformatics*, 23, 950-956.

[71] Borg, M., Mittag, T., Pawson, T., Tyers, M., Forman-Kay, J.D., and Chan, H.S. (2007). Polyelectrostatic interactions of disordered ligands suggest a physical basis for ultrasensitivity. *Proc. Natl. Acad. Sci*, USA, 104, 9650-9655.

[72] Lytle, B.L., Volkman, B.F., Westler, W.M., and Wu, J.H. (2000). Secondary structure and calcium-induced folding of the *Clostridium thermocellum* dockerin domain determined by NMR spectroscopy. *Arch. Biochem. Biophys*, 379, 237-244.

[73] Dyson, H.J., and Wright, P.E. (2002). Coupling of folding and binding for unstructured proteins. *Curr. Opin. Struct. Biol*, 12, 54-60.

[74] Chong, P.A., Ozdamar, B., Wrana, J.L., and Forman-Kay, J.D. (2004). Disorder in a target for the smad2 mad homology 2 domain and its implications for binding and specificity. *J. Biol. Chem*, 279, 40707-40714.

[75] Fuxreiter, M., Simon, I., Friedrich, P., and Tompa, P. (2004). Preformed structural elements feature in partner recognition by intrinsically unstructured proteins. *J. Mol. Biol,* 338, 1015-1026.

[76] Eliezer, D., Barre, P., Kobaslija, M., Chan, D., Li, X., and Heend, L. (2005). Residual structure in the repeat domain of tau: echoes of microtubule binding and paired helical filament formation. *Biochemistry*, 44, 1026-1036.

[77] Oldfield, C.J., Cheng, Y., Cortese, M.S., Romero, P., Uversky, V.N., and Dunker, A.K. (2005). Coupled folding and binding with alpha-helix-forming molecular recognition elements. *Biochemistry*, 44, 12454-12470.

[78] Uversky, V.N., Oldfield, C.J., and Dunker, A.K. (2005). Showing your ID: intrinsic disorder as an ID for recognition, regulation and cell signaling. *J. Mol. Recognit*, 18, 343-384.

[79] Vise, P.D., Baral, B., Latos, A.J., and Daughdrill, G.W. (2005). NMR chemical shift and relaxation measurements provide evidence for the coupled folding and binding of the p53 transactivation domain. *Nucleic Acids Res*, 33, 2061-2077.

[80] Radivojac, P., Vucetic, S., O'Connor, T.R., Uversky, V.N., Obradovic, Z., and Dunker, A.K. (2006). Calmodulin signaling: analysis and prediction of a disorder-dependent molecular recognition. *Proteins*, 63, 398-410.

[81] Baker, J.M., Hudson, R.P., Kanelis, V., Choy, W.Y., Thibodeau, P.H., Thomas, P.J., and Forman-Kay, J.D. (2007). CFTR regulatory region interacts with NBD1 predominantly via multiple transient helices. *Nat. Struct. Mol. Biol*, 14, 738-745.

[82] Eliezer, D., and Palmer, A.G., III. (2007). Biophysics: proteins hunt and gather. *Nature,* 447, 920-921.

[83] Mészáros, B., Tompa, P., Simon, I., and Dosztányi, Z. (2007). Molecular principles of the interactions of disordered proteins. *J. Mol. Biol*, 372, 549-561.

[84] Sugase, K., Dyson, H.J., and Wright, P.E. (2007). Mechanism of coupled folding and binding of an intrinsically disordered protein. *Nature*, 447, 1021-1025.

[85] Vise, P., Baral, B., Stancik, A., Lowry, D.F., and Daughdrill, G.W. (2007). Identifying long-range structure in the intrinsically unstructured transactivation domain of p53. *Proteins,* 67, 526-530.

[86] Moscarello, M.A., Gagnon, J., Wood, D.D., Anthony, J., and Epand, R. (1973). Conformational flexibility of a myelin protein. *Biochemistry*, 12, 3402-3406.

[87] Smith, R. (1992). The basic protein of CNS myelin: its structure and ligand binding. *J. Neurochem*, 59, 1589-1608.

[88] DeBruin, L.S., and Harauz, G. (2007). White matter rafting - Membrane microdomains in myelin. *Neurochem. Res*, 32, 213-228.

[89] Polverini, E., Rangaraj, G., Libich, D.S., Boggs, J.M., and Harauz, G. (2008). Binding of the proline-rich segment of myelin basic protein to SH3-domains - Spectroscopic, microarray, and modelling studies of ligand conformation and effects of post-translational modifications. *Biochemistry,* 47, 267-282.

[90] Cheng, Y., LeGall, T., Oldfield, C.J., Dunker, A.K., and Uversky, V.N. (2006). Abundance of intrinsic disorder in protein associated with cardiovascular disease. *Biochemistry*, 45, 10448-10460.

[91] Uversky, V.N. (2007). Neuropathology, biochemistry, and biophysics of alpha-synuclein aggregation. *J. Neurochem*, 103, 17-37.

[92] DeBruin, L.S., Haines, J.D., Bienzle, D., and Harauz, G. (2006). Partitioning of myelin basic protein into membrane microdomains in a spontaneously demyelinating mouse model for multiple sclerosis. *Biochem. Cell Biol*, 84, 993-1005.

[93] Harauz, G., and Musse, A.A. (2007). A tale of two citrullines - Structural and functional aspects of myelin basic protein deimination in health and disease. *Neurochem. Res*, 32, 137-158.

[94] Musse, A.A., and Harauz, G. (2007). Molecular "negativity" may underlie multiple sclerosis: role of the myelin basic protein family in the pathogenesis of MS. *Int. Rev. Neurobiol*, 79, 149-172.

[95] Ferentz, A.E., and Wagner, G. (2000). NMR spectroscopy: a multifaceted approach to macromolecular structure. *Q. Rev. Biophys*, 33, 29-65.

[96] Wider, G. (2000). Structure determination of biological macromolecules in solution using nuclear magnetic resonance spectroscopy. *Biotechniques*, 29, 1278-1292.

[97] Baleja, J.D. (2001). Structure determination of membrane-associated proteins from nuclear magnetic resonance data. *Anal. Biochem*, 288, 1-15.

[98] Kanelis, V., Forman-Kay, J.D., and Kay, L.E. (2001). Multidimensional NMR methods for protein structure determination. *IUBMB Life*, 52, 291-302.

[99] Pochapsky, T.C., and Sondej Pochapsky, S. (2007). *NMR for Physical and Biological Scientists*. Taylor and Francis, New York, USA.

[100] Bax, A. (1989). Two-dimensional NMR and protein structure. *Annu. Rev. Biochem*, 58, 223-256.

[101] Grage, S.L., Wang, J., Cross, T.A., and Ulrich, A.S. (2002). Solid-state (19)F-NMR analysis of (19)F-labeled tryptophan in Gramicidin A in oriented membranes. *Biophys. J*, 83, 3336-3350.

[102] Gardner, K.H., and Kay, L.E. (1998). The use of H-2, C-13, N-15 multidimensional NMR to study the structure and dynamics of proteins. *Annual Review of Biophysics and Biomolecular Structure*, 27, 357-406.

[103] Markley, J.L., Ulrich, E.L., Westler, W.M., and Volkman, B.F. (2003). Macromolecular structure determination by NMR spectroscopy. *Methods Biochem. Anal*, 44, 89-113.

[104] Torizawa, T., Shimizu, M., Taoka, M., Miyano, H., and Kainosho, M. (2004). Efficient production of isotopically labeled proteins by cell-free synthesis: a practical protocol. *J. Biomol. NMR*, 30, 311-325.

[105] Millet, O., Mittermaier, A., Baker, D., and Kay, L.E. (2003). The effects of mutations on motions of side-chains in protein L studied by 2H NMR dynamics and scalar couplings. *J. Mol. Biol*, 329, 551-563.

[106] Andrew, E.R., Bradbury, A., and Eades, R.G. (1959). Removal of dipolar broadening of nuclear magnetic resonance spectra of solids by specimen rotation. *Nature,* 183, 1802-1803.

[107] Lowe, I.J. (1959). Free induction decays of rotating solids. *Phys. Rev. Lett,* 2, 285-287.

[108] Ramamoorthy, A., Wu, C.H., and Opella, S.J. (1999). Experimental aspects of multidimensional solid-state NMR correlation spectroscopy. *J. Magn. Reson*, 140, 131-140.

[109] De Groot, H.J. (2000). Solid-state NMR spectroscopy applied to membrane proteins. *Curr. Opin. Struct. Biol*, 10, 593-600.

[110] Arora, A., and Tamm, L.K. (2001). Biophysical approaches to membrane protein structure determination. *Curr. Opin. Struct. Biol*, 11, 540-547.

[111] Schwalbe, H., and Bielecki, A. (2001). Recent advances in high-resolution solid-state NMR spectroscopy. *Angew. Chem. (Int. Ed.)*, 40, 2045-2050.

[112] Tycko, R. (2001). Biomolecular solid state NMR: advances in structural methodology and applications to peptide and protein fibrils. *Annu. Rev. Phys. Chem*, 52, 575-606.

[113] Opella, S.J., Nevzorov, A., Mesleb, M.F., and Marassi, F.M. (2002). Structure determination of membrane proteins by NMR spectroscopy. *Biochem. Cell Biol*, 80, 597-604.

[114] Thompson, L.K. (2002). Solid-state NMR studies of the structure and mechanisms of proteins. *Curr. Opin. Struct. Biol*, 12, 661-669.

[115] Drechsler, A., and Separovic, F. (2003). Solid-state NMR structure determination. *IUBMB Life*, 55, 515-523.

[116] Marassi, F.M., and Opella, S.J. (2003). Simultaneous assignment and structure determination of a membrane protein from NMR orientational restraints. *Protein Science*, 12, 403-411.

[117] Straus, S.K. (2004). Recent developments in solid-state magic-angle spinning, nuclear magnetic resonance of fully and significantly isotopically labelled peptides and proteins. *Philos. Trans. R. Soc. Lond. B Biol. Sci*, 359, 997-1008.

[118] Watts, A., Straus, S.K., Grage, S.L., Kamihira, M., Lam, Y.H., and Zhao, X. (2004). Membrane protein structure determination using solid-state NMR. *Methods Mol. Biol*, 278, 403-473.

[119] Hong, M. (2006). Oligomeric structure, dynamics, and orientation of membrane proteins from solid-state NMR. *Structure*, 14, 1731-1740.

[120] Hong, M. (2006). Solid-state NMR studies of the structure, dynamics, and assembly of beta-sheet membrane peptides and alpha-helical membrane proteins with antibiotic activities. *Acc. Chem. Res,* 39, 176-183.

[121] Baldus, M. (2006). *Spectral assignment of (membrane) proteins under magic angle spinning.* in Ramamoorthy, A. (ed.), NMR Spectroscopy of Biological Solids, Taylor and Francis Group, LLC, Boca Raton, FL.

[122] Durr, U.H., Yamamoto, K., Im, S.C., Waskell, L., and Ramamoorthy, A. (2007). Solid-state NMR reveals structural and dynamical properties of a membrane-anchored electron-carrier protein, cytochrome b5. *J. Am. Chem. Soc*, 129, 6670-6671.

[123] Castellani, F., van Rossum, B., Diehl, A., Schubert, M., Rehbein, K., and Oschkinat, H. (2002). Structure of a protein determined by solid-state magic-angle-spinning NMR spectroscopy. *Nature,* 420, 98-102.

[124] Rienstra, C.M., Hohwy, M., Mueller, L.J., Jaroniec, C.P., Reif, B., and Griffin, R.G. (2002). Determination of multiple torsion-angle constraints in U-(13)C,(15)N-labeled peptides: 3D (1)H-(15)N-(13)C-(1)H dipolar chemical shift NMR spectroscopy in rotating solids. *J. Am. Chem. Soc*, 124, 11908-11922.

[125] Rienstra, C.M., Tucker-Kellogg, L., Jaroniec, C.P., Hohwy, M., Reif, B., McMahon, M.T., Tidor, B., Lozano-Perez, T., and Griffin, R.G. (2002). *De novo* determination of peptide structure with solid-state magic-angle spinning NMR spectroscopy. *Proc. Natl. Acad. Sci*, USA, 99, 10260-10265.

[126] Jaroniec, C.P., MacPhee, C.E., Bajaj, V.S., McMahon, M.T., Dobson, C.M., and Griffin, R.G. (2004). High-resolution molecular structure of a peptide in an amyloid fibril determined by magic angle spinning NMR spectroscopy. *Proc. Natl. Acad. Sci*, USA, 101, 711-716.

[127] Straus, S.K., Bremi, T., and Ernst, R.R. (1998). Experiments and strategies for the assignment of fully 13C/15N-labelled polypeptides by solid state NMR. *J. Biomol. NMR*, 12, 39-50.

[128] Hong, M. (1999). Resonance assignment of 13C/15N labeled solid proteins by two- and three-dimensional magic-angle-spinning NMR. *J. Biomol. NMR*, 15, 1-14.

[129] Igumenova, T.I., Wand, A.J., and McDermott, A.E. (2004). Assignment of the backbone resonances for microcrystalline ubiquitin. *J. Am. Chem. Soc*, 126, 5323-5331.

[130] Igumenova, T.I., McDermott, A.E., Zilm, K.W., Martin, R.W., Paulson, E.K., and Wand, A.J. (2004). Assignments of carbon NMR resonances for microcrystalline ubiquitin. *J. Am. Chem. Soc*, 126, 6720-6727.

[131] McDermott, A., Polenova, T., Bockmann, A., Zilm, K.W., Paulson, E.K., Martin, R.W., Montelione, G.T., and Paulsen, E.K. (2000). Partial NMR assignments for uniformly (13C, 15N)-enriched BPTI in the solid state. *J. Biomol. NMR*, 16, 209-219.

[132] Pauli, J., Baldus, M., van Rossum, B., de Groot, H., and Oschkinat, H. (2001). Backbone and side-chain C-13 and N-15 signal assignments of the alpha-spectrin SH3 domain by magic angle spinning solid-state NMR at 17.6 tesla. *Chembiochem*, 2, 272-281.

[133] van Rossum, B.J., Castellani, F., Rehbein, K., Pauli, J., and Oschkinat, H. (2001). Assignment of the nonexchanging protons of the alpha-spectrin SH3 domain by two- and three-dimensional 1H-13C solid-state magic-angle spinning NMR and comparison of solution and solid-state proton chemical shifts. *Chembiochem*, 2, 906-914.

[134] Egorova-Zachernyuk, T.A., Hollander, J., Fraser, N., Gast, P., Hoff, A.J., Cogdell, R., De Groot, H.J., and Baldus, M. (2001). Heteronuclear 2D-correlations in a uniformly [13C, 15N] labeled membrane-protein complex at ultra-high magnetic fields. *J. Biomol, NMR*, 19, 243-253.

[135] Luca, S., Heise, H., and Baldus, M. (2003). High-resolution solid-state NMR applied to polypeptides and membrane proteins. *Acc. Chem. Res*, 36, 858-865.

[136] Park, S.H., Mrse, A.A., Nevzorov, A.A., Mesleh, M.F., Oblatt-Montal, M., Montal, M., and Opella, S.J. (2003). Three-dimensional structure of the channel-forming trans-membrane domain of virus protein "u" (Vpu) from HIV-1. *Journal of Molecular Biology*, 333, 409-424.

[137] Luca, S., Heise, H., Lange, A., and Baldus, M. (2005). Investigation of ligand-receptor systems by high-resolution solid-state NMR: recent progress and perspectives. *Arch. Pharm. (Weinheim)*, 338, 217-228.

[138] Frericks, H.L., Zhou, D.H., Yap, L.L., Gennis, R.B., and Rienstra, C.M. (2006). Magic-angle spinning solid-state NMR of a 144 kDa membrane protein complex: *E. coli* cytochrome bo3 oxidase. *J. Biomol, NMR*, 36, 55-71.

[139] Lange, A., Giller, K., Hornig, S., Martin-Eauclaire, M.F., Pongs, O., Becker, S., and Baldus, M. (2006). Toxin-induced conformational changes in a potassium channel revealed by solid-state NMR. *Nature*, 440, 959-962.

[140] Lange, A., Giller, K., Pongs, O., Becker, S., and Baldus, M. (2006). Two-dimensional solid-state NMR applied to a chimeric potassium channel. *J. Recept. Signal. Transduct. Res*, 26, 379-393.

[141] Etzkorn, M., Martell, S., Andronesi, O.C., Seidel, K., Engelhard, M., and Baldus, M. (2007). Secondary structure, dynamics, and topology of a seven-helix receptor in native membranes, studied by solid-state NMR spectroscopy. *Angew. Chem. Int. Ed. Engl*, 46, 459-462.

[142] Bajaj, V.S., Hornstein, M.K., Kreischer, K.E., Sirigiri, J.R., Woskov, P.P., Mak-Jurkauskas, M.L., Herzfeld, J., Temkin, R.J., and Griffin, R.G. (2007). 250 GHz CW Gyrotron oscillator for dynamic nuclear polarization in biological solid state NMR. *J. Magn. Reson*, 189, 251-279.

[143] Wishart, D.S., and Sykes, B.D. (1994). Chemical shifts as a tool for structure determination. *Methods Enzymol*, 239, 363-392.

[144] Wishart, D.S., and Nip, A.M. (1998). Protein chemical shift analysis: a practical guide. *Biochem. Cell Biol*, 76, 153-163.

[145] Wüthrich, K. (2003). NMR studies of structure and function of biological macromolecules (Nobel Lecture). *J. Biomol. NMR*, 27,13-39.

[146] Wüthrich, K. (2003). NMR studies of structure and function of biological macromolecules. *Biosci. Rep*, 23, 119-168.

[147] Griffin, R.G. (1998) Dipolar recoupling in MAS spectra of biological solids. *Nat. Struct. Biol*, 5 Suppl, 508-512.

[148] Palmer, A.G., III. (1997). Probing molecular motion by NMR. *Curr. Opin. Struct. Biol*, 7, 732-737.

[149] Kay, L.E. (1998) Protein dynamics from NMR. *Biochem. Cell Biol*, 76,145-152.

[150] Palmer, A.G., III. (2001). NMR probes of molecular dynamics: overview and comparison with other techniques. *Annu. Rev. Biophys. Biomol. Struct*, 30,129-155.

[151] Palmer, A.G., III, Kroenke, C.D., and Loria, J.P. (2001). Nuclear magnetic resonance methods for quantifying microsecond-to-millisecond motions in biological macromolecules. *Methods Enzymol*, 339, 204-238.
[152] Palmer, A.G., III. (2004). NMR characterization of the dynamics of biomacromolecules. *Chem. Rev*, 104, 3623-3640.
[153] Kay, L.E. (2005). NMR studies of protein structure and dynamics. *J. Magn. Reson*. 173, 193-207.
[154] Mittermaier, A., and Kay, L.E. (2006). New tools provide new insights in NMR studies of protein dynamics. *Science*, 312, 224-228.
[155] Dyson, H.J., and Wright, P.E. (2001). Nuclear magnetic resonance methods for elucidation of structure and dynamics in disordered states. *Methods Enzymol*, 339, 258-270.
[156] Tompa, P. (2002). Intrinsically unstructured proteins. *Trends Biochem. Sci*, 27, 527-533.
[157] Dyson, H.J., and Wright, P.E. (2002). Insights into the structure and dynamics of unfolded proteins from nuclear magnetic resonance. *Adv. Protein Chem*, 62, 311-340.
[158] Lietzow, M.A., Jamin, M., Jane Dyson, H.J., and Wright, P.E. (2002). Mapping long-range contacts in a highly unfolded protein. *J. Mol. Biol*, 322, 655-662.
[159] Dyson, H.J., and Wright, P.E. (2004). Unfolded proteins and protein folding studied by NMR. *Chem. Rev*, 104, 3607-3622.
[160] Dyson, H.J., and Wright, P.E. (2005). Elucidation of the protein folding landscape by NMR. *Methods Enzymol*, 394, 299-321.
[161] Olson, K.E., Narayanaswami, P., Vise, P.D., Lowry, D.F., Wold, M.S., and Daughdrill, G.W. (2005). Secondary structure and dynamics of an intrinsically unstructured linker domain. *J. Biomol. Struct. Dyn*, 23, 113-124.
[162] Bezsonova, I., Evanics, F., Marsh, J.A., Forman-Kay, J.D., and Prosser, R.S. (2007). Oxygen as a paramagnetic probe of clustering and solvent exposure in folded and unfolded states of an SH3 domain. *J. Am. Chem. Soc*, 129, 1826-1835.
[163] Marsh, J.A., Neale, C., Jack, F.E., Choy, W.Y., Lee, A.Y., Crowhurst, K.A., and Forman-Kay, J.D. (2007). Improved structural characterizations of the drkN SH3 domain unfolded state suggest a compact ensemble with native-like and non-native structure. *J. Mol. Biol*, 367, 1494-1510.
[164] Mittag, T., and Forman-Kay, J.D. (2007). Atomic-level characterization of disordered protein ensembles. *Curr. Opin. Struct. Biol*, 17, 3-14.
[165] Husted, C.A., Goodin, D.S., Hugg, J.W., Maudsley, A.A., Tsuruda, J.S., de Bie, S.H., Fein, G., Matson, G.B., and Weiner, M.W. (1994). Biochemical alterations in multiple sclerosis lesions and normal-appearing white matter detected by *in vivo* 31P and 1H spectroscopic imaging. *Ann. Neurol*, 36, 157-165.
[166] Laule, C., Vavasour, I.M., Kolind, S.H., Li, D.K., Traboulsee, T.L., Moore, G.R., and Mackay, A.L. (2007). Magnetic resonance imaging of myelin. *Neurotherapeutics*, 4, 460-484.
[167] Block, R.E., Brady, A.H., and Joffe, S. (1973). Conformation and aggregation of bovine myelin proteins. *Biochem. Biophys. Res. Commun*, 54, 1595-1602.
[168] Liebes, L.F., Zand, R., and Phillips, W.D. (1975). Solution behavior, circular dichroism and 220 MHz PMR studies of the bovine myelin basic protein. *Biochim. Biophys. Acta*, 405, 27-39.

[169] Liebes, L.F., Zand, R., and Phillips, W.D. (1976). The solution behavior of the bovine myelin basic protein in the presence of anionic ligands. Binding behavior with the red component of trypan blue and sodium dodecyl sulfate. *Biochim. Biophys. Acta*, 427, 392-409.

[170] Chapman, B.E., and Moore, G.J. (1976). Conformation of myelin basic protein in aqueous solution from nuclear magnetic resonance spectroscopy. *Biochem. Biophys. Res. Commun,* 73,758-766.

[171] Chapman, B.E., and Moore, W.J. (1978). NMR studies on myelin basic protein. I. 13C spectra in aqueous solutions. *Aust. J. Chem*, 31, 2367-2385.

[172] Chapman, B.E., Littlemore, L.T., and Moore, W.J. (1978). Conformation of myelin basic protein and its role in myelin formation. *Adv. Exp. Med. Biol*, 100, 207-220.

[173] Pasaribu, S.J. (1977). Synthesis of the major encephalitogenic determinant of myelin basic protein and the 13C NMR of some protected intermediates. *Aust. J. Chem,* 30, 2533-2544.

[174] Littlemore, L.A.T. (1978). N.M.R. studies of myelin basic protein. II. 1H N.M.R. studies of the protein and constituent peptides in aqueous solution. *Aust. J. Chem*, 31, 2387-2398.

[175] Pasaribu, S.J. (1979). Encephalitogenic peptides: Synthesis and 13C NMR of a pentapeptide fragment of myelin basic protein of the (75-74) region. *Aust. J. Chem*, 32, 1575-1583.

[176] Margetson, S.A., and Moore, W.J. (1980). NMR studies on myelin basic protein. 4. 13C NMR of the peptide encephalitogenic in the guinea pig. *Australian Journal of Chemistry*, 33, 2411-2425.

[177] Pasaribu, S.J. (1980). Synthesis and carbon 13 NMR spectrum of the peptide fragment of myelin basic protein (human). *Australian Journal of Chemistry*, 33, 2427-2440.

[178] Margetson, S.A., Moore, W.J., and Gibbons, W.A. (1981). NMR studies of myelin basic protein. 5. 1H NMR of the peptides encephalitogenic in guinea pig. *Australian Journal of Chemistry,* 34, 1373-1392.

[179] Mendz, G.L., Moore, W.J., and Carnegie, P.R. (1982). Proton N.M.R. evidence for secondary and tertiary structure in myelin basic proteins. *Biochem. Biophys. Res. Commun,* 105, 1333-1340.

[180] Mendz, G.L., Moore, W.J., and Carnegie, P.R. (1982). NMR studies of myelin basic protein. 6. Proton spectra in aqueous solutions of proteins from mammalian and avian species. *Australian Journal of Chemistry*, 35,1979-2006.

[181] Moore, W.J. (1982). Interaction of basic protein and peripheral nerve P2 protein with lipids. *Biophys. J,* 37, 59-61.

[182] Sadikot, H., and Moore, W.J. (1983). NMR studies of myelin basic protein. 7. Concentration and temperature dependence of amide resonances of the guinea pig encephalitogen in dimethylsulfoxide. *Australian Journal of Chemistry*, 36, 33-41.

[183] Mendz, G.L., Moore, W.J., and Martenson, R.E. (1983). NMR studies on myelin basic protein. VIII. Complete assignment of the threonine residues by proton NMR of proteins from five species. *Biochim. Biophys. Acta*, 742, 215-223.

[184] Mendz, G.L., Moore, W.J., and Martenson, R.E. (1983). NMR studies of myelin basic protein. IX. Complete assignments of the tyrosine residues by proton NMR of proteins from six species. *Biochim. Biophys. Acta*, 748, 168-175.

[185] Mendz, G.L., and Moore, W.J. (1983). NMR studies of myelin basic protein. X. Conformation of a determinant encephalitogenic in the rabbit. *Biochim. Biophys. Acta*, 748, 176-183.

[186] Mendz, G.L., Moore, W.J., King, G., and Wright, P.E. (1984). NMR studies of myelin basic protein. 11. Conformation of a nonapeptide from a segment encephalitogenic in rhesus monkey. *International Journal of Peptide and Protein Research*, 24, 208-217.

[187] Bremer, J., and Moore, W.J. (1984). NMR studies of myelin basic protein. 12. Spectra in aqueous solution of a synthetic octapeptide encephalitogenic in the Lewis rat. *Australian Journal of Chemistry*, 37, 1427-1435.

[188] Nygaard, E., Mendz, G.L., Moore, W.J., and Martenson, R.E. (1984). NMR of a peptic peptide spanning the triprolyl sequence in myelin basic protein. *Biochemistry*, 23, 4003-4010.

[189] Mendz, G.L., and Moore, W.J. (1985). NMR studies of myelin basic protein. Conformation of a peptide that is an antigenic determinant for B-cell reactivity. *Biochem. J*, 229, 305-313.

[190] Fraser, P.E., and Deber, C.M. (1985). Structure and function of the proline-rich region of myelin basic protein. *Biochemistry*, 24, 4593-4598.

[191] Martenson, R.E., Mendz, G.L., and Moore, W.J. (1985). Conformation of two antigenic regions in myelin basic protein. *Biochem. Biophys. Res. Commun*, 131, 1269-1276.

[192] Mendz, G.L., Moore, W.J., and Martenson, R.E. (1986). NMR studies of myelin basic protein. XIII. Assignment of histidine residues in rabbit, bovine and porcine proteins. *Biochim. Biophys. Acta*, 871, 156-166.

[193] Price, W.S., Mendz, G.L., and Martenson, R.E. (1988). Conformation of a heptadecapeptide comprising the segment encephalitogenic in rhesus monkey. *Biochemistry*, 27, 8990-8999.

[194] Deibler, G.E., Burlin, T.V., and Stone, A.L. (1995). Three isoforms of human myelin basic protein: purification and structure. *J. Neurosci. Res*, 41, 819-827.

[195] Mendz, G.L., Barden, J.A., and Martenson, R.E. (1995). Conformation of a tetradecapeptide epitope of myelin basic protein. *Eur. J. Biochem*, 231, 659-666.

[196] Koshy, K.M., Hashim, G.A., and Boggs, J.M. (1996). Proton NMR study of peptides from myelin basic protein: evidence for Lys74-His77 interaction revealed from histidine line broadening. *Biochim. Biophys. Acta*, 1293, 23-30.

[197] Joffe, S., and Block, R.E. (1972). Nuclear magnetic resonance studies suggestive of a lipid population tightly bound to myelin structural proteins. *Brain Res,* 46, 381-390.

[198] Littlemore, L.A.T., and Ledeen, R.W. (1979). N.M.R. studies of myelin basic protein. III. Interactions of the protein with lipid micelles by ^{1}H and ^{31}P N.M.R. *Aust. J. Chem*, 32, 2631-2636.

[199] Keniry, M.A., and Smith, R. (1980). A 13C NMR spin-lattice relaxation study of the interaction of myelin proteins with lipid vesicles. *Biophys. Chem*, 12, 133-141.

[200] Stollery, J.G., Boggs, J.M., Moscarello, M.A., and Deber, C.M. (1980). Direct observation by carbon-13 nuclear magnetic resonance of membrane-bound human myelin basic protein. *Biochemistry*, 19, 2391-2396.

[201] Hughes, D.W., Stollery, J.G., Moscarello, M.A., and Deber, C.M. (1982). Binding of myelin basic protein to phospholipid micelles. *J. Biol. Chem*, 257, 4698-4700.

[202] Smith, R. (1982). 1H-nuclear magnetic resonance study of the association of the basic protein of central nervous system myelin with lysophosphatidylcholine. *Biophys. Chem*, 16, 347-354.

[203] Smith, R., Cornell, B.A., Keniry, M.A., and Separovic, F. (1983). 31P nuclear magnetic resonance studies of the association of basic proteins with multilayers of diacyl phosphatidylserine. *Biochim. Biophys. Acta*, 732, 492-498.

[204] Fraser, P.E., and Deber, C.M. (1984). Surface accessibility of 13C-labeled lysine residues in membrane-bound myelin basic protein. *J. Biol. Chem*, 259, 8689-8692.

[205] Ong, R.L., and Yu, R.K. (1984). Interaction of ganglioside GM1 and myelin basic protein studied by carbon-13 and proton nuclear magnetic resonance spectroscopy. *J. Neurosci. Res*, 12, 377-393.

[206] Mendz, G.L., Moore, W.J., Brown, L.R., and Martenson, R.E. (1984). Interaction of myelin basic protein with micelles of dodecylphosphocholine. *Biochemistry*, 23, 6041-6046.

[207] Smith, R., and Cornell, B.A. (1985). Myelin basic protein induces hexagonal phase formation in dispersions of diacylphosphatidic acid. *Biochim. Biophys. Acta*, 818, 275-279.

[208] Deber, C.M., Hughes, D.W., Fraser, P.E., Pawagi, A.B., and Moscarello, M.A. (1986). Binding of human normal and multiple sclerosis-derived myelin basic protein to phospholipid vesicles: effects on membrane head group and bilayer regions. *Arch. Biochem. Biophys*, 245, 455-463.

[209] Fraser, P.E., Moscarello, M.A., Rand, R.P., and Deber, C.M. (1986). Spontaneous vesicularization of myelin lipids is counteracted by myelin basic protein. *Biochim. Biophys. Acta*, 863, 282-288.

[210] Gow, A., Winzor, D.J., and Smith, R. (1987). Equilibrium binding of myristoyllysophosphatidylcholine to bovine myelin basic protein: an example of ligand-mediated acceptor association. *Biochemistry*, 26, 982-987.

[211] Forbes, J., Bowers, J., Shan, X., Moran, L., Oldfield, E., and Moscarello, M.A. (1988). Some new developments in solid-state nuclear magnetic resonance spectroscopic studies of lipids and biological membranes, including the effects of cholesterol in model and natural systems. *Journal of the Chemical Society-Faraday Transactions I*, 84, 3821-3849.

[212] Mendz, G.L., Moore, W.J., Kaplin, I.J., Cornell, B.A., Separovic, F., Miller, D.J., and Brown, L.R. (1988). Characterization of dodecylphosphocholine/myelin basic protein complexes. *Biochemistry*, 27, 379-386.

[213] Fraser, P.E., Rand, R.P., and Deber, C.M. (1989). Bilayer-stabilizing properties of myelin basic protein in dioleoylphosphatidylethanolamine systems. *Biochim. Biophys. Acta*, 983, 23-29.

[214] Mendz, G.L., Brown, L.R., and Martenson, R.E. (1990). Interactions of myelin basic protein with mixed dodecylphosphocholine/palmitoyllysophosphatidic acid micelles. *Biochemistry*, 29, 2304-2311.

[215] Mendz, G.L., Miller, D.J., Jamie, I.M., White, J.W., Brown, L.R., Ralston, G.B., and Kaplin, I.J. (1991). Physicochemical characterization of dodecylphosphocholine/palmitoyllysophosphatidic acid/myelin basic protein complexes. *Biochemistry*, 30, 6509-6516.

[216] Mendz, G.L., Jamie, I.M., and White, J.W. (1992). Effects of acyl chain length on the conformation of myelin basic protein bound to lysolipid micelles. *Biophys. Chem*, 45, 61-77.

[217] Hayer-Hartl, M., Brophy, P.J., Marsh, D., and Watts, A. (1993). Interaction of two complementary fragments of the bovine spinal cord myelin basic protein with phosphatidylglycerol bilayers, studied by 2H and 31P NMR spectroscopy. *Biochemistry*, 32, 9709-9713.

[218] Nezil, F.A., Bayerl, S., and Bloom, M. (1992). Temperature-reversible eruptions of vesicles in model membranes studied by NMR. *Biophys. J*, 61, 1413-1426.

[219] Husted, C., Montez, B., Le, C., Moscarello, M.A., and Oldfield, E. (1993). Carbon-13 "magic-angle" sample-spinning nuclear magnetic resonance studies of human myelin, and model membrane systems. *Magn. Reson. Med*, 29, 168-178.

[220] Reinl, H.M., and Bayerl, T.M. (1993). Interaction of myelin basic protein with single bilayers on a solid support: an NMR, DSC and polarized infrared ATR study. *Biochim. Biophys. Acta*, 1151,127-136.

[221] Roux, M., Nezil, F.A., Monck, M., and Bloom, M. (1994). Fragmentation of phospholipid bilayers by myelin basic protein. *Biochemistry,* 33, 307-311.

[222] Mendz, G.L., Miller, D.J., and Ralston, G.B. (1995). Interactions of myelin basic protein with palmitoyllysophosphatidylcholine: characterization of the complexes and conformations of the protein. *Eur. Biophys. J*, 24, 39-53.

[223] Ohler, B., Graf, K., Bragg, R., Lemons, T., Coe, R., Genain, C., Israelachvili, J., and Husted, C. (2004). Role of lipid interactions in autoimmune demyelination. *Biochim. Biophys. Acta*, 1688, 10-17.

[224] Pointer-Keenan, C.D., Lee, D.K., Hallok, K., Tan, A., Zand, R., and Ramamoorthy, A. (2004). Investigation of the interaction of myelin basic protein with phospholipid bilayers using solid-state NMR spectroscopy. *Chem. Phys. Lipids*, 132, 47-54.

[225] Deber, C.M., Cheifetz, S., and Moscarello, M.A. (1983). Microheterogeneity of bovine myelin basic protein studied by nuclear magnetic resonance spectroscopy. *Biopolymers*, 22, 377-380.

[226] Cheifetz, S., Moscarello, M.A., and Deber, C.M. (1984). NMR investigation of the charge isomers of bovine myelin basic protein. *Arch. Biochem. Biophys*, 233, 151-160.

[227] Persaud, R., Fraser, P., Wood, D.D., and Moscarello, M.A. (1988). The glycosylation of human myelin basic protein at threonines 95 and 98 occurs sequentially. *Biochim. Biophys. Acta,* 966, 357-361.

[228] Mendz, G.L., Moore, W.J., Easterbrook-Smith, S., and Linthicum, D.S. (1985). Proton NMR study of interaction of myelin basic protein with a monoclonal antibody. *Biochem. J*, 228, 61-68.

[229] Libich, D.S., Robertson, V.J., Monette, M.M., and Harauz, G. (2004). Backbone resonance assignments of the 18.5 kDa isoform of murine myelin basic protein (MBP). *J. Biomol. NMR*, 29, 545-546.

[230] Farès, C., Libich, D.S., and Harauz, G. (2006). Solution NMR structure of an immunodominant epitope of myelin basic protein. Conformational dependence on environment of an intrinsically unstructured protein. *FEBS J*, 273, 601-614.

[231] Bamm, V.V., Ahmed, M.A., Ladizhansky, V., and Harauz, G. (2007). Purification and spectroscopic characterization of the recombinant BG21 isoform of murine golli myelin basic protein. *J. Neurosci. Res*, 85, 272-284.

[232] Libich, D.S., Monette, M.M., Robertson, V.J., and Harauz, G. (2007). NMR assignment of an intrinsically disordered protein under physiological conditions: the 18.5 kDa isoform of myelin basic protein. *Biomolecular NMR Assignments*, 1, 61-63.

[233] Ahmed, M.A., Bamm, V.V., Harauz, G., and Ladizhansky, V. (2007). The BG21 isoform of Golli myelin basic protein is intrinsically disordered with a highly flexible amino-terminal domain. *Biochemistry*, 46, 9700-9712.

[234] Zhong, L., Bamm, V.V., Ahmed, M.A., Harauz, G., and Ladizhansky, V. (2007). Solid-state NMR spectroscopy of 18.5 kDa myelin basic protein reconstituted with lipid vesicles: spectroscopic characterisation and spectral assignments of solvent-exposed protein fragments. *Biochim. Biophys. Acta (Biomembranes),* 1768, 3193-3205.

[235] Bates, I.R., Matharu, P., Ishiyama, N., Rochon, D., Wood, D.D., Polverini, E., Moscarello, M.A., Viner, N.J., and Harauz, G. (2000). Characterization of a recombinant murine 18.5-kDa myelin basic protein. *Protein Expr. Purif*, 20, 285-299.

[236] Bates, I.R., Libich, D.S., Wood, D.D., Moscarello, M.A., and Harauz, G. (2002). An Arg/Lys→Gln mutant of recombinant murine myelin basic protein as a mimic of the deiminated form implicated in multiple sclerosis. *Protein Expr. Purif*, 25, 330-341.

[237] Kaur, J., Libich, D.S., Campagnoni, C.W., Wood, D.D., Moscarello, M.A., Campagnoni, A.T., and Harauz, G. (2003). Expression and properties of the recombinant murine Golli-myelin basic protein isoform J37. *J. Neurosci. Res*, 71, 777-784.

[238] Bates, I.R., Feix, J.B., Boggs, J.M., and Harauz, G. (2004). An immunodominant epitope of myelin basic protein is an amphipathic α-helix. *J. Biol. Chem*, 279, 5757-5764.

[239] Buck, M. (1998). Trifluoroethanol and colleagues: cosolvents come of age. Recent studies with peptides and proteins. *Q. Rev. Biophys*, 31, 297-355.

[240] Roccatano, D., Colombo, G., Fioroni, M., and Mark, A.E. (2002). Mechanism by which 2,2,2-trifluoroethanol/water mixtures stabilize secondary-structure formation in peptides: a molecular dynamics study. *Proc. Natl. Acad. Sci*, USA, 99, 12179-12184.

[241] Perham, M., Liao, J., and Wittung-Stafshede, P. (2006). Differential effects of alcohols on conformational switchovers in alpha-helical and beta-sheet protein models. *Biochemistry,* 45, 7740-7749.

[242] Povey, J.F., Smales, C.M., Hassard, S.J., and Howard, M.J. (2007). Comparison of the effects of 2,2,2-trifluoroethanol on peptide and protein structure and function. *J. Struct. Biol,* 157, 329-338.

[243] Otzen, D.E., Sehgal, P., and Nesgaard, L.W. (2007). Alternative membrane protein conformations in alcohols. *Biochemistry*, 46, 4348-4359.

[244] Bates, I.R., Boggs, J.M., Feix, J.B., and Harauz, G. (2003). Membrane-anchoring and charge effects in the interaction of myelin basic protein (MBP) with lipid bilayers studied by spin-labelling. *J. Biol. Chem*, 278, 29041-29047.

[245] Musse, A.A., Boggs, J.M., and Harauz, G. (2006). Deimination of membrane-bound myelin basic protein in multiple sclerosis exposes an immunodominant epitope. *Proc. Natl. Acad. Sci*, USA, 103, 4422-4427.

[246] Bates, I.R., and Harauz, G. (2003). Molecular dynamics exposes α-helices in myelin basic protein. *J. Mol. Mod*, 9, 290-297.

[247] Polverini, E., Boggs, J.M., Bates, I.R., Harauz, G., and Cavatorta, P. (2004). Electron paramagnetic resonance spectroscopy and molecular modelling of the interaction of

myelin basic protein (MBP) with calmodulin (CaM)-diversity and conformational adaptability of MBP CaM-targets. *J. Struct. Biol*, 148, 353-369.

[248] Boggs, J.M., Yip, P.M., Rangaraj, G., and Jo, E. (1997). Effect of posttranslational modifications to myelin basic protein on its ability to aggregate acidic lipid vesicles. *Biochemistry,* 36, 5065-5071.

[249] Boggs, J.M., Rangaraj, G., and Koshy, K.M. (1999). Analysis of the membrane-interacting domains of myelin basic protein by hydrophobic photolabeling. *Biochim. Biophys. Acta,* 1417, 254-266.

[250] Seelig, J. (1978). 31P nuclear magnetic resonance and the head group structure of phospholipids in membranes. *Biochim. Biophys. Acta*, 515, 105-140.

[251] Davis, J.H. (1983). The description of membrane lipid conformation, order and dynamics by 2H-NMR. *Biochim. Biophys. Acta*, 737, 117-171.

[252] Dufourc, E.J., Mayer, C., Stohrer, J., Althoff, G., and Kothe, G. (1992). Dynamics of phosphate head groups in biomembranes. Comprehensive analysis using phosphorus-31 nuclear magnetic resonance lineshape and relaxation time measurements. *Biophys. J,* 61, 42-57.

[253] Auger, M. (2000). Biological membrane structure by solid-state NMR. *Curr. Issues Mol. Biol*, 2, 119-124.

[254] Ouellet, M., Bernard, G., Voyer, N., and Auger, M. (2006). Insights on the interactions of synthetic amphipathic peptides with model membranes as revealed by 31P and 2H solid-state NMR and infrared spectroscopies. *Biophys. J*, 90, 4071-4084.

[255] Hallock, K.J., Henzler, W.K., Lee, D.K., and Ramamoorthy, A. (2002). An innovative procedure using a sublimable solid to align lipid bilayers for solid-state NMR studies. *Biophys. J*, 82, 2499-2503.

INDEX

A

B

C

D

E

F

G

H

I

P

Q

R

S

T

U

V

W

X

Y

Z